Transactions on Engineering Technologies

Sio-Iong Ao · Haeng Kon Kim
Oscar Castillo · Alan Hoi-Shou Chan
Hideki Katagiri
Editors

Transactions on Engineering Technologies

International MultiConference of Engineers
and Computer Scientists 2017

 Springer

Editors
Sio-Iong Ao
International Association of Engineers
Hong Kong
Hong Kong

Haeng Kon Kim
Department of Computer
and Communication Engineering
Daegu Catholic University
Daegu
Korea (Republic of)

Oscar Castillo
Instituto Tecnologico de Tijuana
Tijuana
Mexico

Alan Hoi-Shou Chan
Department of Systems Engineering
and Engineering Management
City University of Hong Kong
Hong Kong
Hong Kong

Hideki Katagiri
Department of Industrial Engineering
and Management
Kanagawa University
Yokohama
Japan

ISBN 978-981-10-7487-5   ISBN 978-981-10-7488-2   (eBook)
https://doi.org/10.1007/978-981-10-7488-2

Library of Congress Control Number: 2017962019

Printed on acid-free paper

This Springer imprint is published by Springer Nature
The registered company is Springer Nature Singapore Pte Ltd.
The registered company address is: 152 Beach Road, #21-01/04 Gateway East, Singapore 189721, Singapore

Preface

A large international conference on Advances in Engineering Technologies and Physical Science was held in Hong Kong, March 15–17, 2017, under the International MultiConference of Engineers and Computer Scientists 2017 (IMECS 2017). IMECS 2017 is organized by the International Association of Engineers (IAENG). IAENG is a nonprofit international association for the engineers and the computer scientists, which was founded originally in 1968 and has been undergoing rapid expansions in recent few years. IMECS conference serves as a good platform for the engineering community to meet with each other and to exchange ideas. The conference has also struck a balance between theoretical and application development. The conference committees have been formed with over three hundred committee members who are mainly research center heads, faculty deans, department heads, professors, and research scientists from over 30 countries with the full committee list available at our conference Web site (http://www.iaeng.org/IMECS2017/committee.html). The conference is truly an international meeting with a high level of participation from many countries. The response that we have received for the conference is excellent. There have been more than six hundred manuscript submissions for IMECS 2017. All submitted papers have gone through the peer review process, and the overall acceptance rate is 51%.

This volume contains twenty-eight revised and extended research articles written by prominent researchers participating in the conference. Topics covered include electrical engineering, communications systems, engineering mathematics, engineering physics, and industrial applications. The book offers the state of the art of tremendous advances in engineering technologies and physical science and applications, and also serves as an excellent reference work for researchers and graduate students working with/on engineering technologies and physical science and applications.

Hong Kong   Sio-Iong Ao
Daegu, Korea (Republic of)   Haeng Kon Kim
Tijuana, Mexico   Oscar Castillo
Hong Kong   Alan Hoi-Shou Chan
Yokohama, Japan   Hideki Katagiri

Contents

Transportation Safety Improvements Through Video Analysis: An Application of Obstacles and Collision Detection Applied to Railways and Roads

Hui Wang, Xiaoquan Zhang, Lorenzo Damiani, Pietro Giribone, Roberto Revetria and Giacomo Ronchetti

Abstract Obstacles detection systems are essential to obtain a higher safety level on railways. Such systems has the ability to contribute to the development of automated guided trains. Even though some laser equipments have been used to detect obstacles, short detection distance and low accuracy on curve zones make them not the best solution. In this paper, after an assessment of the risks related to railway accidents and their possible causes, computer vision combined with prior knowledge is used to develop an innovative approach. A function to find the starting point of the rails is proposed. After that, bottom-up adaptive windows are created to focus on the region of interest and ignore the background. The whole system can run in real time thanks to its linear complexity. Experimental tests demonstrated that the system performs well in different conditions.

Keywords Computer vision · Experimental assessment · Obstacles detection Prior knowledge · Transportation · Transport safety video forensic

H. Wang · X. Zhang
Ulster University, Belfast, UK
e-mail: h.ang@ulster.ac.uk

X. Zhang
e-mail: zxq8984@163.com

L. Damiani · P. Giribone · R. Revetria (✉) · G. Ronchetti
Genoa University, via Opera Pia 15, 16145 Genoa, Italy
e-mail: Roberto.Revetria@unige.it

L. Damiani
e-mail: LorenzoDamiani@unige.it

P. Giribone
e-mail: Piero@itim.unige.it

G. Ronchetti
e-mail: giacomoronchetti@live.it

© Springer Nature Singapore Pte Ltd. 2018
S.-I. Ao et al. (eds.), *Transactions on Engineering Technologies*,
https://doi.org/10.1007/978-981-10-7488-2_1

1

1   Introduction

With the rapid development of economy, transportation became more and more important. Train is one of the safest means of transport, however several accidents still happened in recent years. In the development of automated guided trains, it is urgent to set up a system to help figuring out anomaly conditions in front of a running engine.

Computer vision is always more advanced, but changing lighting conditions, background, and running speed make obstacles detection on railways a hard task. In this paper a system implemented in C++ and OpenCV is proposed.

The system [1] has the ability to tolerate changing environment and to find obstacles on or between the rails. It can therefore be a safety component for automated guided trains, but it can even represent a useful tool for the drivers of conventional trains. The paper is structured as follows. In Sect. 2 the topic of railway safety is assessed. In Sect. 3 some related works in subject of obstacles detection are presented. Section 4 describes our assumptions for the problem and the approach developed.

Experimental results are shown in Sect. 5. Conclusions and future works are presented in Sect. 7.

2   Railway Safety

In the railway sector, the risk can be defined in relation to the events that damage safety or transportation stability. As stated above railway science is a complicated subject, which is interdisciplinary and requires competences of different actors.

2.1   Railway Safety Risk Assessment

In this section an overview about current risk assessment methods is presented.

Risk management is defined as the culture, process and structures that are directed towards the effective management of potential opportunities and adverse effects.

The use of a bottom-up risk assessment process yields a higher level of confidence that all of the failure events of a railway system and their respective causes are identified. Therefore, compared with the top-down, the bottom-up approach has the following characteristics:

- Omission of system failure events and their respective causes are less likely;
- It may be more convenient to incorporate into a computer package;
- It may be more suitable to apply to safety risk analysis of a large railway system, with a high level of uncertainty.

2.2   Railway Collision Due to Obstacle

The elements that influence the risk of accident by collision with obstacles on the train track are several.

From the infrastructural point of view a risk is represented by the concrete point on the track (entrance or exit of tunnels, flyover, etc.) which is considered potentially hazardous, concerning the risk of falls of the obstacles. It is characterised by:

- *Falling rate*: it is the estimated rate of appearance of obstacles in that point (average number of falls per unit of time). It is difficult to estimate, being very small and dependent from varied factors.
- *Maximum visibility*: it is the maximum visibility that driver of the train has at the risk point. The effective visibility will be a percentage of the above mentioned, depending on weather conditions.
- *Time of automatic detection*: assuming that an automatic detector of obstacles has been installed in the point of risk, it is the time delay to reach the traffic control centre the signal informing that there is an obstacle.
- *Time of manual detection*: it is the average time that the warning of the obstacles in the way takes to arrive at the traffic control centre b non-automatic means. It will be an estimation.
- Other elements of the infrastructure that are associated with the railroad:
- *Slope of the railroad*: it is imperative to have the profiles of the physical slopes to calculate time and distance when applying the emergency brake.
- *Time of communication*: it is the time that it takes the order of the emergency brake process from the traffic control centre to the train. From rolling stock point of view, vital elements are:
- *Mass in normal shipment*: total of the train in normal conditions.
- *Time in reaction*: it is the time that it takes for the emergency brake t act since the order is received in the train.
- *Strength of braking process*: it is the curve of the effort in the braking process of the train in function of the speed of the train.
- *Air resistance*: it is the curve of air resistance of the train in function o the speed off the train.

To make an analysis of accident risk during a temporal horizon the characteristics of composition and density of the traffic during the horizon have to be defined as well.

3   Related Work

In the subject of obstacles detection on railways, there exist two basic ideas. The first one is based on the use of sensors, such as [1–3]. The other one is based on video processing. Sensors can find obstacles more easily but they cannot avoid the drawbacks of cost emitting devices, short detection distances and low accuracy on curve zones. So, even though there is not yet a perfect solution for the problem, video based methods may be a better choice. The current approaches can be mainly divided into two strategies: some of the ideas use Canny Operator, some others apply prior Knowledge. Using forensic video analysis is also possible to provide detailed reconstruction for accident dynamic and evaluate quantitatively the effects on human bodies as presented in [4, 5].

3.1 Based on Canny Method

Approaches based on Canny method [6] use Canny operator to extract edges from an image. The first step is to smooth the image by a Gaussian filter in order to decrease the influence of the noise. Then the gradient image is obtained using a Sobel operator. Finally non-maximum suppression is applied to sharpen the edges and two thresholds are used to detect the effective edges. Canny [7] shows an adaptive way to find the thresholds for the Canny method. Its basic principle is to split the image's pixels into two classes and to find the best threshold value through the maximum variance value between the two classes. However, it cannot get the edge of the rails because of the fickle background. Fang et al. [8] shows an obstacles detection method on railways based on Canny method. It uses a model train to imitate the real train and a small camera to get the video from the head of the train. Rails and sleepers show obvious geometric lines characteristics after Canny edge detection, obstacles undermine the integrity of these lines. Thus an obstacle on the condition that disconnected length of lines exceeds a certain threshold can be confirmed. The effect on Canny image is very clear, an obstacle can be detected in two aspects: one is through the number of consecutive pixels where the rails lines lie to determine the integrity of the rails, the other is by determining if the distance at both ends of each sleeper is abnormal with regard to what expected.

Yao et al. [9] finds the candidate obstacle areas based on the edge detection. Then it trains a SVM to classify the candidate obstacle areas and to verify whether they represent a real obstacle or not, using features such as dimension and colour information of these areas. Even though using Canny method is a good way to generate rails intuitively, it suffers the drawback of selecting effective thresholds and shadows may make the conditions even worse. Figure 1 shows the different results of two thresholds. The original image of a railway, the result of an adaptive threshold and the result of a low threshold are shown. We can figure out the importance and the difficulty of choosing a proper threshold.

Fig. 1 Result of different thresholds

3.2 Based on Prior Knowledge

Prior knowledge for image analysis on railways is mainly based on two concepts, the first one is that rails are always in the picture and they meet at the top of the image, the second one that rails from bird's-eye view are always parallel. Tong et al. [10] is based on the fact that the rails look like lines projecting from the bottom of the image to the horizon and they seem to converge in a far point. In the vicinity of the driver zone, the rails appearance is quite monotonous, while on the horizon it exists greater variability, especially in curve sections. To delimit a region of interest is the first task to start the search for obstacles and to decide if the way is free and can be safely travelled. In this case the rails, the area between them and the close outer area represent the zone of interest. One strategy for detecting the rails consists in extracting the lines in the image and selecting the most likely candidates to be the rails. They used the Hough transform to detect the rails in each frame and to start the delimitation of the region of interest to check for obstacles. Rodriguez et al. [11] presented a comparison of various techniques for rails extraction. Then they introduced a new generic and robust approach. The algorithm uses edge detection and applies simple geometric constraints provided by the rail characteristics. Espino and Stanciulescu [12] uses Canny method to get the rails. Then it gets the bird's-eye view image by trapezoidal projection and it creates a region of interest moving from the bottom to the top of the image. In the ROI, a function grope_rails (R; ae; ge) is created to find the pair of lines with the maximum score. It restricts the angle ae and the gap ge to make it run in real time. However, there can be problems when there are more than one pair of lines in the region of interest.

4   Novel Approach

After sufficient studies on previous works, we decided to develop an adaptable and less fine-tuned parameters method. We create a novel linear way to detect rails and to check for obstacles, based on four assumptions:

1. The gap between the rails at the initialization stage has to be constant.
2. The starting position of the rails cannot move saltatorial.
3. The pair of rails in the bird's-eye view has to be parallel.
4. The pair of rails has to come to a vanishing point.

The first two assumptions rectify the false conviction that the starting position has to be constant. The bird's-eye view image helps to identify obstacles between the rails and the last assumption is the stop condition for the system. Figure 2 shows the system diagram designed according to the previous assumptions.

Fig. 2 System diagram

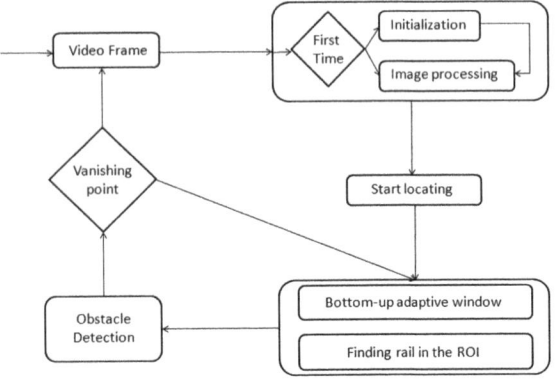

The system collects video information from a camera mounted in front of the train and analyzes each frame. For each frame the methodology works as follows:

1. Pre-processing, it includes initialization and image processing.
2. Locating the start of the rails.
3. Creating bottom-up adaptive windows and finding the rails.
4. Checking for obstacles.

Next frame or back to the step 3.

4.1 Initialization

In order to decrease the human contribution and the number of parameters required, the initialization stage has to be carried out just once for each train. After that, the system will keep a configuration file and will update it automatically. During initialization, to get the gap between the rails and to select the start position of the left rail is necessary. This is the only human action required in the whole system. In this stage Sobel operator is used to get the gradient image of the frame. Gradient image can get the whole information of the edges in the image and it allows to avoid the drawback of using thresholds imported by Canny operator. We get the gradient value by using Euclidean distance (Fig. 3).

Fig. 3 Raw frame and gradient image

4.2   Locating the Start of the Rails

After we get the gradient image, we need to locate the start of the rails in order to create bottom-up windows and to check for obstacles. This method applies the assumption that the gap between the rails is constant and it does not move in a saltatorial way. The idea is shown from a geometrical point of view in Fig. 4 which case(s).

To make it a real time operating system, the algorithm has to be linear. In order to do so, we first create a pair of sliding windows. The gap between the windows is the gap between the rails and it is read from the configuration file. After that, we define a function f(x) such that:

Since we move the sliding windows horizontally, the time complexity of finding their position is $O(w)$. w is the width of the image. After we get the pair of windows, finding the correct position of the rail in the window is $O(L^2)$, and since L is constant, the time cost of this procedure is $O(w + L^2)$. Given that both the width of the window and the width of the image are constant, the computational time complexity of finding the start of the rails is linear and it guarantees the real time property of the system.

4.3   Creating Bottom-up Windows

After finding out the starting position of the rails, adaptive bottom-up windows are created through an iterative process. It is desirable the windows to contain the same amount of information, so the top windows will be smaller than the bottom ones. The adaptive windows perform in two ways: size and position. Since we assume that the rail in the window is a line, the scale ratio between two adjacent windows

Fig. 4 Locating the start of the rails

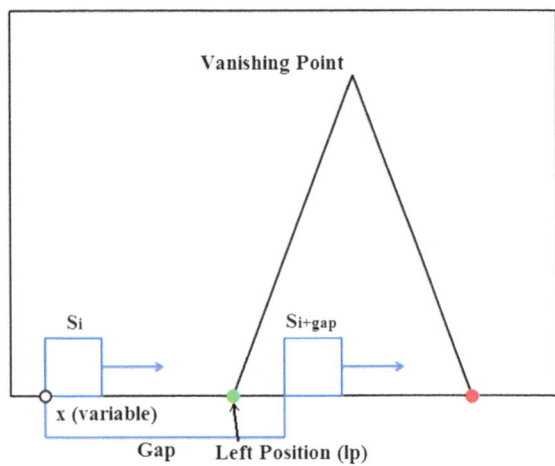

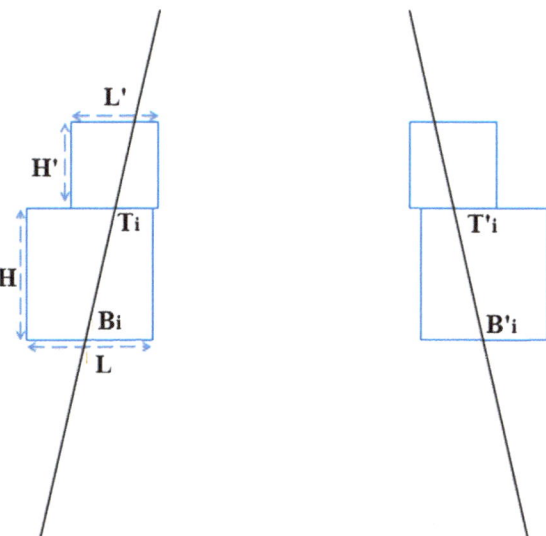

Fig. 5 Creating bottom-up windows

is linear. The upper windows are generated through an iterative process and their size is defined by the following equations (Fig. 5):

$$L' = L \times \frac{T'_i - T_i}{B'_i - B_i}; \; H' = H \times \frac{T'_i - T_i}{B'_i - B_i}$$

The rail is consistent and therefore the top point of the rail in a window is supposed to be the bottom point of the rail in the adjacent one. The position of the window is adaptive since the angle of the rail can change sharply in the small windows. The position of the window changes according to the top position of the rail in the lower window. As shown in Fig. 6a, if, for instance, the position T_i is on the right corner of the window, we claim the rail may be deflected sharply to the right. In such condition, the next window is created at the right corner in order to keep the rail inside. The same happens when T_i is on the left (c), obviously. When T_i is in the middle, the next window is centred on the previous one (b). The position of the adaptive windows is therefore strictly related to T_i and it responds well wherever a sharp deflection appears.

After we fix the position of the next window, we apply local maximum idea to get the maximal average gradient line to be the rail. Subsequently the obstacle detection stage is worked. If there are no obstacles in the window, the next pair of top windows is generated. This process is iterated until the vanishing point is reached. The vanishing point is considered reached when the left window intersects with the right window, as shown in Fig. 6. The time complexity is still linear since, when we get the rail, the bottom point is already set as in the previous stage.

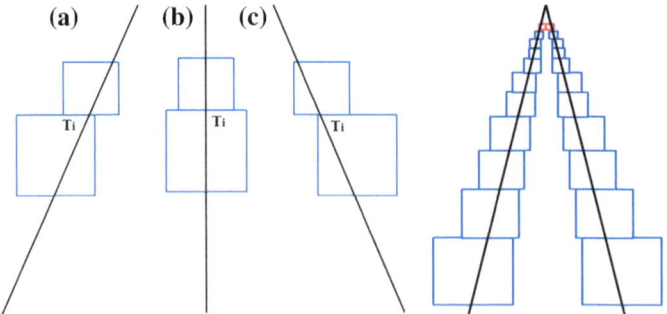

Fig. 6 Adaptability of the windows (left) and Iteration of the process (right)

4.4 Obstacles Detection

Every time a pair of bottom-up windows is created, the obstacle detection stage is worked. Three methods to check for obstacles on the railway are used.

(1) Compare the angle generated by the lines detected between two adjacent windows. If the angle is sharper than a certain value, there might be an obstacle or the rail could be broken. The threshold value is looser for the windows further away because of the zoom rate. Figure 6 shows a gradient image where a brick lies on the left rail generating an anomalous angle between the lines detected by the system. In this case, the algorithm correctly assumes there is an obstacle obstructing the rail.

(2) Get the average gradient value of the rail. If the value is less than the expected, we suppose there is a break point. If we find the maximum consistent break point length is more than 20% of rail in the window, an obstacle may come. Here again the value becomes looser when the windows get higher in the image. If an obstacle is detected here, we check if most of the pixels do not come to an average value. If it is true, a shadow may hide the light.

(3) In order to check for obstacles in the middle of the railway, trapezoidal projection to get the bird's-eye view image is used. We get the projection matrix by trapezoidal mapping. We use T_i, T'_i, B_i, B'_i to get the trapezium and we project it into the rectangle RT_i, RT'_i, RB_i, RB'_i. The same process is applied for the adjacent windows. After that, the background subtraction method is used to check for the contours in the result image and to check for obstacles.

5 Experimental Results

The system has been implemented in C++ in Visual Studio 2015 and OpenCV 2.4.11. The dataset is an open source video captured by a camera in front of a train from Malmo to Angelholm, Sweden.

The frame size is 640 × 360, corresponding to an aspect ratio of 16:9. The frame rate is 25 fps. Initially we have separated the video into four fragments in order to check the ability of the system to solve the problem in different conditions. Since there are very rare cases of obstacles on railways, the videos have been modified adding digital obstacles of different nature, shape and obstruction trajectory.

For this procedure we used Adobe Premiere Pro CC 2015. We have included on the railway bricks, rocks, cars and pedestrians in different positions and at different distances. After that, we tested the system and 10 out of 10 obstacles were successfully identified, but there is still a significant occurrence of false positive. Table 1 summarizes results obtained from the test conducted on a 2 min recording.

At the end of the test, we can state the system shows the following strengths:

(1) Sensitivity equal to 100%. This is a great achievement for safety, because it is essential that false negative do not occur, namely that all the obstacles are identified. Figure 7 shows obstacles of different nature and in different position identified by the system.

Table 1 Experimental results

Object	Total	Success	Failure	Ratio (%)
Brick	3	3	0	100
Car	1	1	0	100
Pedestrian	3	3	0	100
Rock	3	3	0	100
Total	**10**	**10**	**0**	**100**
False positive	3	–	–	–

Fig. 7 Obstacles detected on the rail

(2) Ability of the system dealing with shadows. There are many scenes where the lighting conditions are modified by the presence of trees, bridges, and so on. Dealing with shadows is the main reason why we did not choose to apply the Canny method.
(3) Ability of the system to get the rail even when the train shakes. Many methods which apply prior knowledge assume the position of the rail in the image is constant, however the train will shake during his journey. This methodology allows to create bottom up windows and to find the rail in linear time even when a shake occurs between two consecutive frames.

6 Involving G Sensors in Incident Detection

Computer vision is playing a key role in transportation evolution. Video imaging scientists are providing intelligent sensing and processing technologies for a wide variety of applications and services. There are many interesting technical challenges including imaging under a variety of environmental and illumination conditions, data overload, recognition and tracking of objects at high speed, distributed network sensing and processing, energy sources, as well as legal concerns. This methodology makes full use of computer vision techniques combined with data analysis and it can be applied both for future automated guided trains and as a useful tool for the drivers of conventional trains. It can introduce a distinct improvement in the subject of rail transportation safety. It represents a safety device for reducing the occurrence of accidents or at least their consequences. The potential of computer vision combined with data analysis has been clearly shown in this work. Present work is then link to another one proposed by the same authors but focused on road vehicles rather than on rails. The purpose of both works is to introduce technologies to increase the level of safety in the field of transportation. The second project is focused on video recording accidental events using a camera mounted in front of vehicles for public transportation, especially buses and coaches. When an accident occurs it is often difficult to determine who is guilty and who is not. Introducing a camera to record what has happened would be useful in such cases. The main difference between the two projects is that the obstacle detection system plays an active role picking up an potential danger on the railways and directly applying the emergency brake or warning the driver, while the accident recording system acts indirectly continuous monitoring driver's behaviour. If a bus driver knows its actions are being recorded, indeed, he will behave more carefully and the hazards will be reduced. It is not possible, however, to record all the period while a vehicle is being driven since it would produce a huge amount of data to be stored. The key point of the system is to understand whenever an accident happened in order to store only the video showing few seconds before and after the event. The current system uses two cameras, rear and front, a GPS, and a accelerometer sensor. In the existing version of the system only the GPS and the accelerometer sensor are used to determine if an accident occurred. In the following paragraph a short description of how the method works is presented. Accelerometer sensor data is the main variable which is used to determine if an event has occurred. This is done using a simple algorithm. The main idea is to calculate the differential in kinetic energies between two points.

$$E = \frac{m \times V^2}{2};$$

The mass m of the vehicle is a constant regardless of the fact the vehicle is carrying passengers, luggage or whatever else and it does not influence the variation of energy:

$$\Delta E = (V_1^2 - V_2^2)$$

In order to get the data from the accelerometer the algorithm is implemented following these steps:

(1) The raw data coming from the accelerometer is divided by the three axes x, y, and z. To obtain the acceleration is necessary to sum the three components. The signal is framed into discrete intervals in order to reduce the total amount of data.
(2) The velocity is then obtained by integrating the acceleration within the time interval using definite integral.
(3) The equation is applied to get the variation of energy.

If the system notes a variation of energy higher than a certain threshold an impact due to an accidental event may have occurred. However, when a vehicle stops or departs from a bus stop, for example, a considerable acceleration/deceleration may be observed, and a false accidental event may wrongly be recorded. For this reason, the GPS is used as support to understand if the variation of energy recorded represents a real accident. If the amount of energy changes in the proximity of a bus stop, for example, the system supposes there is nothing abnormal happened and it does not store the video. On the other hand, if something similar happens into an intersection it is supposed an accident occurred. For this purpose, a handy online tool can be used to create a map of the points of interest, such as bus stops, pedestrian crossings, intersections and so on, for any city in the world.

Notwithstanding this measure, the result of the tests carried out on the system presented a significant occurrence of false positive. We find therefore the same drawback of the obstacle detection system and that is why we thought we can apply a similar approach to spot false positive. The idea is to extend the work providing the cameras with processors, in order to make them active devices, able to contribute to the detection of an anomalous situation, not just mere video recorders. We want to analyse a video frame to understand if it shows a traffic accident or not. On the road certainly a wider range of scenarios can be found compared with the railway, where the view is rather monotonous. The algorithm and the classifier we need to apply should be completely different from those used for obstacle detection on railways, but the idea of detecting an anomalous event, in this case a traffic accident, by using computer vision combined with data analysis is very similar.

7 Conclusions

This work updates techniques based on prior knowledge and creates an automatic and real time method for obstacles detection on railways. The key contribution are:

1. Except the initialization stage, the system is completely automatic and it does not require the human contribution.
2. It can find the rails in linear time even in critical conditions, which guarantees the real time property of the system.
3. The bottom-up windows are adaptive in position and size and perform well.

The obstacles detection methods have proved to be effective but we need to manage false positive in a better way. Many objects, indeed, can affect the monotony of the images without posing a real danger to the rail journey. Recognition of these objects is essential for a reliable behaviour of the system. Learning and identification of benign object provides interesting challenges for computer vision algorithms. In our particular case, since we use the contours extraction to check for obstacles between the rails, some big metal plates may lead to false positive problems. A solution we are thinking to apply is to collect a large amount of images and to create a dataset composed of positive and negative samples. After that we can use Local Binary Patter to extract features from images and train a Support Vector Machine classifier to be embedded in the system. Another problem we need to solve is identifying crossing areas. In the current work there is no research in this direction, although there are many crossings during the train journey, especially nearby the stations. We hope to manage this problem using one-dimension projection and Gaussian Mixture Model to find the rails correctly in these critical areas.

References

1. H. Wang, X. Zhang, L. Damiani, P. Giribone, R. Revetria, G. Ronchetti, Video analysis for improving transportation safety: obstacles and collision detection applied to railways and roads, in *Proceedings of the International MultiConference of Engineers and Computer Scientists 2017*. Lecture Notes in Engineering and Computer Science, 15–17 Mar 2017, Hong Kong, pp. 909–915
2. R. Passarella, B. Tutuko, A.P.P. Prasetyo, Design concept of train obstacle detection system in Indonesia. IJRRAS **9**(3), 453–460 (2011)
3. F. Kruse, S. Milch, H. Rohling, Multi sensor system for obstacle detection in train applications. Proc. IEEE Trans. 42–46 (2003)

4. E. Briano, C. Caballini, R. Revetria, The maintenance management in the highway context: a system dynamics approach, in *Proceedings of FUBUTEC* (2009), pp. 15–17; R.W. Lucky, Automatic equalization for digital communication. Bell Syst. Tech. J. **44**(4), 547–588 (1965)
5. E. Briano, C. Caballini, R. Revetria, M. Schenone, A. Testa, Use of system dynamics for modelling customers flows from residential areas to selling centers, in *Proceedings of the 12th WSEAS International Conference on Automatic Control, Modelling and simulation* (World Scientific and Engineering Academy and Society (WSEAS), 2010), pp. 269–273
6. S. Sugimoto, H. Tateda, H. Takahashi, M. Okutomi, Obstacle detection using millimeter-wave radar and its visualization on image sequence, in *2004 Proceedings of the 17th International Conference on Pattern Recognition, ICPR 2004*, vol. 3 (IEEE, 2004) pp. 342–345
7. J. Canny, A computational approach to edge detection. IEEE Trans. Pattern Anal. Mach. Intell. **6**, 679–698 (1986)
8. M. Fang, G.X. Yue, Q.-C. Yu, The study on an application of Otsu method in canny operator, in *International Symposium on Information*
9. T. Yao, S. Dai, P. Wang, Y. He, Image based obstacle detection for automatic train supervision, in *2012 5th International Congress on Image and Signal Processing (CISP)* (IEEE, 2012), pp. 1267–1270; E.H. Miller, A note on reflector arrays (Periodical style— Accepted for publication). Eng. Lett.
10. L. Tong, L. Zhu, Yu. Zujun, B. Guo, Railway obstacle detection using onboard forward-viewing camera. J. Transp. Syst. Eng. Inf. Technol. **4**, 013 (2012)
11. L.F. Rodriguez, J.A. Uribe, J.F.V. Bonilla, Obstacle detection over rails using hough transform, in *2012 XVII Symposium of Image, Signal Processing, and Artificial Vision (STSIVA)* (IEEE, 2012), pp. 317–322
12. J.C. Espino, B. Stanciulescu, Rail extraction technique using gradient information and a priori shape model, in *2012 15th International IEEE Conference on Intelligent Transportation Systems (ITSC)* (IEEE, 2012), pp. 1132–1136

IT-Security Concept Cornerstones in Industrie 4.0

Yübo Wang, Riham Fakhry, Sebastian Rohr and Reiner Anderl

Abstract The majority of enterprises identified the potential and the benefits of Industrie 4.0. However, many companies consider Industrie 4.0 more as a security challenge than an opportunity to optimize their processes or an enabler for new business models. Therefore effective security methods to protect the Industrie 4.0 systems and its associated values and assets are needed. Based on the connectivity infrastructure in the shopfloor, the diversity in the corporate landscape of the global mechanical and plant engineering ultimately causes that every enterprise has to develop its own way of production IT security management. The purpose of the followed sections is to analyze the challenges of IT security in Industrie 4.0 and to identify and combine the requirements from manufacturing automation, mechanical engineering, process engineering and the properties of cyber-physical systems with well-established core elements of IT security descriptions. A process model, which consists of a data model and an algorithm as its core element is developed to consider the challenges and to cope with the requirements. As a prototypical implementation security measures and their properties are presented in a technological Industrie 4.0 Toolbox IT-Security, as the result of the whole process and data model.

Keywords CPS · Defense in depth · Industrie 4.0 · IT security
Security by design · Toolbox

Y. Wang (✉) · R. Anderl
Department of Computer Integrated Design at the Technical University Darmstadt,
Otto-Berndt-Straße 2, 64287 Darmstadt, Germany
e-mail: y.wang@dik.tu-darmstadt.de

R. Fakhry · S. Rohr
GmbH, Marktstraße 47-49, 64401 Groß-Bieberau, Germany
e-mail: fakhry@accessec.com

S. Rohr
e-mail: rohr@accessec.com

© Springer Nature Singapore Pte Ltd. 2018
S.-I. Ao et al. (eds.), *Transactions on Engineering Technologies*,
https://doi.org/10.1007/978-981-10-7488-2_2

17

1 Introduction

The fourth industrial revolution (Industrie 4.0) is distinguished by a growing network and intelligence of machines, products, services and data. The property is reflected in the creation of value-added networks across enterprise boundaries [1, 2]. In addition, the traditional way of engineering and its production system goes through a change. The change manifests in digital enterprises, wherein the vision of Industrie 4.0 establishes, on the one hand, the collaborative product development and virtual engineering, on the other hand, it sets up an integrated and intelligent production planning and controlling [3, 4]. Furthermore, the digital network is an opportunity to improve collaboration, coordination and transparency across all business sectors of an enterprise [5].

Holistic technical and organizational changes accompany the horizontal and vertical integration of Industrie 4.0. These circumstances create new security requirements and implications for the product development and enterprise processes, incurred data, production plants and resources [6]. The successful creation of a security culture is one of the important key prerequisites for Industrie 4.0. Only a holistic view for an Industrie 4.0 security culture leads to trustworthy, resilient and socially accepted Industrie 4.0 systems and processes. Such a culture includes both safety and security, where security is divided into office IT security and production IT security [7]. This paper focuses on the environment of the production IT-Security.

The challenges for the production IT security possess diverse characteristics. The diversity, on the one hand, stems from the increasing use of internet technologies in the production area and the trend that in Industrie 4.0 cyber-physical systems (CPS) and cyber-physical production systems (CPPS) merge with Internet of Things, Internet of Services and Internet of Data [8, 9]. Whenever CPS and internet technologies are adopted into machines, the increasing threat of vulnerable IT systems is automatically transferred to the industrial plant. Regrettably there is no way to build an absolutely secure system, however, there can only be different degrees of protection [10].

On the other hand, the networked production resources, cross-organizational collaboration, coordination and transparency across all business sectors of an enterprise and cross-enterprise networks forms a chain of security measures, that—like any chain—is only as strong as its weakest link. Currently used systems have not been designed for the connected mode of operation they are deployed in, hence they usually do not provide security functionality of even security related information like modern IT components does. The IT sub-systems that have so far been used in production environments are not adequately aligned to current security requirements, never mind under the complex networked circumstances of Industrie 4.0 [7].

Consequently, any Industrie 4.0 initiatives need to be designed IT security as one of the key design criteria. An additional challenge is to address on the technology level the topics of secure networks, secure processes, secure services, and secure data—depending on the requirements each system and process has. Resilient and trusted cyber-physical machines have to be designed and build at the system level. Only based on this foundation, Industrie 4.0 applications will be able to unfold the full potential of its possibilities and benefits.

2 Approaches to IT-Security Strategy and Design in Industrie 4.0

Constantly, new vulnerabilities in IT and CP systems are detected and, consequently, new attack methods are developed to exploit these vulnerabilities. With this insight in mind, it is easy to understand why it is impossible to design a perfectly secure system or to maintain a high-security level over time with no additional effort. IT security is not a goal that is achieved once—it is rather a continuous process based on behavioral change [11].

Security solutions for Industrial Control Systems (ICS) so far have been based only on adding to the complexity of the architecture. This strategy, known as Security by Obscurity, has however only worked well for environments without external communication connections [12]. This single protective measure is bound to create problems on the usability side and primarily does not meet the requirements derived from Industrie 4.0 scenarios, wherein the ICS is not only connected to the internal network but is also extensively connected with external networks and networked nodes under the control of third parties. Therefore, the Defense in Depth strategy has been defined as a model to meet the security requirements for ICS. This layered approach means that in order to secure the entire environment, the use of several protective measures at different points and levels of the system is required [13]. A prerequisite for the Defense in Depth strategy is the division of the system into several separate areas. This separation may be logical and/or physical, e.g. by separating assets, values or different domains [5].

Another important approach is to establish Security by Design as a core design principle. This approach is particularly relevant for the development of future systems. It requires an early assessment of possible threats and the consideration of necessary protective measures in the first phases of product development and in the construction of infrastructures [5]. The result of such an approach is, that only systems with technologies that meet the defined security requirements are allowed to be used [14]. It has to pass the threat analysis and risk assessment for individual components over systems to entire industries [2]. Due to this approach, security is transformed from a subordinate, retrospective issue to an integrated topic in the development process [5].

3 Challenges of IT-Security in Industrie 4.0 in the German Industrial Landscape

In order to cope with the complexity in Industrie 4.0, several organizations have created guidance for different stakeholders. While the leading institution for promoting German Industrie 4.0 activities, the Platform Industrie 4.0, has announced 17 technology development areas and an implementation roadmap for describing the vision of Industrie 4.0 [15], the Mechanical Engineering Industry Association (VDMA), who represents over 3000 mostly medium-sized companies, published the "Guideline Industrie 4.0" to set up the "Guiding principles for the implementation of Industrie 4.0 in small and medium-sized businesses". The core method for this is the Industrie 4.0 Toolbox Product and Industrie 4.0 Toolbox Production [16].

In addition, the "Generic Procedure Model to introduce Industrie 4.0 in Small and Medium-sized Enterprises" (GPMI4.0) performs in four different project formats. First, a holistic company-specific project format over 1 year with a focus on developing specific solutions and to subsequently implement these solutions successfully in a real production environment [17]. Second, a workshop concept to support the enterprise in generating their own framework to implement Industrie 4.0 [16]. Third, regular competence-building events focusing on knowledge transformation from research projects to industrial approach with respect to Industrie 4.0. Fourth, coaching event for trainers, which imparts the methods and procedures to develop a corporate Industrie 4.0 workshop. Accordingly, there is huge supply and demand in the German market.

Since 2011, when Industrie 4.0 was officially presented during the 2011 Hannover Messe Industrie (HMI) fair, both global players and small and medium-sized enterprises (SME) recognize and identify the potential of Industrie 4.0 applications and its possibilities and benefits with the support and assistance by public or private offers. Enterprises always build specific implementation roadmaps after consuming these initial supports and assistance offers. Furthermore, their present production systems or subsystems are analyzed, existing system processes are modeled and ideas to optimize them resiliently or upgrade mechanic systems over mechatronic systems over adaptronic systems to cyber-physical systems are generated. However, IT security issues have been discussed, but have not yet been detailed in secure networks, secure processes, secure services, and secure data. In the context of Industrie 4.0, an integrated concept is needed, which combines the requirements from manufacturing automation, mechanical engineering, process engineering and the properties of cyber-physical systems with well-established core elements of IT security descriptions. These results in:

A. Diversity of Systems and Processes (R1)

Industrie 4.0 includes a large number of different, networked systems with many actors and components. These are all involved in several processes throughout the entire product lifecycle [2]. An approach to creating an Industrie 4.0 Toolbox IT-Security is designed to meet these requirements.

B. Defense in Depth Strategy (R2)

The Defense in Depth strategy is based on the idea of achieving an acceptable level of security for industrial systems not only with a single protective measure but with several adjunct and layered measures. The tendency is to achieve the targeted security level by the combined use of several protective measures of a considered system or component, so that the attack becomes more complicated to execute, and that damage potential of a successful attack is minimized [5, 18].

C. Threat Analysis and Risk Assessment (R3)

Threat analysis and risk assessment must be taken as the basis for the procedure in the development of security solutions. In order to ensure the security of the production and the produced product within the Industrie 4.0, the production processes and applications of the produced product must be taken into account [5, 7, 10].

D. Application-Specific Security Solutions (R4)

The type and level of security measures should vary based on the threat situation, the value of the considered assets, as well as the present process protection requirements. It should not be defined uniformly [5, 7].

E. Usability (R5)

For developing security procedures and measures, the user-friendliness, a holistic, extendable and simple approach and balanced implementation are the key factors. If the measures provide semi-automated assistance and are application-friendly, the user acceptance will be increased [5, 10].

4   Security Concept Cornerstones

Most enterprises in the industrial landscape would agree that they are severely limited to solve the described challenges and the requirements of IT security in Industrie 4.0. Having taken all these factors into account a process model, which consists of a data model and an algorithm as its core element is developed to consider the challenges and cope with the requirements.

After giving a brief overview of the developed process model in Sect. 4.1, the data model will be described first in Sect. 4.2, the algorithm with its main methods will be explained in detail in Sect. 4.3 and a prototypical implementation of an Industrie 4.0 Toolbox IT-Security will be explained in detail in Sect. 4.4 [1].

4.1 Process Model Level

As shown in Fig. 1, on the process model level the developed algorithm accepts
different industrial systems as input. Further threat analysis and risk assessments are
finalized for the input system. The algorithm acquires and filters security measures
from a database. It eliminates security measures that are irrelevant for the given
system based on its identified, significant characteristics.

As an output, the algorithm generates a customized catalog of optional security
measures represented in a comprehensive framework, the so-called Industrie 4.0
Toolbox IT-Security. The algorithm is based on the developed data model and
suggests suitable security measures for the input system. As the core element of the
process model level, the algorithm is designed using pseudo code for generating the
Industrie 4.0 Toolbox IT-Security.

Due to a large number of systems and processes within the Industrie 4.0, creation
of a static Industrie 4.0 Toolbox IT-Security is not sufficient. A procedure oriented
Industrie 4.0 Toolbox IT-Security is expected to provide relevant security measures
depending on the system and application. These security measures and their
properties are presented in a technological Industrie 4.0 Toolbox IT-Security, as the
result of the whole process and data model.

In the following, the components of the process model level will be explained,
starting with the database, followed by the system and ends with the toolbox [1].

A. Database

The database is essential for the whole model as it stores all important data that was
identified by the data model and is significant for the process model itself.

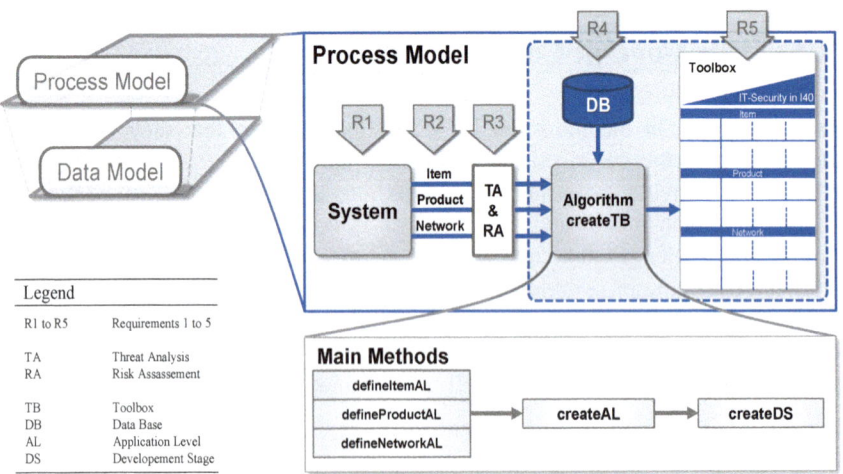

Fig. 1 Process model and data model as security concept cornerstones

Information about system details, security attacks and threats and their correlation to compatible security measures are combined into this database so that the algorithm can use this information to create the expected results. In order to maintain its impact, this database must always be updated with any new systems, arising cyber-attacks and their corresponding security measures. After generating a specific Industrie 4.0 Toolbox IT-Security, it would be likewise stored in the database and therefore can be used or updated for similar systems without running through the whole evaluation again.

B. System as Input

The input in the process model is an arbitrary system or subsystem that needs to be secured. A system in this context consists of one or more components as hardware or/and software that interact together, building one whole system that is integrated into any industrial process. These are mostly CPS but not exclusively. To enable the algorithm to work with this system, it must be analyzed and represented according to the developed data model. Therefore, the system is divided into three main categories: items, products and networks, which will be referring as components. More information about these three categories will be provided in the section Data Model Level.

C. Industrie 4.0 Toolbox IT-Security as Output

The Industrie 4.0 Toolbox IT-Security is an informative framework to represent the results of the foregoing analysis process of the given system as the output of the presented algorithm. It consolidates important information including characteristics of the given system and Industrie 4.0 along with the properties of risk assessment and threat model. The generated toolbox does not recommend the optimal solution for the given system nor can it replace the whole role of an IT expert of implementing a security measure, as it requires mostly more complex analysis and examination of more properties. More likely it is an approach to reduce the efforts and time invested in such a process enabling people with limited knowledge in IT security to take first steps towards designing more secure systems without having to rely on external IT security expertise right from the start.

4.2 Data Model Level

The various properties of Industrie 4.0 and IT security are aggregated in the developed data model and represented in its classes and attributes. This data model is represented as a UML class diagram in Fig. 2. To give a simplified overview of the data model only classes are presented in this chapter. The operations each class are defined in prospect to the algorithm. The attributes are significant characteristics that cover necessary information of each class and prepare them for the operations

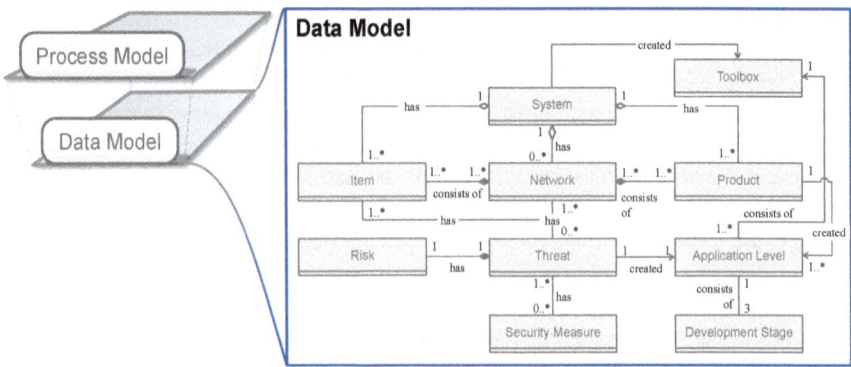

Fig. 2 Detailed data model level

to use them while running the algorithm. Both attributes and operations will not be described any further [1].

The first class is the **Class System**. As explained earlier, a given system is the input for the algorithm and is divided into three groups: item, product and network.

Any given system consists of at least one item, i.e. a server, a machine or simply confidential data. Hence an item object is any item that can't be divided into smaller items or where sub items are not necessarily to be secured individually. This way, not only a system is secured but also different assets are secured to fulfill the requirement of secure data. Most objects of this class are items with known threats and corresponding security measures.

A *product object* is comparable to the *item object* yet it is defined as a separate class. The **Class Product** describes only new and innovative products, wherefore no or only limited security-related information are established. Both, *item* and *product objects*, would both be parts of the same industrial process in case there are more than one item and an innovative product in the given system that should be secured as well.

The reason why it is necessary to divide the system into three categories is to adopt the defense in depth strategy by securing each component standing alone and as part of a network. Therefore, the **Class Network** is necessary to represent the interaction of different items in one system.

Hereby, relevant information about each class are stored to enable the measures-filtering process. Especially, attributes like technical requirements reflecting the industrial process, common threats, security goals and the whole use case of the system, are very important. Furthermore, properties of the networks like for example its topology structure or its transmission medium are also stored. According to these attributes amongst other the algorithm is enabled to search for suitable security measures.

To secure an *item* or *network object* the **Class Threat** is defined. This class holds specific information about a well-known threat or cyber-attack that is related to an *item* or *network object*. Furthermore, it is related to two other classes: **Class risk** and **Class security** measure. The risk resulting from a threat is evaluated so that the component is secured depending on its risk level.

The **Class Security Measure** is likewise related to threats and corresponding security goals, items and networks. Most security measures can be deployed for more than one threat case and thus for different items or networks. Therefore, a *security measure object* has many attributes describing when and how it can be used along with its advantages and disadvantages in a structured and a unified way. In addition to that, each security measure is evaluated in regard to costs, user-friendliness, latency and security strength. These properties are key factors while choosing between many optional security measures as this choice will differ conforming to the use case and preferences of the user. Especially, when talking about Industrie 4.0 the latency factor is very determining. Many industrial processes are critical, real-time processes and hence it's not tolerable that a security measure cause any latency within such critical processes.

The strength of a security measure is defined as its ability to withstand a direct attack [19]. A user can hereby weigh between how secure he wants his system and components to be and the worth of his system in terms of costs and effort. The result of the user's preferences concerning a component is hence added to the object of this component.

With **Class Threat, Class Risk** and **Class Security** measure along with their attributes the requirements resulting from the threat modeling and risk assessment process are fulfilled. Whereas **Class System, Class Item, Class Product** and **Class Network** aim to represent industrial systems and processes. To represent the result of the security process within the *Industrie 4.0 Toolbox IT-Security* the **Class Toolbox, Class Application Level** and **Class Development Stage** are defined. This model, describes a *toolbox object* consisting of horizontally arranged *application level objects*. Each *application level object* is defined by a component, namely an item, a product or a network. Furthermore, it is also defined by a threat and the targeted security goal which together describes an application level of a cyber-attack and its properties. An *application level object* is composed of three security measures each represented as a *development stage object*.

4.3   Algorithm as Core Element

The algorithm in this model is also designed to support the Defense in Depth strategy by securing a system and its components. The input of the algorithm is hence a system categorized into items, products and networks. The output is a collection of suggested security measures and some of its properties represented in an *Industrie 4.0 Toolbox IT-Security* that is conformingly divided into three categories. The algorithm contains three main operations that are responsible for each of

```
1   PROCEDURE createTB (item, product, network)
2       TB := defineItemApplicationLevel (item)
3       TB := defineProductApplicationLevel (item)
4       TB := defineNetworkApplicationLevel (item)
5   END PROCEDURE
```

Listing-1 Main method for creating Industrie 4.0 Toolbox IT-Security

the three categories (Listing-1). To give a brief overview, the scope of describing the algorithm is limited to its important parts of the pseudo code.

The operation *defineProductApplicationLevel* generates the application level for the products of a given system. The approach for securing a product differs from the approach adopted for items or networks due to the lacking experience with this innovation. This algorithm doesn't look for well-known threats or cyber-attacks as they are mostly not available for an innovative product, but it rather tries to categorize the product to a technology level. Determined by the identified technology level the algorithm suggests a suitable and rather general security measure. After gaining more experience with this product, it can then be considered as an *item object*.

The operations *defineItemApplicationLevel* and *defineNetwork-ApplicationLevel* perform similar steps to generate the application level for their components. Each operation goes sequentially through the list of components of a given system. For each component, it identifies its security goals and subsequently it identifies the threats related to a security goal. Based on the assets of the component item or network, the process in which it is involved and the properties of this specific threat the risk associated with this threat can be evaluated. Based on the assets of the component item or network, the process in which it is involved and the properties of this specific threat the risk associated with this threat can be evaluated. To avoid unnecessary efforts and expenses items or networks are secured according to their risk level, so that low-risk threats are sorted out without searching for suitable security measures or any further actions. After filtering the threats, the algorithm starts searching for security measures in the database that matches the user's requirements of the component and the network properties for the case that the component in consideration is a network and not an item.

After identifying suitable security measures for a specific application level, each of the above-described operations calls the operation *createApplicationLevel* given the identified collection of security measures and the user's preferences concerning the component (Listing-2). A collection of security measures is categorized into three different levels analogous to the three levels of security strengths described earlier this section. For each category, the algorithm searches for a security measure that matches the user's preferences. For example, in some cases, the user's priority throughout the security process is to deploy economical or user-friendly measures

```
1    PROCEDURE createApplicationLevel (component, measures, featuresPrio)
2          if component == Product then
3                name := setProductALName (TBProductAL)
4          else
5                name := seALName (component, protectionTarget, threat)
6          end if
7          measuresCategories := sortMeasures (measure)
8          for all (measuresCategories) do
9                optimalMeasure := getOptimalMeasure (measuresCategories, featuresPrio)
10               developmentStage =: createDevelopmentStage (optimalMeasure)
11         end for return applicationLevel
12   END PROCEDURE
```

Listing-2 Main method for creating development stage

and thus the security measures would be selected in correspondence to the user's preferences.

The best suitable security measure is then selected and given to the operation *createDevelopmentStage* as an input value. This operation is responsible for generating a new *development stage object* based on the information derived from the security measure (Listing-3).

As stated earlier in this chapter, a security measure is evaluated by its security strength, so that the development stage is sorted accordingly. Furthermore, the other three evaluation factors of a security measure are explicitly represented in a *development stage object*, so that the user is provided with enough information about the selected security measure. In addition, technical requirements of the component that couldn't be fulfilled by the suggested security measure are also explicitly mentioned.

```
1    PROCEDURE createDevelopmentStage (optimalMeasure)
2          stage := setStage (optimalMeasure)
3          stageName := setStageName (optimalMeasure)
4          if further Measures != EMPTY then
5                Point to further Measures furtherMeasures
6          end if
7          stageRealtime := setRealtime (optimalMeasure)
8          stageCosts := setCosts (optimalMeasures)
9          stageEffort := setEffort (optimalMeasures)
10         if missingRequirements != EMPTY then
11               Do not point to fulfilled Requirements missingRequierements
12         end if return developmentStage
13   END PROCEDURE
```

Listing-3 Main method for creating development stage

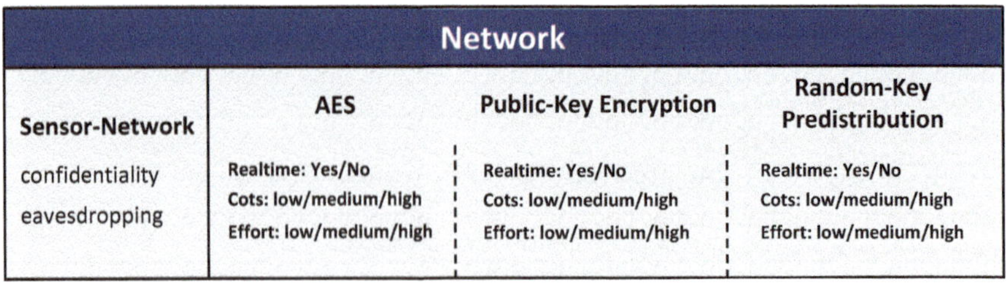

Network			
Sensor-Network	**AES**	**Public-Key Encryption**	**Random-Key Predistribution**
confidentiality eavesdropping	Realtime: Yes/No Cots: low/medium/high Effort: low/medium/high	Realtime: Yes/No Cots: low/medium/high Effort: low/medium/high	Realtime: Yes/No Cots: low/medium/high Effort: low/medium/high

Fig. 3 Example of an application level: Sensor-Network

4.4   A Prototypical Implementation of an Application Level in the Industrie 4.0 Toolbox IT-Security

The basic concept to generate an Industrie 4.0 Toolbox IT-Security for a given system is developed. This approach can be classified as an informative intermediate solution for users with no or little IT security knowledge by representing the solution and the relevant characteristics in an illustrative way. The structure of the Industrie 4.0 Toolbox IT-Security is defined in application levels arranged vertically and development stages arranged horizontally. An application level displays IT security themes in Industrie 4.0, where every application level is broken down into different technological and sequential stages. The highest stage represents the highest IT security measures. As an example, for one variation of application levels, which results from the call of the operation *createApplicationLevel* the application level a network includes the component *Sensor-Network*, the protection Target *confidentiality* and the threat *eavesdropping* as is shown in Fig. 3. The development stages *AES*, *Public-Key Encryption* and *Random-Key Predistribution* with their properties *realtime*, *costs* and *effort* will be returned by calling the operation *createDevelopmentStage*. In addition, these properties will rate as *yes/no* or *low/medium/high*.

5   Conclusion and Future Work

The change of the industrial landscape is described by introducing the horizontal and vertical integration of Industrie 4.0 and the digital enterprises. Consequently, a security culture is one of the key prerequisites for Industrie 4.0. The production IT faces many forms of challenges like the diversity of its characteristics, the enterprise chain of security measures, and the coming complex networked circumstances of Industrie 4.0.

After characterizing the approaches to IT security strategy and design in Industrie 4.0, different aspects were considered and analyzed: Defense in Depth for present systems, Security by Design for new systems as well as the Challenges of

IT security in Industrie 4.0 for the German industrial landscape. These resulted in five requirements. To cover these requirements a process model, consisting of a data model and an algorithm as its core element is developed. As a prototypical implementation and a result of the whole process and data model, security measures and their properties are presented in a technological Industrie 4.0 Toolbox IT-Security.

References

1. Y. Wang, R. Fakhry, R. Anderl, Combined secure process and data model for IT-Security in Industrie 4.0, in *Proceedings of the International MultiConference of Engineers and Computer Scientists 2017*. Lecture Notes in Engineering and Computer Science, (Hong Kong, 15–17 Mar 2017), pp. 846–852
2. H. Kagermann, W. Wahlster, J. Helbig, Recommendations for implementing the strategic initiative Industrie 4.0—securing the future of German manufacturing industry, Apr 2013
3. E. Abele, R. Anderl, J. Metternich, A. Wank, O. Anokhin, A. Arndt, T. Meudt, M. Sauer, Effiziente Fabrik 4.0—Einzug von Industrie 4.0 in bestehende Produktionssysteme, published in ZWF—Zeitschrift für wirtschaftlichen Fabrikbetrieb, pp. 150–153 (2015)
4. BMBF, Zukunftsbild Industrie 4.0 (2014). http://www.bmbf.de/pubRD/Zukunftsbild_Industrie_40.pdf. Accessed 17 Nov 2016
5. BITKOM, VDMA, ZVEI, Umsetzungsstrategie Industrie 4.0—Ergebnisbericht der Plattform Industrie 4.0, April 2015
6. R. Anderl, Industrie 4.0—technological approaches, use cases, and implementation, published in at—Automatisierungstechnik, vol. 63, pp. 753–765 (2015)
7. BMWi, IT-Sicherheit für Industrie 4.0 (2016). http://www.bmwi.de/BMWi/Redaktion/PDF/Publikationen/Studien/it-sicherheit-fuer-industrie-4-0-. Accessed 30 Nov 2016
8. T. Bauernhansl, M. ten Hompel, B. Vogel-Heuser, *Industrie 4.0 in Produktion, Automatisierung und Logistik. Wiesbaden* (Springer Fachmedien Wiesbaden, 2014)
9. BMWi, Autonomik für die Industrie 4.0 (2016), http://www.autonomik.de/de/1003.php. Accessed 10 July 2016
10. K. Böttinger, B. Filipovic, M. Hutle, S. Rohr, Leitfaden Industrie 4.0 Security—Handlungs-empfehlungen für den Mittelstand. Frankfurt am Main, VDMA (2016)
11. T. Phinney, IEC 62443: Industrial Network and System Security, Published in ISA (2016). http://www.isa.org/autowest/pdf/Industrial-Networking-and-Security/Phinneydone.pdf. Accessed 10 Nov 2016
12. D. Kuipers, M. Fabro, *Control Systems Cyber Security: Defense in Depth Strategies—Recommended Best Practice: Defense in Depth, INL-Idaho National Laboratory* (2006)
13. Homeland Security, Recommended Practice: Improving Industrial Control Systems Cybersecurity with Defense-In-Depth Strategies—Control Systems Security: National Cyber Security Division (2009)
14. A. Cardenas, S. Amin, B. Sinopoli, A. Giani, A. Perrig, S. Sastry, Challenges for Securing Cyber Physical Systems (2016), https://chess.eecs.berkeley.edu/pubs/601/. Accessed 10 Nov 2016
15. Plattform Industrie 4.0, Neue Chancen für unsere Produktion—17 Thesen des Wissenschaftlichen Beirats der Plattform Industrie 4.0 (2016), http://www.its-owl.de/fileadmin/PDF/Industrie_4.0/Thesen_des_wissenschaftlichen_Beirats_Industrie_4.0.pdf. Accessed 19 Nov 2016

16. R. Anderl, A. Picard, Y. Wang, J. Fleischer, S. Dosch, B. Klee, J. Bauer, *Guideline Industrie 4.0—Guiding principles for the implementation of Industrie 4.0 in small and medium sized businesses*, *VDMA Forum Industrie 4.0*, Frankfurt (2015). ISBN: 978-3-8163-0687-0
17. Y. Wang, G. Wang, R. Anderl, Generic Procedure Model to Introduce Industrie 4.0 in Small and Medium-sized Enterprises, in *Proceedings of the World Congress on Engineering and Computer Science 2016*. Lecture Notes in Engineering and Computer Science (San Francisco, USA, 19–21 Oct 2016), pp. 971–976. ISBN: 978-988-14048-2-4
18. K. Stouffer, J. Falco, K. Scarfone, Guide to Industrial Control Systems (ICS) Security—NIST Special Publication 800–82. Supervisory Control and Data Acquisition (SCADA) systems, Distributed Control Systems (DCS), and other control system configurations such as Programmable Logic Controllers (PLC): Recommendations of the National Institute of Standards and Technology. http://citeseerx.ist.psu.edu/viewdoc/summary?doi=10.1.1.353.2376. Accessed 15 Nov 2016
19. Bundesamt für Sicherheit in der Informationstechnik, IT-Sicherheitskriterien (1989). https://www.bsi.bund.de/DE/Themen/ZertifizierungundAnerkennung/Produktzertifizierung/ZertifizierungnachCC/ITSicherheitskriterien/DeutscheITSicherheitskriterien/dtitsec.html. Accessed 19 Nov 2016

Further In-Depth Investigation of Precuing Effect on Tracking and Discrete Dual-Task Performance with Different Spatial Mappings

Ken Siu Shing Man, Steve Ngai Hung Tsang, Coskun Dizmen and Alan Hoi-Shou Chan

Abstract The objective of this study was to establish data for making ergonomics recommendations for human-machine interface design in control consoles. The results showed significant precuing effect and spatial S-R compatibility effect, leading to different response times of the participants. Compatibility in both orientations (BC condition) resulted in a better reaction time. The performance of the primary tracking task was subject to its level of difficulty. For secondary discrete task, there was a significant effect of S-R mapping condition. There was also a significant interaction between S-R mapping conditions and the position of signal and that of precue. The same position between signal and precue under BC and TC S-R mapping led to faster reaction times, while the opposite position between signal and precue under BI and LC S-R mapping exhibited faster reaction times. Tactile precue and auditory precue can always improve the performance in reaction time for four S-R mapping conditions with visual stimulus while the visual precue worsens the performance.

Keywords Auditory · Dual-task paradigm · Precue · Spatial stimulus-response compatibility · Stimulus onset asynchronies Tactile · Visual

K. S. S. Man (✉) · S. N. H. Tsang · C. Dizmen · A. H.-S. Chan
Department of Systems Engineering and Engineering Management,
City University of Hong Kong, 83 Tat Chee Avenue, Kowloon, Hong
Kong SAR, China
e-mail: ssman6-c@my.cityu.edu.hk

S. N. H. Tsang
e-mail: nhtsang2-c@my.cityu.edu.hk

C. Dizmen
e-mail: coskun_dizmen@hotmail.com

A. H.-S. Chan
e-mail: alan.chan@cityu.edu.hk

© Springer Nature Singapore Pte Ltd. 2018
S.-I. Ao et al. (eds.), *Transactions on Engineering Technologies*,
https://doi.org/10.1007/978-981-10-7488-2_3

31

1 Introduction

Displays and controls provide a fundamental communication medium in man-machine system. In certain situations, people always choose one preferred response to a stimulus. This well-established phenomenon can be described as the term 'population stereotype'. It is proposed that an excellent pairing of response sets (control devices) and stimulus sets (displays) exists. The probability with which a response is chosen can be used to express population stereotypes. However, the speed and accuracy of response can be used to describe stimulus-response (S-R) compatibility. Faster reaction times (RTs) and lower error rates result from compatible pairings in general. In other words, the human performance can be improved by using compatible pairings. It is believed that lower recoding demands and higher information transfer rates during responses lead to the higher efficiency and accuracy of compatible S-R combination [1]. For a reduction of human error and response time, the design of control and display configurations should conform to the expectations of the relevant user population. It has proven that precuing the spatial location of an impending target stimulus can facilitate response performance, and such precuing benefit exists for both intra-modal and cross-modal precue-target pairings [2–4]. It was generally expected that when the precue and target occur on the same side (valid/ipsilateral precuing), faster and more accurate responses were obtained as opposed to their occurrence on the opposite side (invalid/contralateral precuing) [5]. Such precuing effects were anticipated to exist between all possible combinations of visual, auditory and tactile precue-target pairings [6, 7]. It is also important to note that the precue-target stimulus onset asynchronies (SOAs) may pose a significant impact on the precuing effects. Past studies had shown that when the SOAs were longer than 300 ms, inhibition of return (IOR) may be resulted and attention is inhibited from returning to the previously attended (cued) location [6, 8], leading to an increase in response times. Most of the previous studies on precuing effects however were in a single-task paradigm that participants were asked to fixate to a central fixation cross and respond to a target stimulus after the presentation of an endogenous or an exogenous precue. There have been few studies examining the precuing effects in a dual-task paradigm in which participants have to primarily focus to a dynamic moving target while eliciting responses to a side task [9], let alone the inclusion of spatial compatibility effects in the side task. Given the importance of spatial compatibility in human-machine interface design, it is crucial to understand how the location of a precue and the different precue-target combinations affect the subsequent responses under different spatial stimulus-response mapping conditions. With the ever-increasing complexity of real-world working environments where operators normally require multitasking and shifting attention among different modalities, it is of paramount importance to advance human-machine system design to facilitate the task performance of operators in such complex environments. Thus, it is necessary to have a thorough understanding of the cross-modal attention shifts among visual, auditory and tactile modalities, so as to design a sophisticated multimodal warning system that can effectively capture the operator's attention and hence to improve overall task performance.

2   Method

2.1   Participants

Thirty six Chinese students (21 males and 15 females) aged between 20 and 24 (median = 20) from City University of Hong Kong participated in this experiment. Only one student was left-handers, while the others were right-handers. All of them had normal or corrected-to-normal vision and normal color vision and did not report any physical or health problems involving their hands. Informed consent was obtained from each participant before the start of the experiment.

2.2   Apparatus and Design

In this experiment, a dual-task paradigm with a primary pursuit manual tracking task and a secondary spatial S-R compatibility (SRC) task was employed for testing, using a computer with a 17-in. LCD monitor. Because of the limitation of space, readers may refer to reference [10] for the details of the design.

3   Results

3.1   Primary Task—Root Mean Square Tracking Error (RMSTE)

For three tracking difficulty levels, 36 (36 participants × 1 conditions) RMSTEs were obtained. The range of the values was between 16.31 and 70.22 pixels. The average RMSTEs and standard deviations under the three difficulty levels are summarized as Table 1. Of the three difficulty levels, the average RMSTE for low difficulty level was the lowest (24.74 pixels), while that for high difficulty level was the highest (45.95 pixels). The RMSTE for low difficulty level was 32.43% and 85.75% lower than that for medium difficulty level and high difficulty level respectively.

For three stimulus onset asynchronies, 108 (36 participants × 3 conditions) RMSTEs were obtained. The range of the values was between 14.245 and 80.10

Table 1 Average root mean square tracking errors (RMSTEs) and standard deviations (SDs) for different difficulty levels

Tracking difficulty levels	Average RMSTE (pixel)	SD (pixel)
Low	24.74	7.53
Medium	32.76	6.00
High	45.95	11.45

pixels. One (0.93%) outlier RMSTEs beyond the upper control limit ($+3\sigma$) of 72.58 pixels were excluded from analysis. The average RMSTEs and standard deviations under the three stimulus onset asynchronies are summarized as Table 2. Of the three stimulus onset asynchronies, the average RMSTE for SOA of 600 ms was the lowest (32.83 pixels), while that for SOA of 1000 ms was the highest (35.51 pixels). The RMSTE for SOA of 600 ms was 0.652% and 8.15%, lower than that for SOA of 200 ms and SOA of 1000 ms respectively.

For precues on the ipsilateral and contralateral sides, 72 (36: ipsilateral and 36: contralateral) RMSTEs were obtained. The range of the values was between 15.91 and 71.82 pixels. The average RMSTEs and standard deviations under the two position between signal and precue are summarized as Table 3. Of the two positions between signal and precue, the average RMSTE for the ipsilateral precue was lower (33.72 pixels) than that for the contralateral precue (34.89 pixels), contributing to 3.45% variations between the two precue positions.

For three types of precue, 36 (36 participants $\times$ 1 conditions) RMSTEs were obtained. The range of the values was between 16.31 and 70.22 pixels. The average RMSTEs and standard deviations under the three types of precue are summarized as Table 4. Of the three types of precue, the average RMSTE for tactile precue was the lowest (32.16 pixels), while that for auditory precue was the highest (36.54 pixels). The RMSTE for tactile precue was 13.61% and 6.94% lower than that for auditory precue and visual precue respectively.

Table 2 Average root mean square tracking errors (RMSTEs) and standard deviations (SDs) for different stimulus onset asynchronies

Stimulus onset asynchronies (ms)	Average RMSTE (pixel)	SD (pixel)
200	33.05	12.70
600	32.83	10.43
1000	35.51	12.04

Table 3 Average root mean square tracking errors (RMSTEs) and standard deviations (SDs) for different positions between signal and precue

Positions between signal and precue	Average RMSTE (pixel)	SD (pixel)
Ipsilateral	33.72	12.72
Contralateral	34.89	12.67

Table 4 Average root mean square tracking errors (RMSTEs) and standard deviations (SDs) for different types of precue

Types of precue	Average RMSTE (pixel)	SD (pixel)
Auditory	36.54	13.79
Tactile	32.16	10.72
Visual	34.40	13.60

Further examination of RMSTEs was performed with analysis of variance (ANOVA). The main factors considered were S-R mapping condition (BC, BI, LC and TC), tracking difficulty level (low, medium and high), stimulus onset asynchrony (200, 600 and 1000 ms), position between signal and precue (ipsilateral and contralateral), and type of precue (Auditory, Tactile, Visual). The results (Table 5) showed significant tracking difficulty level effect [$F(2, 70) = 18.46$, $p < 0.0001$]. The stimulus onset asynchrony ($p > 0.05$), position between signal and precue ($p > 0.05$), and type of precue ($p > 0.05$) were not significant for RMSTE. There was no significant two-way interaction.

Table 5 Result of ANOVA performed on the average root mean square tracking errors (RMSTEs)

Sources	Sum of squares	df	Mean square	F	Sig.
Main effect					
Mapping_condition	241.99	3	80.66	0.51	0.68
Tracking difficulty level	2753.41	2	1376.71	18.46	0.00*
SOA	158.33	2	79.17	0.54	0.58
Position between signal and precue	24.39	1	24.39	0.15	0.70
Precue	114.95	2	57.47	0.35	0.71
Two-factor interaction					
Mapping_condition × SOA	269.17	6	44.86	1.05	0.39
Mapping_condition × precue	138.41	6	23.07	0.66	0.69
Mapping_condition × difficulty level	178.21	6	29.70	0.91	0.50
Mapping_condition × position between signal and precue	21.28	3	7.092	0.29	0.83
Precue × SOA	51.53	4	12.88	0.75	0.57
Precue × difficulty level	166.31	4	41.58	0.44	0.78
Precue × position between signal and precue	15.88	2	7.94	0.89	0.43
Difficulty level × position between signal and precue	16.29	2	8.14	1.53	0.24
Difficulty level × SOA	52.03	4	13.01	0.95	0.44
SOA * position between signal and precue	114.75	2	57.38	2.98	0.06

Mapping_condition, Spatial stimulus-response compatibility; SOA, stimulus onset asynchrony; precue, type of precue; *$p < 0.0001$

3.2 Secondary Task—Reaction Time (RT)

For three tracking difficulty levels, Table 6 summarizes the mean reaction times computed for different tracking difficulty levels. The shortest mean reaction time (566 ms) was obtained from the low tracking difficulty level, while the longest mean reaction time (580 ms) was to medium tracking difficulty level. The mean reaction time for the low tracking difficulty level is smaller than that for medium tracking difficulty level (2.47%) and that for high tracking difficulty level (0.71%).

For three stimulus onset asynchronies, Table 7 summarizes the mean reaction times computed for different stimulus onset asynchronies. The shortest mean reaction time (568 ms) was obtained from stimulus onset asynchrony of 200 ms, while the longest mean reaction time (575 ms) was to stimulus onset asynchrony of 1000 ms. The mean reaction time for stimulus onset asynchronies of 200 ms is smaller than that for stimulus onset asynchrony of 600 ms (0.88%) and that for stimulus onset asynchrony of 1000 ms (1.23%).

For two positions between signal and precue, Table 8 summarizes the mean reaction times computed for different positions between signal and precue. The shorter mean reaction time (570 ms) was obtained from the signal and precue on the same side, while the longer mean reaction time (574 ms) was to the between signal and precue on the opposite side. The mean reaction time for the signal and precue

Table 6 Mean reaction times (RTs) and standard deviations (SDs) for different tracking difficulty levels

Tracking difficulty levels	Mean RT (ms)	SD (ms)
Low	566	131
Medium	580	123
High	570	127

Table 7 Mean reaction times (RTs) and standard deviations (SDs) for different stimulus onset asynchronies

Stimulus onset asynchronies (ms)	Mean RT (ms)	SD (ms)
200	568	123
600	573	129
1000	575	130

Table 8 Mean reaction times (RTs) and standard deviations (SDs) for position between signal and precue

Positions between signal and precue	Mean RT (ms)	SD (ms)
Ipsilateral	570	128
Contralateral	574	127

Table 9 Mean reaction times (RTs) and standard deviations (SDs) for different types of precue

Types of precue	Mean RT (ms)	SD (ms)
Auditory	563	130
Tactile	541	116
Visual	613	125

on the same side is smaller than that for the signal and precue on the opposite side (0.70%).

For three types of precue, Table 9 summarizes the mean RTs computed for different types of precue. The shorter mean reaction time (541 ms) was obtained from tactile precue, while the longer mean reaction time (613 ms) was to visual precue. The mean reaction time for tactile precue is smaller than that for auditory (4.07%) and that for visual precue (13.31%), while the mean reaction time for auditory precue is smaller than that for visual (8.88%).

Further examination of reaction times was performed with analysis of variance (ANOVA). The main factors considered were spatial stimulus-response mapping condition (BC, BI, LC and TC), tracking difficulty level (low, medium and high), stimulus onset asynchrony (200, 600 and 1000 ms), position between signal and precue (ipsilateral and contralateral), and type of precue (Auditory, Tactile, Visual). The results (Table 10) showed significant spatial stimulus-response mapping condition effect [$F(3, 6871) = 108.54$, $p < 0.0001$], tracking difficulty level effect [$F(2, 6872) = 7.91$, $p < 0.0001$] and type of precue effect [$F(2, 6872) = 203.88$, $p < 0.0001$]. The stimulus onset asynchrony and position between signal and precue ($p > 0.05$) were not significant. The significant two-way interaction were spatial stimulus-response mapping condition × type of precue [$F(6, 3450) = 2.81$, $p < 0.05$], spatial stimulus-response mapping condition × position between signal and precue [$F(3, 2589) = 2.99$, $p < 0.05$], type of precue × stimulus onset asynchrony [$F(4, 3068) = 3.47$, $p < 0.01$], type of precue × tracking difficulty level [$F(3.822, 2931.474) = 44.01$, $p < 0.001$] and tracking difficulty level × stimulus onset asynchrony [$F(3.954, 3015.16) = 7.41$, $p < 0.0001$].

Figure 1 shows the interaction plot of type of precue and S-R mapping condition. Among the twelve combinations, tactile precue under BC S-R mapping condition resulted in the fastest reaction time (501 ms), while the reaction time for visual precue under TC S-R mapping condition was the slowest (645 ms). From the interaction plot, the four combinations with tactile precue always exhibited faster reaction times than other combinations, implying that if tactile precue was used, the reaction times elicited were always the shortest (Fig. 1). Among the three incompatible cases, incompatibility in the longitudinal orientation (TC) always led to the longest reaction time. The relatively slower performance in this condition indicated that a reverse in front-rear position of the stimulus-key relation was the most confusing to the participants and additional translation time was required for them to recognize the correct responses. Comparing with the previous study [11], this study showed that the mean reaction times for BC, BI, LC and TC S-R mapping

Table 10 Result of ANOVA performed on the average root mean square tracking errors (RMSTEs)

Sources	Sum of squares	df	Mean square	F	Sig.
Main effect					
Mapping_condition	5.03	3	1.68	108.54	0.00*
Tracking difficulty level	0.26	2	0.13	7.91	0.00*
SOA	0.06	2	0.03	1.82	0.16
Position between signal and precue	0.02	1	0.02	1.48	0.22
Precue	6.23	2	3.12	203.88	0.00*
Two-factor interaction					
Mapping_condition × SOA	0.11	5.92	0.02	1.66	0.13
Mapping_condition × precue	0.20	5.94	2.81	2.81	0.01
Mapping_condition × difficulty level	0.15	6	0.03	2.09	0.51
Mapping_condition × position between signal and precue	0.10	3	0.03	2.99	0.03
Precue × SOA	0.16	4	0.04	3.47	0.01
Precue × difficulty level	2.41	3.82	0.63	44.01	0.00*
Precue × position between signal and precue	0.05	2	0.03	2.41	0.09
Difficulty level × position between signal and precue	0.01	2	0.01	0.53	0.59
Difficulty level × SOA	0.33	3.95	0.08	7.41	0.00*
SOA × position between signal and precue	0.02	2	0.01	0.97	0.38

Mapping_condition, Spatial stimulus-response compatibility; SOA, stimulus onset asynchrony; precue, type of precue; $*p < 0.0001$

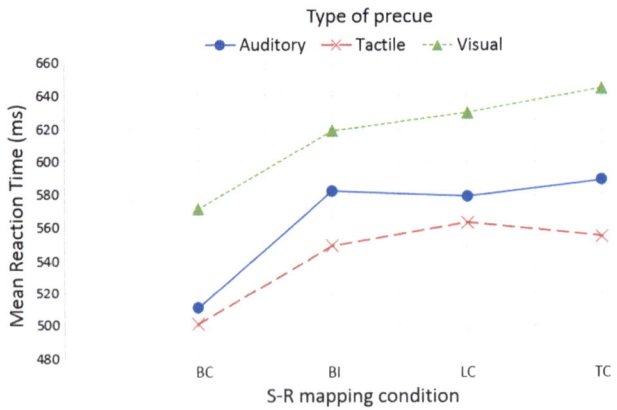

Fig. 1 Interaction plot of mean reaction times (RTs) for the type of precue and S-R mapping condition

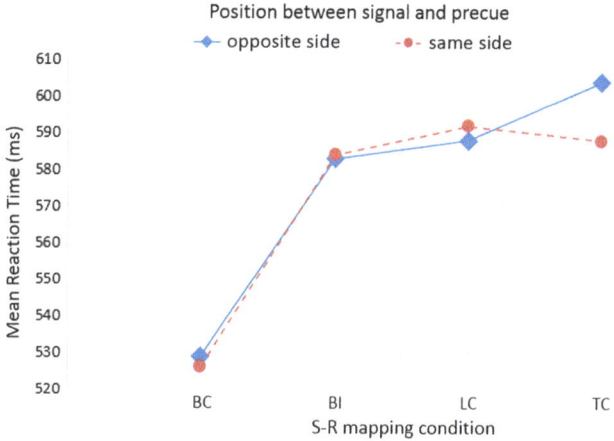

Fig. 2 Interaction plot of mean reaction times (RTs) for the position between signal and precue and S-R mapping condition

conditions with tactile precue was 10.978%, 14.572%, 11.723% and 16.216% faster than that for BC, BI, LC and TC S-R mapping conditions without tactile precue respectively. Moreover, the mean reaction times for BC, BI, LC and TC S-R mapping conditions with auditory precue were 8.806%, 8.076%, 8.636% and 9.508% faster than that for BC, BI, LC and TC S-R mapping conditions without auditory precue respectively.

Figure 2 shows the interaction plot of position between precue and signal and S-R mapping condition. Among the eight combinations, the same position between signal and precue under BC S-R mapping condition resulted in the fastest reaction time (526 ms), while the reaction time for the opposite position between signal and precue under TC S-R mapping condition was the slowest (604 ms). From the interaction plot, the same position between signal and precue led to faster reaction times than the contralateral precue under BC and TC S-R mapping, while the opposite position between signal and precue exhibited faster reaction times than the ipsilateral precue under BI and LC S-R mapping, implying that the position between signal and precue under different S-R mappings results in different reaction times (Fig. 2). Among the four S-R mapping cases, both transverse and longitudinal compatible often led to the fastest reaction times than others. Among the three incompatible cases, both transverse and longitudinal incompatible often led to faster reaction times than others.

3.3 Secondary Task—Response Error (RE)

For three tracking difficulty levels, the mean error percentages (EPs) computed for different tracking difficulty levels is summarized in Table 11. Among the three

Table 11 Mean error
percentages (EPs) and
standard deviations (SDs) for
different tracking difficulty
levels

Tracking difficulty levels	Mean (%)	SD (%)
Low	4.97	4.70
Medium	5.62	4.31
High	5.41	5.41

tracking difficulty levels, low tracking difficulty level exhibited the smallest EP (4.97%), while medium tracking difficulty level was the highest (5.62%). Because of the non-normal distribution of the EP data, the non-parametric Kruskall-Wallis test was conducted for further analysis. The main factor examined was tracking difficulty level. The results showed that the main factor of tracking difficulty level ($p > 0.05$) was not significant.

For three stimulus onset asynchronies, the mean error percentages (EPs) computed for different stimulus onset asynchronies is summarized in Table 12. Among the three stimulus onset asynchronies, stimulus onset asynchrony of 600 ms exhibited the smallest EP (4.39%), while stimulus onset asynchrony of 200 ms was the highest (5.49%). The non-parametric Kruskall-Wallis test was conducted for further analysis because the EP data was a non-normal distribution. The main factor examined was stimulus onset asynchrony. The results showed that the main factor of stimulus onset asynchrony ($p > 0.05$) was not significant.

For two positions between signal and precue, the mean error percentages (EPs) computed for positions between signal and precue is summarized in Table 13. Among the two positions between signal and precue, the opposite position between signal and precue exhibited smaller EP (3.51%), while the same position between signal and precue was greater (5.53%). Because the EP data was a non-normal distribution, the non-parametric Mann-Whitney test was conducted for further analysis. The main factor examined was position between signal and precue. The results showed that the main factor of position between signal and precue ($p > 0.05$) was not significant.

Table 12 Mean error percentages (EPs) and standard deviations (SDs) for different stimulus onset asynchronies

Stimulus onset asynchronies (ms)	Mean (%)	SD (%)
200	5.49	5.93
600	4.39	4.42
1000	4.84	4.84

Table 13 Mean error percentages (EPs) and standard deviations (SDs) for different positions between signal and precue

Positions between signal and precue	Mean (%)	SD (%)
Ipsilateral	5.53	5.53
Contralateral	3.51	2.66

Table 14 Mean error percentages (EPs) and standard deviations (SDs) for different types of precue	Types of precue	Mean (%)	SD (%)
	Auditory	7.18	5.67
	Tactile	2.86	2.95
	Visual	5.95	6.03

For three types of precue, the mean error percentages (EPs) computed for different types of precue is summarized in Table 14. Among the three types of precue, tactile precue exhibited the smallest EP (2.86%), while auditory precue was the highest (7.18%). Because the EP data was a non-normal distribution, the non-parametric Kruskall-Wallis test was conducted for further analysis. The main factor examined was type of precue. The results showed that the main factor of type of precue ($p > 0.05$) was not significant.

4 Discussion

4.1 Primary Task—Root Mean Square Tracking Error (RMSTE)

In previous special compatibility study of Tsang [11], the average root mean square tracking error (RMSTE) for responses to the compatible (BC) condition was significantly lower than that of the other three incompatible conditions. In contrast, the same result cannot be obtained in this study with the presence of precue. It was surprising that there was no significant difference among four mapping conditions with the presence of precue in this study. This diverse result between the two studies was speculated to result from the precuing effects. There was no obvious difference in root mean square tracking error (RMSTE) with different types of precue. This indicated that the presence of precue has no contributions to the performance of primary tracking task due to the function of precue reminding participant that the signal needed a response will appear very soon, accordingly it made participants utilize their spatial resource capacity for the secondary task.

4.2 Secondary Task—Reaction Time (RT)

The interaction between the S-R mapping condition and the position between precue and signal was significant. Figure 2 shows the interaction plot of position between precue and signal, and S-R mapping condition. It was an interesting result that the BC and TC S-R mapping conditions with the signal and precue on the same side resulted in the faster reaction time than those with signal and precue on the opposite side, while BI and LC S-R mapping conditions with signal and precue on the opposite side led to the faster reaction time than those with the signal and precue on the same side. In accordance with multiple resource model [12], it was believed that participant needs fewer mental resources used for perceptual activity and response selection under the BC and TC S-R mapping conditions with the signal and precue on the same side to improve the reaction time because of the same-side positions of signal, precue and response key. Likewise, it was believed that participant needs fewer mental resources used for perceptual activity and response selection under the BI and LC S-R mapping conditions with the opposite position between signal and precue to improve the reaction time due to the same-side positions of precue and response key were on the same side. It was obvious that there was a common condition where the positions of precue and response key were on the same side. The implication of this result was that the positions of precue and response key being the same side lead to a faster reaction time due to fewer mental resources used for perceptual activity and response selection.

Regarding the effects of the type of precue, the result showed that the performance in reaction time with tactile precue was the best (541 ms), followed by the auditory precue and then visual precue. This result was in line with the previous study [13]. Brebner and Welford [14] indicated that the variation of response across sensory modalities might be resulted from differences in the peripheral mechanisms such as some sensory systems are more sensitive than others. It was found that the temporal sensitivity of the skin was very high which was close to that of the auditory system and larger than that of the visual system. The tactile and auditory response time here were faster than the visual response time because the tactile and auditory stimuli are more sensitive than visual stimuli. It was interesting that there was a significant interaction between the type of precue and S-R mapping condition in reaction time. Figure 1 shows among the twelve combinations, tactile precue under BC S-R mapping condition resulted in the fastest reaction time (501 ms), while the reaction time for visual precue under TC S-R mapping condition was the slowest (645 ms). From the interaction plot, the four S-R mapping conditions with tactile precue always exhibited faster reaction times than those with auditory and visual precue, implying that if tactile precue was used, the reaction times elicited were always the shortest. In comparing with the previous study [11], it implicates that using tactile precue and auditory precue can always improve the performance in reaction time for four S-R mapping conditions when the stimulus was visual. According to multiple resource model [12], it was believed that tactile precue and auditory precue did not increase the usage of visual perception resources, thus leaving more visual perception resources for secondary task while visual precue does so to increase the reaction time.

4.3 Secondary Task—Response Error (EP)

In this experiment, there were only a few response errors and these were found to be non-significant with regard to the mapping conditions, tracking difficulty level, position between signal and precue, stimulus onset asynchrony and type of precue indicating participants did not respond to the spatial mapping task at the expense of accuracy, or, it may be that they emphasized response accuracy more than response time despite of being told to respond as fast and accurately as they could.

5 Conclusion

As a result of this study, the following recommendations were made with the aim of improving the design of human-machine interfaces requiring concurrent manual operation for a continuous task and a discrete choice response task.

(a) For the performance of reaction time in the secondary spatial S-R compatibility (SRC) task, using ipsilateral precue is beneficial under BC (both transverse and longitudinal compatible) and TC (transverse compatible and longitudinal incompatible), while using contralateral precue is advantageous under BI (both transverse and longitudinal incompatible) and LC (longitudinal compatible and transverse incompatible).

(b) In the secondary spatial S-R compatibility (SRC) task with visual stimulus, using tactile precue leads to a better performance in reaction time. If tactile precue is not feasible, auditory precue can also be used to improve the reaction time of participants.

Acknowledgements This work was supported by a grant from City University of Hong Kong (SRG7004663).

References

1. C. Umiltá, R. Nicoletti, Spatial stimulus-response compatibility, in *Stimulus-Response Compatibility: An Integrated Perspective*, ed. by R.W. Proctor, T.G. Reeve (Amsterdam, North-Holland, 1990), pp. 89–116
2. J. Driver, C. Spence, Crossmodal spatial attention: evidence from human performance, in *Cross-modal Space and Crossmodal Attention*, ed. by C. Spence, J. Driver (Oxford University Press, New York, 2004), pp. 179–220
3. A.B. Chica, D. Sanabria, J. Lupiáñez, C. Spence, Comparing intramodal and crossmodal cuing in the endogenous orienting of spatial attention. Exp. Brain Res. **179**(3), 353–364 (2007)
4. M.K. Ngo, R.S. Pierce, C. Spence, Using multisensory cues to facilitate air traffic management. Hum. Fact. **54**(6), 1093–1103 (2012)
5. M.I. Posner, Orienting of attention. Quart. J. Experiment. Psychol. **32**(1), 3–25 (1980)
6. R.M. Klein, Inhibition of return. Trends Cogn. Sci. **4**(4), 138–147 (2000)
7. C. Spence, Crossmodal spatial attention. Ann. N.Y. Acad. Sci. **1191**(1), 182–200 (2010)
8. J. Lupiáñez, Inhibition of return, in *Attention and Time*, ed. by A.C. Nobre, J.T. Coull (Oxford University Press, UK, 2010), pp. 17–34

9. T.K. Ferris, N.B. Sarter, Cross-modal links among vision, audition, and touch in complex environments. Hum. Fact. **50**(1), 17–26 (2008)

10. K.S.S. Man, C. Dizmen, S.N.H. Tsang, A.H.S. Chan, Influence of Precue on Spatial Stimulus-Response Compatibility Effect in a Dual-Task Paradigm, Lecture Notes in Engineering and Computer Science,in *Proceedings of The International MultiConference of Engineers and Computer Scientists* (2017), 15–17 March, 2017, Hong Kong, pp. 921–926

11. S.N.H. Tsang, Multi-task Performance in Processing Four-choice Spatial Stimulus-Response (S-R) Mappings: Implications for Multimodal Human-machine Interface Design, P.h.D dissertation, Dept. S.E.E.M., City University Hong Kong, Hong Kong (2014)

12. C.D. Wickens, Multiple resources and performance prediction. Theor. Iss. Ergonom. Sci. **3** (2), 159–177 (2002)

13. A.Y. Ng, A.H.S. Chan, Finger response times to visual, auditory and tactile modality stimuli, in *Proceedings of the International MultiConference of Engineers and Computer Scientists 2012*, IMECS 2012, 14–16 March, 2–12, Hong Kong, pp. 1449–1454

14. J.M.T. Brebner, A.T. Welford, Introduction: an historical background sketch, in *Reaction Times*, ed. by A.T. Welford (Academic Press, London, 1980), pp. 1–23

Improvement of Maximum Production in the Batch Transesterification Reactor of Biodiesel by Using Nonlinear Model Based Control

Arphaphon Chanpirak and Weerawun Weerachaipichasgul

Abstract To achieve a maximum production of biodiesel in the batch transesterification, an optimal operating condition and an effective control strategy are needed to improve the quality of product. An off-line optimization is prior determined by maximizing productivity for the batch transesterification to modify optimal temperature set point. Model based control, model predictive control (MPC) with an estimator has been implemented to drive the reactor temperature tracking to the desired profile. An extended Kalman filter (EKF) has been designed to estimate the uncertain parameter and unmeasurable states variable. In this work, improvement of batch transesterification process under uncertain parameters on the overall heat transfer coefficient has been proposed. Simulation results demonstrate that the EKF can still provide good estimates of the overall heat transfer coefficient and heat of reaction. The control performance of MPC is better than that of PID. Moreover, MPC with the EKF estimator can control the transesterification according to the optimal trajectory and then can achieve maximum product as determined. As a result, the MPC with EKF is still robust and applicable in real plants.

Keywords Batch reactor · Biodiesel production · Extended Kalman filter
Model predictive control · Non-linear model based control · Optimization
Transesterification reaction

A. Chanpirak · W. Weerachaipichasgul (✉)
Faculty of Engineering, Department of Industrial Engineering, Division of Chemical
Engineering, Naresuan University, Phitsanulok 65000, Thailand
e-mail: weerawunw@nu.ac.th

A. Chanpirak
e-mail: arphaphonc@nu.ac.th

© Springer Nature Singapore Pte Ltd. 2018
S.-I. Ao et al. (eds.), *Transactions on Engineering Technologies*,
https://doi.org/10.1007/978-981-10-7488-2_4

1 Introduction

In the current world, the energy security and environmental safety have been concerned because the demand for fuels is rising, on the other hand petroleum reserves are inversed. In addition, the burning of fossil fuels is environment problems that causes climate change. Biodiesel is one of the most promising alternatives to fossil fuel, which can be produced form vegetable oils, waste cooking oils, and animal fats [1–3]. Transesterification is a process to produce biodiesel (fatty acid methyl esters, FAME) between triglycerides and methanol in presence of a catalyst [4–6].

Transesterification process is established of three sequential reversible reactions [6, 7] that can be run in a batch reactor because it can adapt to small volume production and it can provide to scale-up processes from laboratory experiment to industrial manufacturing. Moreover, it is also suitable for the manufacturing of difficult processes to convert to continuous operations. The yield of biodiesel affected by four main factor; the molar ratio of reactants, reaction time, catalyst, and reaction temperature [7, 8]. The molar ratio between alcohol and triglycerides is 6:1 in alkali catalyst that is reported to guarantee the complete conversion to product esters in short time. An excess time in the reaction can give the falling conversion of triglycerides because fatty acids will form to soap. Increasing concentration of catalyst helps to improve conversion of triglycerides (increasing the yield of bio-diesel). The most commonly used catalyst is sodium hydroxide. An increase in reaction temperature as an increase reaction rate and reaction time is shortened. However, the yield of biodiesel is possible to decrease when temperature increase because of the saponification reaction of triglycerides appearing. Then temperature is considered as a control parameter to improve biodiesel yield [6–11].

In the batch transesterification process, the dependence of the reaction kinetics and temperature can be categorized by nonlinear behavior, time-varying system and non-stationary. As a result that, control of transesterification process is more challenging.

Model based control, model predictive control (MPC), has been found to be successful control strategy in several manufacturing applications; to handle non-linear processes, multivariable interactions, constraints, and optimization requirements. Therefore, MPC controller have been applied to control in batch processes in many cases [12–14]. However, batch transesterification process has not been much attended [9–11]. It is well-known that model based control needs the state variables and/or parameters but some of variables cannot know exactly or sometime the measured variables with time delay will be introduced. An extended Kalman filter (EKF) is one of several techniques to estimate the state and parameters by available temperature data [12–14, 18].

The aim of this work is the implementation of a maximum production in the batch transesterification reactor of biodiesel by using MPC controller is studied in this work. An optimization technique based on a sequential optimization approach is applied to moderate the set point of the optimal temperature. The optimal reactor temperature is determined by following the objective function; maximum the amount of product and minimum batch time. The EKF has been designed to estimate the uncertain parameters and unmeasurable states variable. Model based control, model predictive control (MPC) is applied to tracking the desire profile. The control performances of MPC are compared with a conventional control technique under the nominal and mismatch case.

2 Batch Transesterification Reactor

A batch reactor system in this work is presented in Fig. 1 which consists of a reactor and jacket system [12–14]. Transesterification reaction between lipids (animal fats or plants), and short chain of alcohol (methanol or ethanol) in a base or acid catalyst is a process to produce biodiesel. Vegetable oil, methanol and a homogenous alkaline catalyst [15] are studied in this work.

2.1 Kinetic Model

Biodiesel, free fatty acid methyl ester (FAME), can be formed by the transesterification reaction of triglyceride (TG) with three moles of methanol (A) as presented in the overall reaction. However, the mechanism reaction of transesterification reaction consist of three sequential reversible reactions. In each reaction step, FAME can be produced. The sequence converting of substances in this process consists of triglyceride to diglycerides (DG), diglycerides to monoglycerides (MG), and monoglycerides to glycerol (GL).

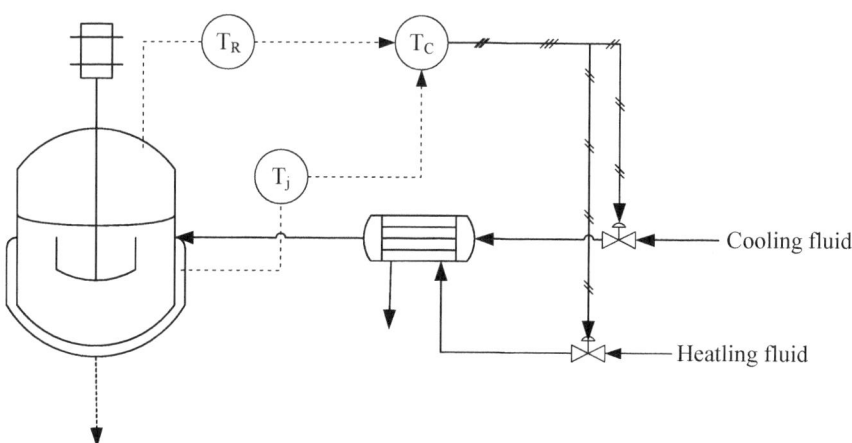

Fig. 1 Schematic diagram of a batch reactor with control system

Overall reaction;

$$TG + 3A \leftrightarrow GL + 3FAME \tag{1}$$

The reaction steps are given;

$$TG + A \underset{k_{-1}}{\overset{k_1}{\leftrightarrow}} DG + FAME \tag{2}$$

$$DG + A \underset{k_{-2}}{\overset{k_2}{\leftrightarrow}} MG + FAME \tag{3}$$

$$MG + A \underset{k_{-3}}{\overset{k_3}{\leftrightarrow}} GL + FAME \tag{4}$$

The assumption of the mathematical model can be obtained by using mass balance of each components in batch reactor following by Noureddini and Zhu [16] and Richard et al. [17]; the reactions are first-order, reactor is operated under atmospheric pressure. Material balances are as follows

$$\frac{dC_{TG}}{dt} = -k_1 C_{TG} C_A + k_{-1} C_{DG} C_E \tag{5}$$

$$\frac{dC_{DG}}{dt} = k_1 C_{TG} C_A - k_{-1} C_{DG} C_E - k_2 C_{DG} C_A + k_{-2} C_{MG} C_{FAME} \tag{6}$$

$$\frac{dC_{MG}}{dt} = k_2 C_{DG} C_A - k_{-2} C_{MG} C_{FAME} - k_3 C_{MG} C_A + k_{-3} C_{GL} C_{FAME} \tag{7}$$

$$\frac{dC_{GL}}{dt} = k_3 C_{MG} C_A - k_{-3} C_G C_{FAME} \tag{8}$$

$$\frac{dC_{FAME}}{dt} = k_1 C_{TG} C_A - k_{-1} C_{DG} C_{FAME} + k_2 C_{DG} C_A - k_{-2} C_{MG} C_{FAME} \cdots$$
$$+ k_3 C_{MG} C_A - k_{-3} C_{GL} C_{FAME} \tag{9}$$

$$\frac{dC_A}{dt} = -\frac{dC_{FAME}}{dt} \tag{10}$$

where C_{TG}, C_{DG}, C_{MG}, C_{FAME}, C_A, C_{GL} are concentrations of triglycerides, diglycerides, monoglycerides, methyl ester, methanol, and glycerol, respectively. The reaction rate constant (k_i) can be expressed by the Arrhenius equation;

$$k_i = k_{0,i} \exp\left(\frac{E_i}{RT_R}\right) \quad for \ i = 1, -1, 2, -2, 3, -3 \tag{11}$$

Table 1 Values of kinetic parameters

Reaction	$k_{0,i}$	$(m^3/kmol\ s)$	E_i/R	(K)
Triglyceride to diglycerides	$k_{0,1}$	3.92×10^7	E_1/R	6614.83
Diglycerides to triglyceride	$k_{0,-1}$	5.77×10^5	E_{-1}/R	4997.98
Diglycerides to monoglycerides	$k_{0,2}$	5.88×10^{12}	E_2/R	9993.98
Monoglycerides to diglycerides	$k_{0,-2}$	9.80×10^9	E_{-2}/R	7366.64
Monoglycerides to glycerol	$k_{0,3}$	5.35×10^3	E_3/R	3231.18
Glycerol to monoglycerides	$k_{0,-3}$	2.15×10^4	E_{-3}/R	4824.87

where $k_{0,i}$ is pre-exponential factor and E_i is activation energy of component "i". Moreover, R is universal gas constant and T_R is reactor temperature. The process parameters in this work are presented in Table 1 [9].

2.2   Reactor Model

Batch reactor in this work is operated under atmospheric pressure, therefore the aim of this work focuses on temperature control of the reactor. The model assumptions for the energy balance include the heat of reaction is considered by using the overall reaction [10], the matters in the reactor are homogeneous. Moreover, molar mass, density and heat capacity of the reactor are each approximated by average constants, and the heat loss is negligible. The energy balances around the reactor are as follows

$$\frac{dT_R}{dt} = \frac{M_R(Q_R + Q_j)}{V\rho_R c_{m,R}} \tag{12}$$

$$\frac{dT_j}{dt} = \frac{F_j\rho_j c_w(T_{jsp} - T_j) - Q_j}{V_j\rho_j c_w} \tag{13}$$

$$Q_R = -V\Delta H_R r \tag{14}$$

$$Q_j = UA(T_j - T_R) \tag{15}$$

$$r = \frac{dC_{FAME}}{dt} \tag{16}$$

where T_R, T_j, and T_{jsp} are reactor, jacket, and set point of jacket temperatures, respectively, moreover the molar mass, density, and molar heat capacity of the reactor contents are M_R, ρ_R and $c_{m,R}$, respectively and c_w is specific heat capacity of water. r and ΔH_R are rate of reaction and heat of reaction, respectively. The surface of heat exchange is A and the overall heat transfer coefficient is U. The initial concentration of triglyceride is 0.3226 mol/L, and methanol is 1.9356 mol/L. Furthermore, the process parameters in this work is presented in Table 2 [9].

Table 2 Process parameters

Parameter	Value	Unit
V	1	m^3
ρ_R	860	kg/m^3
M_R	391.40	kg/kmol
$c_{m,R}$	1277	kJ/kmol K
ΔH_R	-1850	kJ/kmol
UA	450	kJ/min K
F_j	0.348	kg/min
V_j	0.6812	m^3
ρ_j	1000	kg/m^3
c_w	4.12	kJ/kg K

2.3   Optimal Problem Formulation

For batch transesterification process, an optimization problem is transformed into a nonlinear programming (NLP) problem by a sequential approach and uses the control vector parameterization (CVP) technique to ask this problem which is solved using a SQP-based optimization technique and process model is integrated by using the explicit Runge-Kutta Fehlberg (RKF) method. The optimal product is determined by maximizing productivity that is subject to the process model Eqs. (5)–(16) and specified product purity is greater than 0.83 mol/L. Moreover, the range of reactor temperature is between 293.15 and 363.15 K. To solve an optimal problem, the reactor temperature and bath time are selected to be the decision variable into a finite set in, which a piecewise constant function is utilized.

3   Model Predictive Control with Extended Kalman Filter (EKF) Estimator

Due to its high performance, model based control method has received a great deal of attention to control chemical processes. The formulation of the model based controller, model predictive controller (MPC) bases on solving an on-line optimal control problem to minimize the problem referred to an objective function which is the sum of squares of deviation of the set point and predicted values on outputs and inputs over the prediction horizon (P). The optimization decision variables are control variable (T_j) for M time steps.

$$\min_{u(k),\,...,\,k(k+M-1)} \sum_{i=k=1}^{k+P} \left[\left(y_{sp,\,i} - y_{pred,\,i}\right)^2 W_1 + \left(\Delta u_i\right)^2 W_2 \right] \qquad (17)$$

Subject to process models Eqs. (5)–(16) and a constraint on the manipulated variables. W_1 is a weighting matrix and W_2 is a weighting matrix on outputs and inputs, respectively. In general, the process model can be arranged in a discrete time model in state space form as

$$x_{k+1} = A_k x_k + B_k u_k \qquad (18)$$

$$y_k = C_k x_k \qquad (19)$$

where x, u, y are state, input, mad output variables, respectively. In this work, the discrete sate space form is T_R and T_j.

For temperature control, the knowledge of heat of reaction (Q_R) is necessary to control algorithm and control performance. As a result, the EKF estimator is applied to estimate the heat evolution term. In addition, the EKF estimator is also used to provide the estimates of process parameters to handle process-model mismatches [12–14, 18]. In order to develop the EKF estimator, the state estimation based on simplified mathematical models are given in Appendix A.

4 Simulation Results

The application of MPC with the extended Kalman filter to control the reactor temperature, an optimal reactor temperature is prior determined by solving the optimization problem based on an objective function to maximize productivity

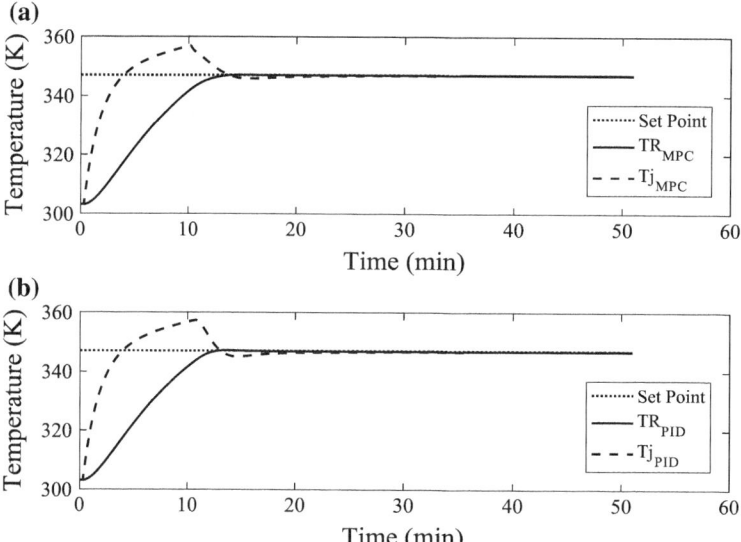

Fig. 2 Control response in nominal case; **a** MPC, **b** PID controllers

(0.848 mol/L in 51 min). Every sampling time (4 s), the temperature of reactor and jacket are introduced to estimator and controller to control the reactor temperature tracking the desired values by manipulating the jacket temperature. Tuning parameters of the MPC controller consist of prediction horizon, control horizon, weighting matrix on outputs, and inputs that are $P = 10$, $M = 10$, $W_1 (1, 1) = 10$, $W_1 (2, 2) = 0.1$, and $W_2 = 5$, respectively. For PID controller, the tuning parameters is tuned to provide identical control response of $K_c = 10$, $\tau_i = 60$, and $\tau_D = 1 \times 10^{-5}$.

The control responses of MPC and PID controllers in the nominal case, all parameters correctly specified, are shown in Fig. 2, and controllers performance are summarized in Table 3. It has been observed that the control performance of MPC and PID controllers are slightly different.

Absolutely, the design of the model based controller has to be concerned about parameter uncertainty. In the actual plant, plant/model mismatches and uncertainty

Table 3 Performance of controllers

Case studies	IAE
PID in nominal case	257.15
MPC in nominal case	252.59
PID in missed math case	326.03
MPC in missed math case	310.43
MPC with EKF in missed math case	254.06

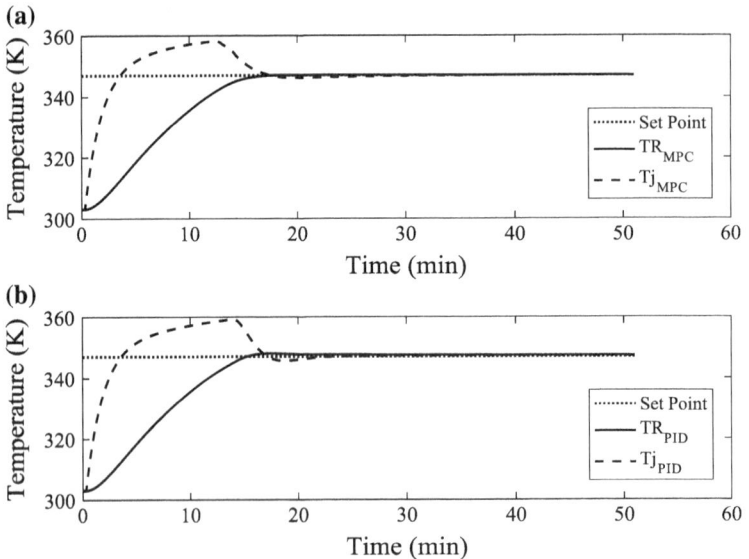

Fig. 3 Control response in mismatch case of UA-30%; **a** MPC, **b** PID controllers

in process parameters subsist. The robustness tests regarding plant/model mismatches and uncertainty must be carried out. The presence of plant/model mismatch in the overall heat transfer coefficient decreased 30% from its real value has been studied in this work. Figure 3 shows the control response of reactor temperature under the MPC and PID controllers. It can be seen that the MPC controller gives a better than the PID does.

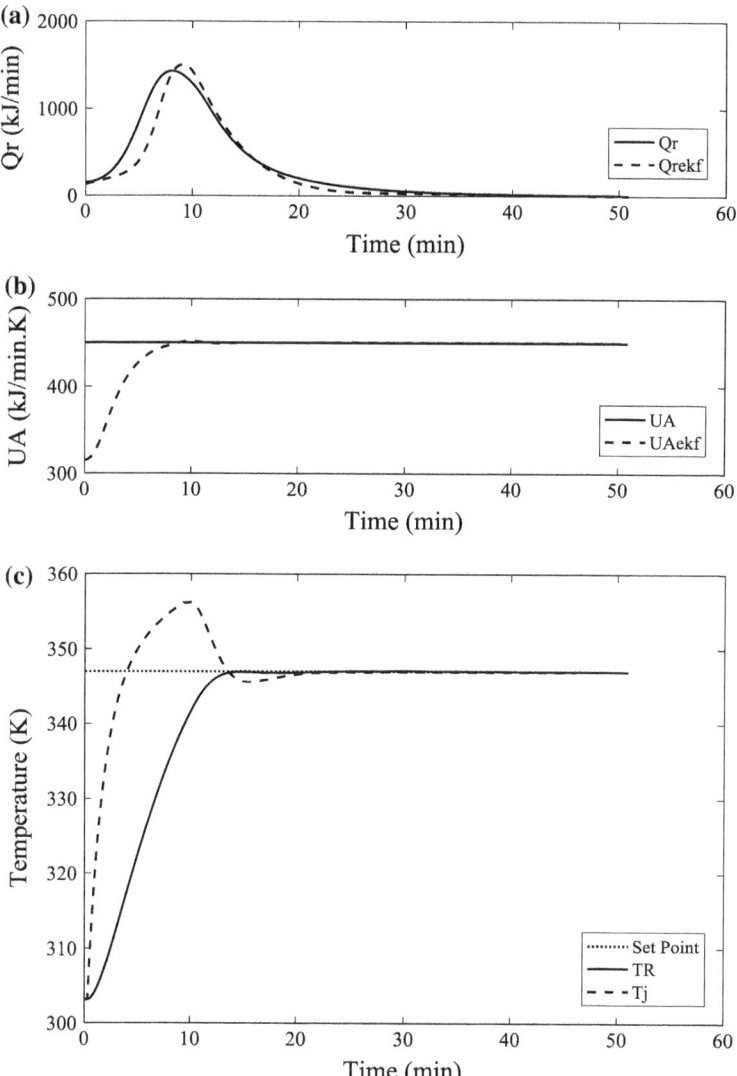

Fig. 4 **a** Estimate of Qr, **b** estimate of UA, **c** control response of MPC with EKF in mismatch case of UA-30%

To improve MPC, the EKF needs to estimate the heat of reaction and the overall heat transfer coefficient that is practiced in the MPC formular. The estimated response of the state (Q_r) and parameter (UA) are showed in Fig. 4a and b, respectively. Figure 4c shows the control response of the MPC with EKF in the mismatch case of the overall heat transfer coefficient. The control performances of the MPC with the EKF for state and parameter estimation are presented in Table 3.

5 Conclusion and Future Work

In this work, a maximum product achievement of biodiesel in the batch transesterification by using model based control with EKF has been proposed. There are two strategies to obtain the objective of this work; optimization to modify the optimal temperature set point, and nonlinear model based control, model predictive control (MPC), has been implemented to drive the reactor temperature tracking to the desired profile. The EKF is incorporated into the MPC algorithm in order to estimate the heat of reaction by measurement of reactor and jacket temperatures. For the comparison of the control performance, the control performance of the MPC is better than that of PID and the MPC is more robust than the PID. In the presence of unknown/uncertain parameters, the estimator is still able to provide accurate value. As a result, the MPC with the EKF is still robust and applicable in real plants.

Appendix A

The model assumptions to design EKF estimator based on simplified mathematical models of batch reactor [18]. In this work, the heat of reaction and the overall heat transfer coefficient are the state and parameter that have to be study. The design of the state estimation bases on simplified mathematical models are given by

$$\frac{dT_R}{dt} = \frac{M_R Q_R + UA(T_j - T_R)}{V \rho_R c_{m,R}} \tag{A.1}$$

$$\frac{dT_j}{dt} = \frac{F_j \rho_j c_j (T_{jsp} - T_j) - UA(T_j - T_R)}{V_j \rho_j c_j} \tag{A.2}$$

$$\frac{dN}{dt} = -bNT_R \tag{A.3}$$

$$\frac{dQ_R}{dt} = N \frac{dT_R}{dt} + T_R \frac{dN}{dt} \tag{A.4}$$

$$\frac{dQ_R}{dt} = N\frac{dT_R}{dt} + T_R\frac{dN}{dt} \tag{A.5}$$

$$\frac{db}{dt} = 0 \tag{A.6}$$

$$\frac{dUA}{dt} = 0 \tag{A.7}$$

The initial condition and parameters to support in the EKF estimator is presented by Chanpirak, and Weerachaipichasgul [18]. Detailed regarding the EKF algorithm is given in [12–14, 18].

References

1. Y. Zhang, M.A. Dube, D.D. McLean, M. Kates, Biodiesel production from waste cooking oil: 2. Economic assessment and sensitivity analysis. Bioresour. Technol. **90**, 229–240 (2003)
2. E.H.S. Moecke, R. Feller, H.A. Santos, M.M. Machado, A.L.V. Cubas, A.R.A. Dutra, L.L.V. Santos, S.R. Soares, Biodiesel production from waste cooking oil for use as fuel in artisanal fishing boats: integrating environmental, economic and social aspects. J. Cleaner Product. **135**, 679–688 (2016)
3. N. Nasir, W. Daud, S. Kamarudin, Z. Yaakob, Process system engineering in biodiesel production: a review. Renew. Sustain. Energy Rev. **22**, 631–639 (2013)
4. K. Khalid, K. Khalid, Transesterification of palm oil for production of biodiesel. Am. J. Appl. Sci. **8**, 804–809 (2011)
5. X. Liu, H. He, Y. Wang, S. Zhu, Transesterification of soybean oil to biodiesel using SrO as a solid base catalyst. Catal. Commun. **8**, 1107–1111 (2007)
6. R. Aliakbar, S. Iman, Modeling the effects of cosolvent on biodiesel production. Fuel **186**, 779–786 (2016)
7. M. Thirumarimurugan, V. Sivakumar, A. Xavier, D. Prabhakaran, T. Kannadasan, Preparation of biodiesel from sunflower oil by transesterification. Int. J. Biosci. Biochem. Bioinfo. **2**, 441–445 (2012)
8. B. Freedman, E.H. Pryde, T.L. Mounts, Variables affecting the yields of fatty esters from transesterified vegetable oils. J. Am. Oil Chem. Soc. **61**, 1638–1643 (1984)
9. P.T. Benavides, U. Diwekar, Optimal control of biodiesel production in a batch reactor. Part I: deterministic control. Fuel **94**, 211–217 (2012)
10. R. Kern, Y. Shastri, Advanced control with parameter estimation of batch transesterification reactor. J. Process Control **33**, 127–139 (2015)
11. R. De, S. Bhartiya, Y. Shastri, Dynamic optimization of a batch transesterification process for biodiesel production, in *Indian Control Conference (ICC) Indian Institute of Technology Hyderabad* (Hyderabad, India, 4–6 Jan 2016), pp. 117–122
12. A. Saengchan, P. Kittisupakorn, W. Paengjuntuek, A. Arpornwichanop, Improvement of batch crystallization control under uncertain kinetic parameters by model predictive control. J. Ind. Eng. Chem. **17**, 430–438
13. W. Paengjuntuek, P. Kittisupakorn, A. Arpornwichanop, Optimization and nonlinear control of a batch crystallization process. J. Chinese Inst. Chem. Eng. **39**, 249–256 (2008)
14. W. Weerachaipichasgul, P. Kittisupakorn, Integrating dynamic composition estimation with model based control for ethyl acetate production, in *Lecture Notes in Electrical Engineering*, vol. 275 (LNEE, 2014), pp. 231–245

15. D.Y.C. Leung, X. Wu, M.K.H. Leung, A review on biodiesel production using catalyzed transesterification. Appl. Energy **87**, 1083–1095 (2010)
16. H. Noureddini, D. Zhu, Kinetics of transesterification of soybean oil. J. Am. Oil Chem. Soc. **74**, 1457–1463 (1997)
17. R. Richard, S. Thiebaud-Roux, L. Prat, Modeling the kinetics of transesterification reaction of sunflower oil with ethanol in microreactors. Chem. Eng. Sci. **87**, 258–269 (2013)
18. A. Chanpirak, W. Weerachaipichasgul, Improvement of biodiesel production in batch transesterification process, in *Lecture Notes in Engineering and Computer Science: Proceedings of The International MultiConference of Engineers and Computer Scientists 2017* (Hong Kong, 15–17 Mar 2017), pp. 806–810

Production Scheduling Tools to Prevent and Repair Disruptions in MRCPSP

Angela Hsiang-Ling Chen, Yun-Chia Liang and Jose David Padilla

Abstract Companies invest countless hours in planning project execution because it is a crucial component for their growth. However, regardless of all the considerations taken in the planning stage, uncertainty inherent to project execution leads to schedule disruptions, and even renders projects unfeasible. There is a vast amount of studies for generating baseline (predictive) schedules, yet, the literature regarding *reactive* scheduling for the Multi-Mode Resource Constrained Project Scheduling Problem (MRCPSP) is scant with only two previous studies found at the time of writing. In contrast, schedule disruption management has been thoroughly studied in the mass production environment, and regardless of the difficulties encountered, they will almost certainly be required to meet the levels planned. With this in mind, this study proposes an integrative (*proactive and reactive*) scheduling framework that uses the experience and methodologies developed in the production scheduling environment and apply it to the MRCPSP. The purpose of this framework is to be used on further empirical research.

Keywords MRCPSP · Proactive—Reactive scheduling · Proactive scheduling
Project management · Project scheduling framework · Reactive scheduling

A. H.-L. Chen (✉)
Department of Business Administration, Nanya Institute of Technology,
Taoyuan, Taiwan
e-mail: achen@nanya.edu.tw

Y.-C. Liang · J. D. Padilla
Department of Industrial Engineering and Management, Yuan Ze University,
Taoyuan, Taiwan
e-mail: ycliang@saturn.yzu.edu.tw

J. D. Padilla
e-mail: s1028909@mail.yzu.edu.tw

© Springer Nature Singapore Pte Ltd. 2018
S.-I. Ao et al. (eds.), *Transactions on Engineering Technologies*,
https://doi.org/10.1007/978-981-10-7488-2_5

1 Introduction

Project execution has a direct impact on firms' performance and this leads companies to collectively invest countless man-hours in careful project planning. Nonetheless, the initial planning and feasibility evaluations are performed well before the project is even proposed to upper management for approval. Then, during the formal planning stage, further assumptions will be made about the possible future conditions in which the project will be developed. In this stage, a baseline (predictive) schedule is usually generated with the assumed future conditions about resource availability and, best case scenario, some considerations for possible uncertain events. And, even though this schedule is predictive and based on assumptions it is necessary because it allows for: the allocation of resources [1]; planning of external activities such as material procurement, maintenance planning, and subcontracting [2]; and cash flow projections, and a measure for the efficiency of the management team and the actual project executioners [3]. Considering its importance, mounds of research have been done on the development of baseline schedules. However, despite the amount of effort and significant advances accomplished, in practice, projects are still being affected because of disrupted schedules.

Even after all the planning is completed, the actual project execution will undoubtedly be subject to uncertainty and we find this from the very definition of a project:

> A temporary organization to which resources are assigned to undertake a unique, novel and transient endeavor managing the *inherent uncertainty* and need for integration in order to deliver beneficial objectives of change [4].

Uncertainty inherent to projects can arise from both unknown and known sources, regardless of the industry, and causes schedule disruptions during the project execution. Failing to handle such disruptions properly, and in time, can result in rapid and poorly evaluated changes to the planned budget, make-span, scope, or quality. In turn, a change in any of these areas can potentially render the entire project unfeasible.

In contrast, disruptions within the production scheduling environment seem to be managed rather efficiently. This is probably because even though their impact may be less significant when compared to that of a firm's project, the probability of occurrence is much higher and this has led to much more studies in this area. Using this vantage point, we intend to extend the work presented in [5] and develop a scheduling framework for the MRCPSP by integrating some of the reactive measures found in production scheduling with a proactive baseline scheduling method developed previously. The remainder of this chapter is organized as follows: Sect. 2 reviews approaches used to handle uncertainty and disruptions in production scheduling. In Sect. 3 we present the proposed framework using the concepts introduced in Sect. 2 and evaluate the feasibility of applying these concepts to the MRCPSP. Finally, Sect. 4 presents our conclusions and directions for future research.

2 Uncertainty and Disruption Management in Production Scheduling

2.1 Stochastic Scheduling

Instead of long term planning of activities, stochastic scheduling handles uncertainty by viewing scheduling as a multi-stage decision process. This requires making dynamic scheduling decisions at stochastic decision points corresponding to the start of the project and the completion of activities based on the observed past, along with a priori knowledge about the activity processing time distributions [3]. A dynamic scheduling procedure determines the beginning of each activity over time and the key objective is to minimize the makespan through the application of scheduling policies.

Stochastic scheduling is especially useful in areas where variations are high such as power system demands [6, 7], as well as network controlled systems which use this model to open and close the required communication throughout a network [8], and even in defensive surveillance of public areas exposed to adversarial attacks [9].

2.2 Rolling Horizon, Decomposition, and Hierarchy

In mass production, parameters can interact in real time and have a severe effect on the production schedule as well as on future planning; therefore, short term scheduling periods (12 h) and long term planning (1 month horizon) need to be considered simultaneously [10].

Rolling Horizon. One approach to dealing with different time scales is to use a rolling horizon, where only a subset of the planning periods include the detailed scheduling decisions with shorter time increments. In this approach, the first planning period is often a detailed scheduling model while the future planning periods include only planning decisions. A rolling horizon approach is used in [11] to resolve a two-level hierarchical planning scheduling problem where uncertainty is explicitly included on the planning level. An example of integrated planning-scheduling can be found in [12], where the authors propose a novel energy management system (EMS) based on a rolling horizon (RH) strategy for a renewable-based micro-grid. And more recently, this approach has been applied in production planning in [13].

Decomposition. A second approach for the integration of different time scales is the decomposition of the horizon into two levels: a higher level for planning which passes information to a more detailed model for scheduling. Examples using multi-level decompositions can be found in [14] which uses decomposition to solve the General Lot Sizing and Scheduling Problem for Parallel Production Lines. Then Ghaddar et al. present a Mixed-Integer Non-linear Programming (MINLP) formulation of the optimal pump scheduling problem and solve it using decomposition [15];

and Blom et al. consider the multiple-time-period, short-term production scheduling problem for a network of multiple open-pit mines and ports [16].

Hierarchy. Lastly, taking decomposition as basis, another approach for uncertainty handling used in job-shops was presented in [17]. The author presents a three-level hierarchy scheduling agent: an operational level agent in charge of solving temporal constraints; a tactical level agent which solves technological constraints by balancing the load between resources; and finally a strategic level agent which supervises and can re-sequence decisions made through a job's process plan or, as last resort, relax the temporal constraints. Recent applications include [18, 19], where they try to minimize makespan by using semi-online hierarchical scheduling problems on two identical machines and [20] where they seek to allocate resources for diverse data-center workloads using multi-resource hierarchical schedulers.

2.3 Knowledge Based Systems

Manufacturing processes are often highly complex and are frequently hampered by the unavailability of required resources. However, reactive scheduling essentially reduces to quick logical thinking, taking into account a series of parameters when evaluating different scenarios and selecting a scenario to achieve the performance goals or the best possible solution. Using this premise and taking advantage of advances in computing, O'Kane proposed the creation of an Expert System (ES) or Knowledge Based System (KBS) combined with a simulator-based adviser in a Flexible Manufacturing System (FMS) [21]. The ES would initially be fed by rules of thumb followed by the experience of schedulers and later be capable of *learning* by adding new information and deleting obsolete information from its own knowledge base. In turn, the simulator would help evaluate possible options and present the ES the result of each scenario. Nonetheless, one of the biggest drawbacks of this technology is the amount and/or quality of the rules required at the outset, which ultimately limits the speed at which the system can learn. Also, two definitions are crucial: *learning*, in terms of how and what to learn; and *adaptation*, using statistics to evaluate the success of the rules in particular circumstances.

A recent example of the use of a KBS can be found in [22], where they apply it to develop an integrated system to capture information and knowledge of building maintenance operations when/after maintenance is carried out to understand how a building is deteriorating and to support preventive/corrective maintenance decisions.

2.4 Sensitivity Analysis

One of the less-frequently used approaches for reactive scheduling involves trying to answer *what if* type questions. In Sensitivity Analysis (SA), the researchers try to answer questions such as: What is the effect of every parameter on the objective function? Which are the parameters that affect the schedule the most? What are the limits for a given parameter while still ensuring schedule feasibility? Research is still scarce in this area; however, if reliable global indicators and thresholds for a schedule could be determined, they would help management predict when a change in the current schedule has become absolutely necessary without risking the project's feasibility.

Within this field, an early example includes Sotskov et al. which discussed the extent of deviations in activity duration wherein remaining optimal [23]. Some studies perform a sensitivity analysis of statistically computed schedules for the problem when actual communication delays differ from estimated delays [24]; others handle uncertainty in short-term scheduling using the idea of inference-based sensitivity analysis with the resulting information being used to determine: (a) the importance of different parameters and constraints, (b) the range of parameters where the optimal solution remains unchanged [25]. Recent applications of SA in scheduling include one performed in [26] which uses SA to determine the best possible parameter combination to achieve optimal or near optimal solutions for the job shop scheduling problem. In [27], the authors' objective was to determine their simulation models robustness to be further used in applications. And in [28], SA was used to study power grid imbalances in terms of five independent variables.

However, Hall and Posner pointed out a number of issues associated with the application of sensitivity analysis in scheduling problems, including: the applicability of SA to special classes of scheduling problems; the efficiency of SA approaches when simultaneous parameter changes occur; the selection of a schedule with minimum sensitivity; and the computational complexity of answering SA questions for intractable scheduling problems [29]. Despite the difficulties it presents, if done correctly, SA could effectively help management teams either timely recover distressed projects or raise flags for projects which should be abandoned. Both actions would be performed early enough in the project life-cycle to allow for an effective course of action.

2.5   Proactive-Reactive Scheduling

The purpose of scheduling is to optimally allocate limited resources to processing tasks over time. However, as aforementioned, uncertainty is ever present and one of the most common solutions to manage uncertainty is to create proactive baseline schedules. This means that the schedule will not only minimize the make-span but also will include a buffer capable of handling certain variations. Nevertheless, one of the biggest limitations of such schedules is that they rely on parameters and other conditions that are not exactly known and this exposes them to disruptions. Should a disruption that can't be handled by the proactive schedule occur, a reactive schedule takes the updated progress and conditions and revises the baseline schedule.

Here we first discuss some methodologies used to generate proactive baseline schedules and then, how reactive scheduling is applied.

2.5.1 Fuzzy Set Theory

Given that scheduling relies heavily on the prediction of unknown parameters and duration, one alternative is to use fuzzy set theory and interval arithmetic to describe the imprecision and uncertainty. Though the uncertainty in process scheduling is generally described through probabilistic models, over the past decades, fuzzy set theory has been applied to schedule optimization using heuristic search techniques. The main difference between the stochastic and fuzzy optimization approaches is in the way uncertainty is modeled. Here, fuzzy programming considers random parameters as fuzzy numbers, and constraints are treated as fuzzy sets [30]. The output of a fuzzy scheduling pass will normally be a fuzzy schedule, which indicates fuzzy starting and ending times for the activities. Such fuzzy time instances may be interpreted as start or completion to a certain extent only [31].

The study of a fuzzy model of resource-constrained project scheduling was initiated by [32]. Fuzzy set theory is used when solving parallel machine scheduling in [33] as well as in [34]. In [35], fuzzy set theory is used to model uncertainties associated with different input parameters for optimized scheduling of repetitive construction projects under uncertainty. Fuzzy activity duration and lead times in procurement scheduling are used in [36] and recently Xu et al. used, the fuzzy set theory to solve the flexible job shop scheduling problem and project scheduling problem [37, 38].

2.5.2 Robust Optimization and Robust Scheduling

The basic idea of robust optimization is that by reformulating the original problem, or by solving a sequence of problems, we may find a solution which is robust to the uncertainty in the data. A solution to an optimization is considered to be solution robust if it remains close to the optimum for all scenarios, and model robust if it remains feasible for most scenarios. Applying this concept into scheduling produces *schedule robustness* as introduced by [39] which defined it as *the ability (of a schedule) to cope with small increases in the time duration of some activities that may result from uncontrollable factors.*

Among the first efforts to produce robust schedules we find [40] who introduces the concept of temporal protection, which extends the duration of activities based on the uncertainty statistics of the resources used for their execution. Later, one of the most popular approaches was Critical Chain Scheduling/Buffer Management (CC/BM) which builds a baseline schedule using aggressive median or average activity duration estimates as presented in [41]. In [42] they propose improvements of the temporal protection technique with their time window slack and focused time window slack approaches which do not include slack into activity duration, but rather explicitly compute available slack time per activity in solution schedules. The study in [43] proposed a modification on the Al-Fawzan and Haouari model and established that a robust schedule maximizes the free slack/activity duration ratio. Chtourou and Haouari proposed 12 different robustness measures based on activity slack [44]. More recent applications of robust baseline scheduling include Chen et al. [45] where they used an entropy function to determine the upper bound or *minimized maximum* makespan of a robust baseline schedule. Rezaeian et al. solved the MRCPSP as a bi-objective problem, optimized robustness and make-span [46]. Meanwhile, Lamas and Demeulemeester defined a new robustness measure and introduce a branch-and-cut method to solve a sample average approximation of the RCPSP [47].

2.5.3 Reactive Scheduling

Reactive scheduling focuses on repairing the schedule, accounting for disruptive incidents and updated status of the conditions. There are two main strategies with which reactive scheduling responds: the first one achieves the quick restoration of the schedule through what is generally called *schedule repairing actions*, moves forward in time all the activities that are affected by the disruption. The actions which follow this strategy usually lead to poor results because they do not consider the re-sequencing of activities but merely push them until the resources become available.

The second strategy involves the re-sequencing of the affected activities, possibly all of them. The actions under this strategy are known as rescheduling and although the solved model has some differences with the original, it is very similar to the generation of a new baseline schedule. As such, it may adopt any performance measure used by the baseline schedule or any other deterministic baseline schedule. Among these measures we find some within the minimum perturbation strategy, which seeks to produce a new schedule that deviates as little as possible from the original schedule. An example of this strategy can be found in [48]. In pursuit of rescheduling stability, algorithms have also been proposed that use a match-up point, i.e., the time instance where the state reached by the revised schedule is the same as the initial schedule [2, 49]. The goal is to match up with the baseline schedule at a certain time in the future, whenever a deviation from the initial parameter values arises.

A different advent within reactive scheduling considers the manual changes made by management to the execution of a project or contingent scheduling. To assist in this decision, a solution proposes group sequence, a totally or partially ordered set of groups of operations, and considers all the schedules obtained by an arbitrary choice of the ordering of the operations inside each group. This provides the decision maker with several feasible schedules, making it possible to react to disruptions by switching among solutions without incurring any loss in performance. An example of this methodology is presented in [50].

Finally, activity crashing is a form of reactive scheduling which can be applied to some or perhaps all activities. This is the execution of activities with an increase in the amount of resources used in order to accelerate the completion of such activities.

Although the current trend in scheduling is proactive scheduling, recent examples of reactive scheduling can be found in [51, 52].

As aforementioned, the model solved for the proactive baseline schedule can very well be used for the reactive schedule. However, there are some issues that need to be considered. First of all, after a disruption, timely decisions need to be made, usually resulting in a trade-off between making well-considered decisions and speeding up the recovery process. Second, new constraints may emerge. For example, if a resource breaks down, executing activities in the pre-selected modes may no longer be feasible. Also, new modes may become available. And finally, the objective function solved could be the same as the original, but it could also change to a different measure (for example, minimizing the total cost of the deviation) or it could become a multi-objective optimization problem. The reactive scheduling model must include all these considerations to apply a re-scheduling policy.

3   An Integrative Framework

3.1   Why Do We Need It?

As we discussed previously, projects will inevitably be subject to uncertainty; and, considering the potential impact of their success or failure, it's crucial for practitioners to know how to react to disruption threats. The biggest setback, however, is that unlike in production scheduling, to the best of our knowledge there are only two previous studies regarding reactive scheduling for MRCPSP. Some work has been done in the single-mode RCPSP such as considering two types of disruptions: activity duration variability and resource availability. In [53] and [54, 55] they dealt with the problem of coping with activity duration variability while in [56, 57] they tackled the problem of uncertainty with respect to resource availability. However, proactive-reactive scheduling policies in the multi-mode RCPSP, which is a generalization of the single-mode RCPSP, have been largely overlooked. In [58], the author proposed a branch-and-cut procedure for a general class of reactive scheduling problems and in [59] reactive scheduling procedures for repairing a disrupted schedules are proposed and evaluated.

More recently, and in the only attempt we have found to integrate proactive and reactive scheduling in project management, Herroelen [60], defines a methodology for planning under uncertainty using both quantitative and qualitative analysis. The author develops an iterative procedure that follows guidelines developed by the Project Management Institute (PMI) and which are currently the main guidelines for practitioners. Even so, the main downside to this procedure, is that it relies heavily on probabilities and assumptions: the probability of an event occurring and the impact we think it will have. In the 2-phase methodology, Phase 1 deals with decisions about the amount of regular and irregular renewable resource capacities to be allocated in the project. Phase 2 defines a robust baseline schedule based on decisions made in Phase 1 and the output of a quantitative schedule risk analysis. The reactive scheduling, is also part of Phase 2 and used only to *repair* the initial schedule which is simulated a number of times under uncertainty. Therefore, the output of this procedure is still only a baseline schedule for the project and does not consider repairs if disruptions occurred during the actual execution.

3.2 The Framework

The lack of methodology that follows through the project leaves us with the need to define a framework that guides practitioners not only through the initial baseline schedule generation stage, but also through the actual execution of the project. Also, given that PMI procedures rely mainly on expert's judgment and in contrast there are already tested procedures within the production scheduling environment, we use these to create our integrative framework. However, there is a significant difference between projects and mass production. The Project Management Institute defines a project as *a temporary endeavor undertaken to create a unique product, service or result*. This means that it has a defined beginning and end in time, and therefore defined scope and resources. Also, it is unique in that a specific set of operations are designed to accomplish a singular goal or *beneficial objectives of change* as stated by [4].

Even so, being produced only one of the units at the time, this single unit will have a specific start and completion time; it will require specific resources for its production; the activities required to produce it must follow precedents; and in the end it will be one unit meaning that it will be, even if only for a while, a unique result. Using this analogy, we could consider mass production as a project that is repeated indefinitely and this is a key point to apply mass production's reactive scheduling methods to MRCPSP in the proposed framework. Keeping this in mind, we now present the main contribution of this research: a framework to create a baseline schedule and handle disruptions in the MRCPSP based on the methods applied in the production schedule environment.

Let's analyze the decision and steps made in this framework. First of all, notice how the framework is divided in two different but connected stages as shown in Fig. 1; transitioning from the predictive stage to the reactive as the project is executed. First, the framework begins by asking if there is reliable information about the parameters and activity duration; the purpose is to reduce the degree of uncertainty coming from known sources. The second decision point is regarding the scope and impact of the project. If a company has high investment in a project or is expecting a crucial result from it, a detailed plan should be formulated. Furthermore, if they expect it to last for a long period of time, they should plan in advance and schedule for shorter terms either by *rolling horizon, hierarchy or decomposition*. In this way, they can get more current information as time passes and plan accordingly, therefore reducing the need for reactive measures and keeping the executed schedule closer to the baseline schedule. However, if the project is expected to be executed in a short

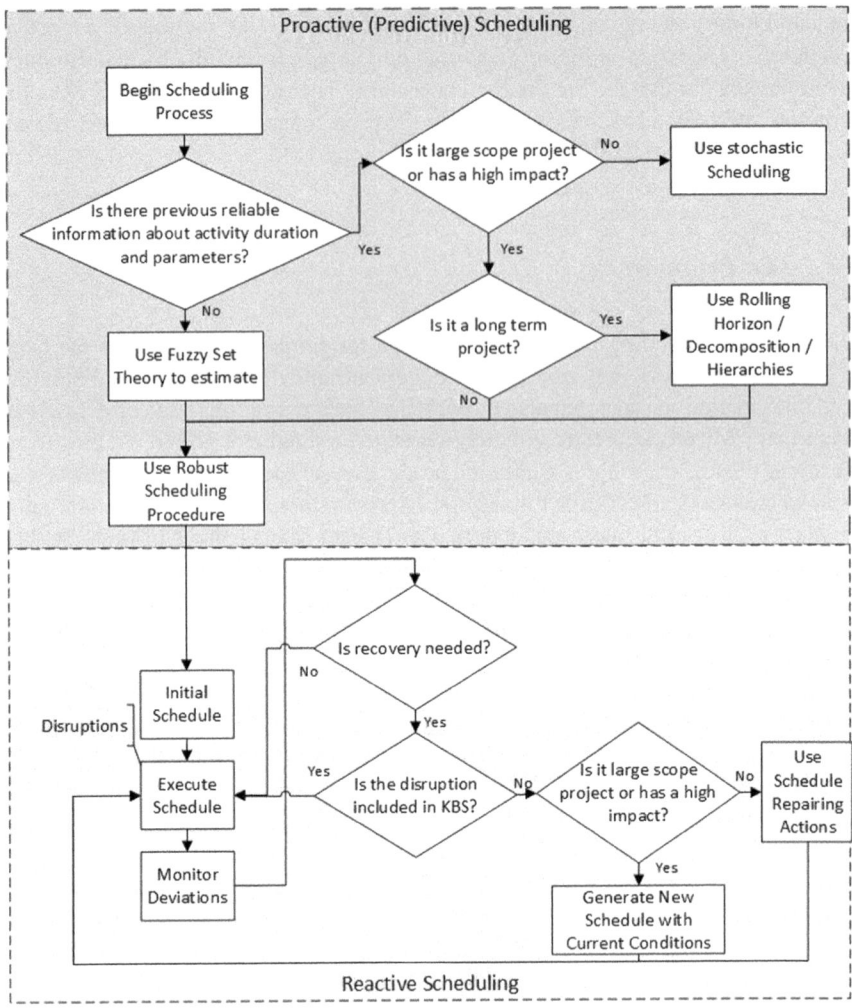

Fig. 1 Production-scheduling based framework for mrcpsp scheduling and disruption handling

time or if its success is not decisive for a company's growth, then it can be planned as it goes by *stochastic scheduling*, since it can make full use of the opportunities arising but without the risk of putting the company in danger.

If there is no reliable information regarding the parameters or activity duration, a safer choice would be the use of *fuzzy set theory* as it allows for scheduling under known uncertainty sources. After this, a *robust scheduling* procedure should be used to create the baseline schedule. For this, we recommend the methodology developed in [45] which uses an entropy function to account for uncertainty coming mainly from the activities precedents and their duration. Recall that the purpose of this schedule is to be capable of absorbing some of the variation without risking the success of the project.

Once the project is actually being executed, the schedule execution should be monitored at all times and if deviations are larger than expected in the baseline schedule the manager should decide if recovery actions are needed. If we count with a reliable *Knowledge Based System*, the system could handle some of the events and provide suggestions as to how to proceed, but if we don't, then the manager should decide again if the project has a high impact or large scope. If not, *schedule repairing actions* can be applied, such as the right shift rule. However, if the project does have high impact or large scope, management will want to consider creating a new baseline schedule in order to minimize the deviation of the project and reduce its overall cost.

Throughout the framework, the main objective is to reduce the need for reactive measures and their inherent cost. For this, we need to look for reliable information, wait until a schedule is absolutely necessary and finally create a robust schedule when it's needed. The initial goal is to create a schedule that can include as much of the expected variation as possible so as to avoid major changes during the execution. However, if the schedule is disrupted and reactive measures are needed, we want to reduce their cost and overall impact. The most inexpensive way could be to simply right-shift the activities until the required resources become available. But if the cost (whether in time, budget or quality) of right-shifting is high, or if the project is crucial to the company, then we need to reschedule with a multi-objective optimization in mind.

4 Conclusion and Future Research

Given the absence of a methodology to guide project management practitioners through their activities this study proposes the creation of an integrative scheduling framework for the MRCPSP. Unlike current procedures which rely heavily on probabilities, expert's judgment and/or simulations, this framework adopts measures from the production scheduling environment because it has been thoroughly studied given their higher likelihood of disruptions. Furthermore, in contrast to the existing reactive procedures for MRCPSP, we consider that taking reactive measures not only affects the project's make-span but also its scope, budget and/or outcome quality. For this reason, this framework accounts for uncertainty from the outset, prior to schedule generation. Then, it includes as much of the updated information as possible in order to reduce the need for reactive measures. And if reactive measures are still required, they should be based on the project scope or impact to reduce overall costs.

As noted in [21] *the challenge is no longer how to generate schedules but rather how schedules can be maintained to ensure that the goals are achieved*. For this reason, despite the difficulties it presents, in further studies we are developing what we call *threshold parameters* based on the complexity of the precedence relationships and other selected parameters that can help us monitor and predict changes to project feasibility. This way we can effectively help management teams either recover distressed projects or raise flags for projects which should be abandoned; in either of these, actions would be performed early enough in the project life-cycle to allow for an effective reaction.

Acknowledgements This work was partially supported by the Ministry of Science and Technology Taiwan grants: [MOST103-2221-E-253-005 and MOST104-2221-E-253-002].

References

1. H. Aytug, M.A. Lawley, K. McKay, S. Mohan, R. Uzsoy, Executing production schedules in the face of uncertainties: a review and some future directions. Eur. J. Operation. Res. **161**(1), 86–110 (2005)
2. S.D. Wu, R.H. Storer, P.C. Chang, One machine rescheduling heuristics with efficiency and stability as criteria. Comput. Operation. Res. **20**, 1–14 (1993)
3. W. Herroelen, R. Leus, Project scheduling under uncertainty-survey and research potentials. Eur. J. Operation. Res. **165**(2), 289–306 (2005)
4. J.R. Turner, R. Müller, On the nature of the project as a temporary organization. Int. J. Project Manag. **21**:1–8 (2003)
5. A.H.L. Chen, Y.C. Liang, J.D. Padilla, An experimental reactive scheduling framework for the multi-mode resource constrained project scheduling problem, in *Lecture Notes in Engineering and Computer Science: Proceedings of the International MultiConference of Engineers and Computer Scientists 2017*, pp. 853–858, Hong Kong, 15–17 March 2017
6. L. Ju, Z. Tan, J. Yuan, Q. Tan, H. Li, F. Dong, A bi-level stochastic scheduling optimization model for a virtual power plant connected to a windphotovoltaicenergy storage system considering the uncertainty and demand response. Appl. Energy **171**, 184–199 (2016)
7. H. Wu, M. Shahidehpour, Z. Li, W. Tian, Chance-constrained day-ahead scheduling in stochastic power system operation. IEEE Trans. Power Syst. **29**(4), 1583–1591 (2014)
8. K. Liu, E. Fridman, K.H. Johansson, Networked control with stochastic scheduling. IEEE Trans. Autom. Control **60**(11), 3071–3076 (2015)
9. T. James, *Control of multiclass queueing systems with abandonments and adversarial customers* (Lancaster University, Thesis, 2016)
10. D.E. Shobrys, D.C. White, Planning, scheduling and control systems: why can they not work together? Comput. Chem. Eng. **24**, 163–173 (2000)
11. G. Sand, S. Engell, A. Mrkert, R. Schultz, and C. Schulz. Approximation of an ideal online scheduler for a multiproduct batch plant. Comput. Chem. Eng. **24**, 361–367 (2000)
12. R. Palma-Behnke, C. Benavides, F. Lanas, B. Severino, L. Reyes, J. Llanos, D. Saz, A microgrid energy management system based on the rolling horizon strategy. IEEE Trans. Smart Grid **4**(2), 996–1006 (2013)
13. P.C. Lin, R. Uzsoy, Chance-constrained formulations in rolling horizon production planning: an experimental study. Int. J. Product. Res. **54**(13), 3927–3942 (2016)
14. H. Meyr, M. Mann, A decomposition approach for the general lotsizing and scheduling problem for parallel production lines. Eur. J. Operation. Res. **229**(3), 718–731 (2013)
15. B. Ghaddar, J. Naoum-Sawaya, A. Kishimoto, N. Taheri, B. Eck, A lagrangian decomposition approach for the pump scheduling problem in water networks. Eur. J. Operation. Res. **241**(2), 490–501 (2015)
16. M.L. Blom, A.R. Pearce, P.J. Stuckey, A decomposition-based algorithm for the scheduling of open-pit networks over multiple time periods. Manag. Sci. **62**(10), 3059–3084 (2016)

17. P. Prosser, A reactive scheduling agent. Eleventh Int. Joint Conf. Artificial Intelligen. **89**, 1004–1009 (1989)
18. X. Chen, Z. Xu, G. Dsa, X. Han, and H. Jiang. Semi-online hierarchical scheduling problems with buffer or rearrangements. Informat. Process. Lett. **113**(4), 127–131 (2013)
19. X. Chen, N. Ding, G. Dsa, X. Han, and H. Jiang. Online hierarchical scheduling on two machines with known total size of low-hierarchy jobs. Int. J. Comput. Mathemat. **92**(5), 873–881 (2015)
20. A.A. Bhattacharya, D. Culler, E. Friedman, A. Ghodsi, S. Shenker, I. Stoica. Hierarchical scheduling for diverse datacenter workloads. in *Proceedings of the 4th Annual Symposium on Cloud Computing*, p. 4. ACM (2013)
21. J.F. O'Kane, A knowledge-based system for reactive scheduling decision-making in FMS. J. Intelligen. Manufact. **11**(5), 461–474 (2000)
22. I. Motawa, A. Almarshad, A knowledge-based BIM system for building maintenance. Automat. Construct. **29**, 173–182 (2013)
23. Y. Sotskov, N.Y. Sotskova, F. Werner, Stability of an optimal schedule in a job shop. Omega **25**(4), 397–414 (1997)
24. F. Guinand, A. Moukrim, E. Sanlaville, Sensitivity analysis of tree scheduling on two machines with communication delays. Parallel Comput. **30**(1), 103–120 (2004)
25. Z. Jia, M.G. Ierapetritou, Short-term scheduling under uncertainty using MILP sensitivity analysis. Indust. Eng. Chem. Res. **43**, 3782–3791 (2004)
26. S. Maqsood, S. Noor, M.K. Khan, A. Wood, Hybrid genetic algorithm (GA) for job shop scheduling problems and its sensitivity analysis. Int. J. Intelligen. Syst. Technol. Appl. **11**(1–2), 49–62 (2012)
27. J.C. Thiele, W. Kurth, V. Grimm, Facilitating parameter estimation and sensitivity analysis of agent-based models: a cookbook using NetLogo and 'R'. J. Artif. Soc. Social Simulat. **17**(3), 11 (2014)
28. A. Muzhikyan, A.M. Farid, K. Youcef-Toumi, An enterprise control assessment method for variable energy resource-induced power system imbalances 2014;Part II: Parametric sensitivity analysis. IEEE Trans. Indust. Electron. **62**(4), 2459–2467 (2015)
29. N.G. Hall, M.E. Posner, Sensitivity analysis for scheduling problems. J. Schedul. **7**(1), 49–83 (2004)
30. Z.K. Li, M. Ierapetritou, Process scheduling under uncertainty: review and challenges. Comput. Chem. Eng. **32**, 715–727 (2008)
31. J. Dorn, R. Kerr, G. Thalhammer, Reactive scheduling: improving robustness of schedules and restricting the effects of shop floor disturbances by fuzzy reasoning. Int. J. Human Comput. Studies **42**, 687–704 (1995)
32. M. Hapke, A. Jaskievicz, R. Slowinski, Fuzzy project scheduling system for software development. Fuzzy Sets Syst. **21**, 101–117 (1994)
33. S.A. Torabi, N. Sahebjamnia, S.A. Mansouri, M. Aramon, Bajestani. A particle swarm optimization for a fuzzy multi-objective unrelated parallel machines scheduling problem. Appl. Soft Comput. **13**(12), 4750–4762 (2013)
34. W.C. Yeh, P.J. Lai, W.C. Lee, M.C. Chuang, Parallel-machine scheduling to minimize makespan with fuzzy processing times and learning effects. Informat. Sci. **269**, 142–158 (2014)
35. I. Bakry, O. Moselhi, T. Zayed, Fuzzy dynamic programming for optimized scheduling of repetitive construction projects. in *IFSA World Congress and NAFIPS Annual Meeting (IFSA/NAFIPS), 2013 Joint*, pp. 1172–1176 (2013)
36. V. Dixit, R.K. Srivastava, A. Chaudhuri, Procurement scheduling for complex projects with fuzzy activity durations and lead times. Comput. Indust. Eng. **76**, 401–414 (2014)
37. J. Xu, Y. Ma, X. Zehui, A bilevel model for project scheduling in a fuzzy random environment. IEEE Trans. Syst. Man Cybernet. **45**(10), 1322–1335 (2015)
38. Y. Xu, L. Wang, S.Y. Wang, M. Liu, An effective teaching-learning-based optimization algorithm for the flexible job-shop scheduling problem with fuzzy processing time. Neurocomputing **148**, 260–268 (2015)

39. M.A. Al-Fawzan, M. Haouari, A bi-objective model for robust resource-constrained project scheduling. Int. J. Product. Econom. **96**, 175–187 (2005)
40. H. Gao, *Building robust schedules using temporal protection - an empirical study of constraint based scheduling under machine failure uncertainty* (University of Toronto, Thesis, 1995)
41. E. Goldratt. *Critical Chain*. The North River Press (1997)
42. A. Davenport, C. Gefflot, C. Beck, Slack-based techniques for robust schedules. in *Sixth European Conference on Planning*, pp. 43–49 (2014)
43. P. Kobyalanski, D. Kutcha, A note on the paper by M. A. Al-Fawzan and M. Haouari about A bi-objective model for robust resource-constrained project scheduling. Int. J. Product. Econom. **107**, 496–501 (2007)
44. H. Chtourou, M. Haouari, A two stage priority rule based algorithm for robust resource constrained project scheduling. Comput. Indust. Eng. **55**(1), 183–194 (2008)
45. A.H.L. Chen, Y.C. Liang, J.D. Padilla, An entropy-based upper bound methodology for robust predictive multi-mode RCPSP schedules. Entropy **16**, 5032–5067 (2014)
46. J. Rezaeian, F. Soleimani, S. Mohaselafshary, A. Arab, Using a meta-heuristic algorithm for solving the multi-mode resource-constrained project scheduling problem. Int. J. Operation. Res. **24**(1), 1–16 (2015)
47. P. Lamas, E. Demeulemeester, A purely proactive scheduling procedure for the resource-constrained project scheduling problem with stochastic activity durations. J. Sched. **19**(4), 409–429 (2016)
48. O. Alagz, M. Azizoglu, Rescheduling of identical parallel machines under machine eligibility constraints. Eur. J. Operation. Res. **149**(3), 523–532 (2003)
49. M.S. Akturk, E. Gorgulu, Match-up scheduling under a machine breakdown. Eur. J. Operation. Res. **112**, 81–97 (1999)
50. P. Mauguire, J.C. Billaut, C. Artigues, Grouping jobs on a single machine with heads and tails to represent a family of dominant schedules. In *8th Workshop on Project Management and Scheduling, Valencia*, pp. 3–5 (2002)
51. F. Quesnel, A. Lbre, M. Sdholt, Cooperative and reactive scheduling in large-scale virtualized platforms with DVMS. Concurr. Comput. Pract. Exp. **25**(12), 1643–1655 (2013)
52. L. Nie, L. Gao, P. Li, X. Li, A GEP-based reactive scheduling policies constructing approach for dynamic flexible job shop scheduling problem with job release dates. J. Intelligen. Manufact. **24**(4), 763–774 (2013)
53. S. Van de Vonder, *Proactive-reactive procedures for robust project scheduling* (Katholieke Universiteit Leuven, Thesis, 2006)
54. S. Van de Vonder, F. Ballestn, E. Demeulemeester, and W. Herroelen. Heuristic procedures for reactive project scheduling. Comput. Indust. Eng. **52**(1), 11–28 (2007)
55. S. Van de Vonder, E. Demeulemeester, W. Herroelen, Proactive heuristic procedures for robust project scheduling: an experimental analysis. Eur. J. Operation. Res. **189**(3), 723–733 (2008)
56. O. Lambrechts, E. Demeulemeester, W. Herroelen, Proactive and reactive strategies for resource-constrained project scheduling with uncertain resource availabilities. J. Sched. **11**(2), 121–136 (2008)
57. O. Lambrechts, E. Demeulemeester, W. Herroelen, A tabu search procedure for developing robust predictive project schedules. Int. J. Product. Econom. **111**(2), 496–508 (2007)
58. G. Zhu, J. Bard, G. Yu, Disruption management for resource-constrained project scheduling. J. Operation. Res. Soc. **56**(4), 365–381 (2005)
59. F. Deblaere, E. Demeulemeester, W. Herroelen, Reactive scheduling in the multi-mode RCPSP. Comput. Operation. Res. **38**(1), 63–74 (2011)
60. W. Herroelen. *A Risk Integrated Methodology for Project Planning Under Uncertainty*, book section 9, pp. 203–217. Springer, New York (2014)

A Supply Chain Design of Perishable Products Under Uncertainty

Himanshu Shrivastava, Pankaj Dutta, Mohan Krishnamoorthy and Pravin Suryawanshi

Abstract Most supply chain (SC) studies often consider conventional products and assign little importance to the product perishability. In addition, most SC models in the literature assume that transportation routes are disruption-free. However, in reality, transportation routes are subject to various sorts of disruptions. In this chapter, we develop a stochastic mathematical model for a perishable product under conditions of route disruption and demand uncertainty. We investigate optimal facility location and distribution strategies that minimise the total cost of the SC. We propose two policies for decision-making under uncertainty. The first one is the risk-neutral policy in which the expected cost of the SC is minimised. The second policy is the risk-averse policy. The risk-averse policy is proposed through conditional value-at-risk (CVaR) approach in which the worst-case cost is minimised. The effectiveness of our model is demonstrated through an illustrative example. We observe that a resilient SC and a disruption-free SC have different designs. Finally, the effect of disruption uncertainties is presented through a statistical analysis.

Keywords Conditional value-at-risk (CVaR) · Distribution planning · Network design · Perishable products · Resilient supply chains · Uncertainty

H. Shrivastava
IITB-Monash Research Academy, IIT Bombay, Powai, Mumbai 400076,
Maharashtra, India
e-mail: himanshu.shrivastava@iitb.ac.in

P. Dutta (✉) · P. Suryawanshi
Shailesh J. Mehta School of Management, Indian Institute of Technology
Bombay, Powai, Mumbai 400076, Maharashtra, India
e-mail: pdutta@iitb.ac.in

P. Suryawanshi
e-mail: pravinsuryawanshi@iitb.ac.in

M. Krishnamoorthy
Department of Mechanical and Aerospace Engineering, Monash University,
Melbourne, VIC 3180, Australia
e-mail: m.krishnamoorthy@uq.edu.au

M. Krishnamoorthy
School of Information Technology and Electrical Engineering, The University
of Queensland, Brisbane, QLD 4072, Australia

© Springer Nature Singapore Pte Ltd. 2018
S.-I. Ao et al. (eds.), *Transactions on Engineering Technologies*,
https://doi.org/10.1007/978-981-10-7488-2_6

1 Introduction

Supply chain risk management is an important part of organisational strategy [1]. In most firms, one of the major concerns in the current competitive environment is SC disruptions [1, 2]. With the increase in the complexity of SCs in recent years, the occurrence and severity of SC disruption has also increased [3]. As per the World Economic Forum's report, 'Building Resilience in Supply Chains', SC disruptions cause a reduction of 7% in the share prices of the affected companies [3]. Additionally, 80% of global companies considers SC protection as their top priority [3]. Authors such as [2, 4] provide ample examples of SC disruptions. A review article by [2] presents an overview of a study that was carried out in the aspect of SC disruptions; it also discusses the various modeling approaches in this context and provide insights into 180 research articles under the four disruption mitigating categories: "(a) mitigating disruption through inventories; (b) mitigating disruptions through sourcing and demand flexibility; (c) mitigating disruptions through facility location; and (d) mitigating disruptions through interaction with external partners." A global SC comprises the sourcing of raw materials from (and distributing goods to) other countries. This, inevitably, gives rise to various disruptions. Enterprises must manage SC disruptions and reduce this vulnerability [5]. Therefore, managing and mitigating disruptions has become an important research issue in the recent past [2, 5, 6].

Most of the studies about SC have been limited to 'regular products' for which perishability is not a major concern. In India, the perishable product market is one of the fast developing markets, for example, India stands second in the production of fruits and vegetables in the world, after China[1]. Also, the SC challenges in perishable goods are more complex when compared to regular products, because the value of the product deteriorates over time [7]. Furthermore, economic shocks, product varieties and management issues are unavoidable in the overall working of a perishable product's SC [8].

There are many approaches to the mitigating, quantifying and analysing of risk problems. These problems can be demonstrated by using fuzzy concepts and also through stochastic modeling using distribution functions [4]. The stochastic modeling approach has been used in this chapter. In normal unbiased risk conditions, mathematical modeling could throw light on the quality of the SC system. However, due to the involvement of various risk factors, it would be favourable to optimise for worst case SC scenarios too. Moreover, different organisations would have different risk-tackling behaviours. We deploy the Conditional-Value-at-Risk (CVaR) [5, 9–12] approach to manage SC risks in the worst case.

This chapter examines an integrated SC network for perishable products under transportation route disruptions and uncertain demand. We aim to determine an optimal network structure and a suitable distribution strategy for perishable products under uncertain environments which minimises the SC total cost. The overall objective is to help the firm with decision-making under disruption risks. We have

---

[1]http://mofpi.nic.in/documents/reports/annual-report.

proposed two types of policies for decision-making. One is a risk-neutral policy in which the expected total cost of the SC is minimised, while the second policy is risk-averse decision-making in which the CVaR concept is used to analyse and quantify the risk and the worst case SC cost is optimised. Finally, we present an extensive statistical analysis of the disruption uncertainties that are present in the SC.

2 Problem Description and Model Formulation

The research objectives are (a) to minimise the total cost of the SC under uncertain environments, and (b) to quantify and analyse the uncertainties that are present in the SC. In this section, we develop a mixed-integer programming based mathematical model. We consider a two-echelon, single-product and single-period SC system that comprises several manufacturers and retailing outlets under random disruptions. The manufacturers produce a single product which is perishable in nature and are also responsible for processing, packaging and labeling of the products. The potential locations of manufacturers are known in advance. Manufacturers have different limited capacities. The transportation routes between each manufacturer and each retailer may face different amounts of disruptions. If the disruption occurs along the transportation routes, then the shipment that is enroute may experience either full or partial loss.

We assume that the product becomes obsolete at the end of the period, and the retailer has to bear the excess cost of dumping the product. Similarly, if the retailer is unable to meet the demand then an opportunity is lost, and the cost equivalent to it (that is, understocking/opportunity cost) will be incurred. Hence the retailer has to create a balance between its supply and the demand in order to minimise the cost. The retailers foresee their demand and orders it to the manufacturers at the beginning of the period. The lead time for retailers and manufacturers is assumed to be small and constant so that there is no deterioration of the product along the distribution route and at the facilities itself.

We first formulate the cost minimisation model, which determines the optimal network structure under random disruptions. This model minimises the expected total cost of the SC and assumes a completely risk-neutral approach. However, decision makers may have different approaches to undertaking risk. Therefore, we also formulate and analyse a risk-averse model through the CVaR approach.

2.1 Risk Neutral Model

We use a mixed integer stochastic programming approach to formulate a two-stage cost optimisation model. In two-stage stochastic programming, the decision variables are divided into two sets. The first-stage decision variables are those that have to be identified before the realisation of uncertain events, while the second-stage decision variable involves the recourse decisions and are decided after the realisation of uncertain events [13]. In this chapter, the decisions for opening the facilities are modeled at the first stage and are considered as binary decisions. The second-stage decisions enable us to determine the optimal distribution flow in the SC network and are considered here as continuous decision variables. Therefore, the first stage involves strategic decisions, while the second stage involves tactical decisions, which are decided after the realisation of uncertain disruptions. To incorporate uncertainty in the second stage, a scenario-based modeling approach is used [9–11]. The stochastic formulation for the minimisation problem is [10]:

$$min \quad \sum_{w \in W} x_w \cdot H_w + \Omega[Q(x, s)]$$

$$subject \ to \ x_w \in \{0, 1\} \quad \forall \quad w \in W$$

(1)

where x_w (where $w \in W$ is index for facilities) is the facility location decision at the first stage, H_w is the fixed cost associated with the facility w and $Q(x, s)$ is the second-stage optimal solution that corresponds to first-stage decision x and scenario s; $\Omega[Q(x, s)]$ is the expected cost assumed with respect to scenario, s, which indicates the realisation of a facility's state. For a finite set of scenarios, the deterministic equivalent of $E[Q(x, s)]$ is given as, $\sum_{s \in S} \rho_s \cdot Q(x, s)$ [10], where ρ_s is the probability of the occurrence of scenario, s. The details of the scenario and the probability of its occurrence is explained later. We have used the following notations to formulate mathematical model:

Indices:

$$m \in M \longrightarrow \text{The set of potential locations for manufacturers;}$$
$$r \in R \longrightarrow \text{The set of retailers that need to be serviced;}$$
$$s \in S \longrightarrow \text{The set of disruption scenarios;}$$

Decision variables:

$$y_m \longrightarrow \text{Equals 1 if manufacturer is open at location } m \text{ and 0 otherwise;}$$
$$x_{mrs} \longrightarrow \text{Quantity of final product shipped from } m \text{ to } r \text{ in scenarios;}$$

Parameters:

$$F_m \longrightarrow \text{Manufacturer's fixed opening cost at candidate location } m;$$
$$D_r \longrightarrow \text{Demand at retailer } r;$$
$$E(D_r) \longrightarrow \text{Expected demand at retailer } r;$$
$$F(D_r) \longrightarrow \text{Cumulative distribution function of } D_r;$$
$$O_r \longrightarrow \text{Handling cost per unit at retailer } r;$$
$$K_m \longrightarrow \text{Capacity of manufacturer } m;$$

$P_m \longrightarrow$ Sum of unit production and unit holding cost at manufacturer m;

$B \longrightarrow$ Budget limit of opening manufacturer's facilities;

$C_{mr} \longrightarrow$ Unit cost of shipping final product from manufacturer m to retailer r;

$\beta_{mrs} \longrightarrow$ Fraction of supply disruption in the link between m and r in scenario s;

$\sigma_{mr} \longrightarrow$ Unit penalty cost of disruption;

$Z \longrightarrow$ Desired level of fill rate;

$C_S \longrightarrow$ Unit shortage cost to retailer;

$C_E \longrightarrow$ Unit excess cost to retailer;

We assume that β_{mrs} follows a certain known distribution whose mean and standard deviation are known in advance. In order to formulate the mathematical model, we first compute the SC cost at each echelon in scenario, s.

The total cost of the SC from manufacturer m to retailer r:

$$F_m \cdot y_m + P_m \cdot x_{mrs} + C_{mr} \cdot x_{mrs} + \beta_{mrs} \cdot x_{mrs} \cdot \sigma_{mr} \qquad (2)$$

The first term in Eq. (2) indicates the fixed opening cost of the manufacturer's facilities and the second term denotes the production and holding costs at manufacturer m while the third term indicates the transportation cost from manufacturer m to retailer r. The last term in the above equation denotes the penalty cost of disruption as the transportation link is assumed to have an associated risk of disruption.

If disruption occurred $\beta_{mrs}\%$ of supply is assumed to be disrupted. Hence the quantity arriving at the retailer r is $(1 - \beta_{mrs}) \cdot x_{mrs}$.

The total cost at retailer r (T_r):

$$T_r = \sum_{m \in M} O_r \cdot (1 - \beta_{mrs}) \cdot x_{mrs} + C_E \left(\sum_{m \in M} (1 - \beta_{mrs}) \cdot x_{mrs} - D_r \right)^+$$
$$+ C_S \left(D_r - \sum_{m \in M} (1 - \beta_{mrs}) \cdot x_{mrs} \right)^+ \qquad (3)$$

where, $A^+ = \max\{A, 0\}$.

In other words, the total cost of the retailer is comprised of the handling cost (which is a combination of holding cost and processing/packaging cost) plus excess cost (of overstocking plus shortage cost (of unfulfilled demand). Due to the perishable nature of the product and single period SC distribution planning, we are deploying a *newsvendor style* model [14, 15] for managing and calculating the inventory of the retailer. Equation (3) is simplified to following equation:

$$T_r = \sum_{m \in M} O_r \cdot (1 - \beta_{mrs}) \cdot x_{mrs} + C_E \left(\int_0^{\sum_{m \in M}(1-\beta_{mrs}) \cdot x_{mrs}} F\left(D_r\right) dD_r \right)$$

$$+ C_S \left(\int_0^{\sum_{m \in M}(1-\beta_{mrs}) \cdot x_{mrs}} F\left(D_r\right) dD_r - \sum_{m \in M} (1 - \beta_{mrs}) \cdot x_{mrs} + E(D_r) \right) \quad (4)$$

For finite set of scenarios , the deterministic equivalent of stochastic formulation for minimisation of total SC's is given below:

Objective function:

$$Min \ U = \sum_{m \in M} F_m \cdot y_m + \sum_{s \in S} \rho_s \left(\sum_{m \in M} \sum_{r \in R} P_m \cdot x_{mrs} + \sum_{r \in R} \sum_{m \in M} C_{mr} \cdot x_{mrs} \right.$$

$$+ \sum_{r \in R} \sum_{m \in M} \beta_{mrs} \cdot x_{mrs} \cdot \sigma_{mr} + \sum_{r \in R} \sum_{m \in M} O_r \cdot (1 - \beta_{mrs}) \cdot x_{mrs}$$

$$+ \left(C_E + C_s \right) \left[\sum_{r \in R} \left(\int_0^{\sum_{m \in M}(1-\beta_{mrs}) \cdot x_{mrs}} F\left(D_r\right) dD_r \right) \right]$$

$$\left. - C_S \left[\sum_{r \in R} \left(\sum_{m \in M} (1 - \beta_{mrs}) \cdot x_{mrs} - E(D_r) \right) \right] \right) \quad (5)$$

Subject to:

$$\sum_{r \in R} x_{mrs} \leq K_m \cdot y_m \ ; \quad \forall \ m \in M \ ; \quad \forall \ s \in S \quad (6)$$

$$\sum_{m \in M} F_m \cdot y_m \leq B \quad (7)$$

$$Z \leq \frac{\sum_{m \in M}(1 - \beta_{mrs}) \cdot x_{mrs}}{E(D_r)} \ ; \quad \forall \ r \in R \ ; \quad \forall \ s \in S \quad (8)$$

$$x_{mrs} \geq 0 \ ; \quad \forall \ r \in R \ ; \quad \forall \ m \in M \ ; \quad \forall \ s \in S \quad (9)$$

$$y_m \in \{0,1\} \ ; \quad \forall \ m \in M \quad (10)$$

The objective function minimises the total cost of the SC network. Constraint Eq. (6) imposes capacity constraints on manufacturers. Constraint Eq. (7) represents the budget constraint. Constraint Eq. (8) ensures that service level should be greater or equal to $Z\%$. Constraint Eq. (9) and Eq. (10) respectively impose the non-negativity and binary restrictions.

The decision variables addresses the optimal network structure. The decision variable in our model includes binary variables that represents the existence of manufacturers and the continuous variable that represent the material flow from manufacturers to retailers.

We have considered demand to be uniformly distributed. However, the model can be used for other distributions too. The uniform demand distribution, $F(D)$, in the interval $[a, b]$ is given as:

$$F(D) = \frac{D - a}{b - a} \qquad a \leq D \leq b \tag{11}$$

Substituting $F(D)$ in the objective function, the resulting expression is:

$$
\begin{aligned}
Min \quad U = &\sum_{m \in M} F_m \cdot y_m + \sum_{s \in S} \rho_s \left(\sum_{m \in M} \sum_{r \in R} P_m \cdot x_{mrs} + \sum_{r \in R} \sum_{m \in M} C_{mr} \cdot x_{mrs} \right. \\
&+ \sum_{r \in R} \sum_{m \in M} \beta_{mrs} \cdot x_{mrs} \cdot \sigma_{mr} + \sum_{r \in R} \sum_{m \in M} O_r \cdot (1 - \beta_{mrs}) \cdot x_{mrs} + \left(C_E + C_s \right) \\
&\left[\sum_{r \in R} \left(\frac{\left(\sum_{m \in M} (1 - \beta_{mrs}) \cdot x_{mrs} \right)^2}{2 \cdot (b_r - a_r)} - a_r \cdot \frac{\left(\sum_{m \in M} (1 - \beta_{mrs}) \cdot x_{mrs} \right)}{b_r - a_r} \right) \right] \\
&\left. - C_s \left[\sum_{r \in R} \left(\sum_{m \in M} (1 - \beta_{mrs}) \cdot x_{mrs} - E(D_r) \right) \right] \right)
\end{aligned}
\tag{12}
$$

subject to: Eqs. (6)–(10). The above is a quadratic expression and hence we have mixed integer quadratic model.

2.1.1 Scenario Generation and Scenario Reduction

As stated earlier, to deal with the uncertainty in the second stage, a scenario based modeling approach is used. To determine the probability of scenario s (ρ_s), we need to define a scenario. To realise a scenario, we follow the method that is used by [10, 11].

Let State-1 indicate that there is chance of no disruption in the route and State-0 indicate the chance that the route is partly or fully disrupted. Let the total number of routes be F. A scenario is defined thus: the subset of transportation routes (say, F') are in State-1 and the transportation routes in the other set, (say, F/F') are in State-0. Hence there are a total of 2^F possible scenarios. Let Θ be the probability of being in State-1. Hence, the probability of scenario s can be realised is:

$$\rho_s = \prod_{F \in F'} \Theta_{F'} \prod_{F \in F/F'} (1 - \Theta_{F/F'}) \tag{13}$$

We have observed that total number of possible scenarios depends on the total number of transportation routes. If there are three manufacturers and four retailers, then the total number of routes from manufacturers to retailer will be 12. Therefore, it is possible to have as many as 2^{12} scenarios which is a huge number. To overcome this we have deployed the scenario reduction technique that was proposed by [16] and recently deployed by [17].

Let θ_i ($i = 1$ to I) be the vector of uncertain parameters. There is finite set of values that each uncertain parameter, that is, θ_i, can have, and is given by $\theta_i^{j_i}, j_i = 1$ to J_i, $i = 1$ to I and hence there are a total of $\prod_{i=1}^{I} J_i$ scenarios. The probability associated with $\theta_i^{j_i}$ is $p_i^{j_i}$. Assuming each uncertain event is independent, the probability of a scenario is given by $\prod_{i=1}^{I} p_i^{j_i}$. Karuppiah et al. [16] have proposed a heuristic to obtain a minimum number of scenarios from the original set of scenarios in which the sum of the probabilities of the new scenarios with the uncertain parameters value, $\theta_i^{j_i}$ appears is equal to $p_i^{j_i}$. Let $\hat{p}_{1j_1, 2j_2, \ldots, Ij_I}$ be the new probability that is associated with a scenario. The relaxed mathematical formulation for scenario reduction is given below:

$$min \ \sum_{j_1=1}^{J_1} \sum_{j_2=1}^{J_2} \cdots \sum_{j_1=i}^{J_I} (1 - p_1^{j_1} p_2^{j_2} \cdots p_I^{j_I}) \cdot \hat{p}_{1j_1, 2j_2, \ldots, Ij_I} \tag{14}$$

Subject to:

$$\sum_{j_2=1}^{J_2} \sum_{j_3=1}^{J_3} \cdots \sum_{j_1=i}^{J_I} \hat{p}_{1j_1, 2j_2, \ldots, Ij_I} = p_1^{j_1} \quad j_1 = 1, \ldots, J_1 \tag{15}$$

$$\sum_{j_1=1}^{J_1} \sum_{j_3=1}^{J_3} \cdots \sum_{j_1=i}^{J_I} \hat{p}_{1j_1, 2j_2, \ldots, Ij_I} = p_2^{j_2} \quad j_2 = 1, \ldots, J_2 \tag{16}$$

$$\vdots$$

$$\sum_{j_1=1}^{J_1} \sum_{j_3=1}^{J_3} \cdots \sum_{j_1=i-1}^{J_{I-1}} \hat{p}_{1j_1, 2j_2, \ldots, Ij_I} = p_i^{j_I} \quad j_I = 1, \ldots, J_I \tag{17}$$

$$\sum_{j_1=1}^{J_1} \sum_{j_3=1}^{J_3} \cdots \sum_{j_1=i}^{J_I} \hat{p}_{1j_1, 2j_2, \ldots, Ij_I} = 1 \tag{18}$$

$$0 \leq \hat{p}_{1j_1, 2j_2, \ldots, Ij_I} \leq 1 \quad \forall \ j_1, j_2, \ldots, j_I \tag{19}$$

2.2 Risk Averse Model

The model developed in Sect. 2.1 minimises the expected total cost, which is a risk-neutral condition. This approach helps the organisation to evaluate the performance of the SC. However, in the long run, minimising the expected cost may not be very useful, especially, when the SC is subjected to various sorts of risks and uncertainties. In such conditions, minimising downside risk would provide more insight to the SC and would be more beneficial to firms. In other words, due to the various risk factors involved, it would be beneficial to minimise the worst-case total cost of the SC. In this section we present the worst-case risk measure through the CVaR concept.

Value-at-Risk (VaR) provides the basis for CVaR. The VaR is defined as the lowest amount ζ, such that, with a specified probability, λ, the loss will not exceed ζ [12]. This means that we allow $(1 - \lambda)100\%$ of the outcome to exceed ζ. To overcome this, we use the CVaR approach, which is defined as the expected loss that exceeds VaR. In other words, this is the expectation of losses above the amount, ζ.

In our model, the confidence level $\lambda \in (0, 1)$ will fixed by the decision maker to control the risk of losses due to supply and route disruptions and demand fluctuations. A risk-neutral decision maker may choose the low value of λ while a decision-maker who is risk-averse would select a high value of λ. We have presented an extensive sensitivity analysis with respect to λ later on.

Let $\Psi(\mathbf{u}, \mathbf{v})$ be the cost function (which is to be minimised) with decision vector $\mathbf{u}$ and the random vector $\mathbf{v}$. For each $\mathbf{u}$, the cost $\Psi(\mathbf{u}, \mathbf{v})$ is a random variable having a distribution induced by $\mathbf{v}$. Let the probability density function (PDF) of $\mathbf{v}$ be $p(\mathbf{v})$. The probability of $\Psi(\mathbf{u}, \mathbf{v})$ not exceeding the threshold level ζ is given by [12]:

$$\Phi(\mathbf{u}, \zeta) = \int_{\Psi(\mathbf{u},\mathbf{v}) \leq \zeta} p(\mathbf{v})d\mathbf{v} \tag{20}$$

The VaR value at any specified probability level λ is given by [12]:

$$\zeta_\lambda(\mathbf{u}) = min\{\zeta : \Phi(\mathbf{u}, \zeta) \geq \lambda\} \tag{21}$$

Mathematically, CVaR is given as [12]:

$$CVaR_\lambda(\mathbf{x}, \zeta) = \zeta + (1 - \lambda)^{-1} \int_v [f(\mathbf{u}, \mathbf{v}) - \zeta]^+ p(\mathbf{v})d\mathbf{v} \tag{22}$$

where, $[t]^+ = \max \{t, 0\}$.

For a pre-defined set of discrete scenarios, [12] also studied Eq. (22) as a linear optimisation problem (with CVaR minimisation as the objective) by introducing the auxiliary variable τ_s. Here, τ_s is the tail cost for scenario s, where tail cost is defined as the amount by which the cost in scenario s exceeds ζ. Following the study of [9, 10, 12], the risk averse SC model which minimises the expected worst case cost is defined as:

$$min \quad CVaR = \zeta + \frac{1}{1 - \lambda} \sum_{s \in S} \rho_s \cdot \tau_s \tag{23}$$

Subject to: 6–10 and,

$$
\begin{aligned}
\tau_s \geq &\sum_{m \in M} F_m \cdot y_m + \sum_{m \in M} \sum_{r \in R} P_m \cdot x_{mrs} + \sum_{r \in R} \sum_{m \in M} C_{mr} \cdot x_{mrs} \\
&+ \sum_{r \in R} \sum_{m \in M} \beta_{mrs} \cdot x_{mrs} \cdot \sigma_{mr} + \sum_{r \in R} \sum_{m \in M} O_r \cdot (1 - \beta_{mrs}) \cdot x_{mrs} \\
&+ (C_E + C_s) \left[\sum_{r \in R} \left(\int_0^{\sum_{m \in M}(1 - \beta_{mrs}) \cdot x_{mrs}} F(D_r) \, dD_r \right) \right] \\
&- C_S \left[\sum_{r \in R} \left(\sum_{m \in M} (1 - \beta_{mrs}) \cdot x_{mrs} - E(D_r) \right) \right] - \zeta , \ \forall \, s \in S
\end{aligned}
\tag{24}
$$

$$
\tau_s \geq 0 , \ \forall \, s \in S
\tag{25}
$$

Equation (24), in the formulation defines the tail cost for scenario s which exceeds ζ. Constraint Eq. (25) imposes a non negativity condition on the tail cost. By using a suitable demand distribution function the risk-averse model is simplified in a similar way as described in Sect. 2.1.

3   Results and Discussions

We have implemented our formulation in order to design a small SC that has the risk of being disrupted at transportation links. In our example, we have considered three manufacturers and four retailers. Thus, there are a total of twelve transportation routes in the SC. We have solved our model by using the default settings of the CPLEX optimisation software (version 12.6) on an Intel(R) 2.4 GHz computer with 4 GB of RAM.

We first analyse the results of the resilient design model and compare the results with that obtained from the disruption-free design (with no disruptions). In the resilient design, we consider that disruptions can occur in any of the twelve routes between the manufacturing unit to the retailer, which leads to many scenarios. The scenario-reduction technique led us to choose five scenarios. Table 1 presents the overview of the chosen scenario. In all these five scenarios, the transportation route between m_1 and r_1 are in State-0, which signifies that there is a chance that this route will always be disrupted. Similarly, the routes that are in State-1 signify that there is a possibility that these routes will not be disrupted. In the first scenario, seven out of twelve routes have a chance of disruption and there is a high probability of the occurrence of this scenario. The possibility of the occurrence of the first scenario is higher than that of the second and the third scenarios. The fourth and fifth scenarios occur rarely. The major contributor to cost is the first scenario because of its high probability of occurrence. The disruption in the route is characterised by β, which is shown in Table 2 (scenarios 1, 2 and 3) and Table 3 (scenarios 4 and 5). It should be

Table 1 Route disruption scenarios and probability of their occurrence

	$m_1 - r_1$	$m_1 - r_2$	$m_1 - r_3$	$m_1 - r_4$	$m_2 - r_1$	$m_2 - r_2$	$m_2 - r_3$	$m_2 - r_4$	$m_3 - r_1$	$m_3 - r_2$	$m_3 - r_3$	$m_3 - r_4$	Probability of occurrence (ϕ_s)
Scenario 1	0	1	0	1	0	1	0	1	0	1	0	0	0.55
Scenario 2	0	0	1	1	0	0	0	0	1	0	1	0	0.2
Scenario 3	0	0	0	1	0	1	1	1	1	0	1	1	0.15
Scenario 4	0	0	0	0	1	0	0	0	0	1	0	0	0.07
Scenario 5	0	1	0	0	0	0	0	0	0	0	0	1	0.03

Table 2 Supply disruption probabilities (β) in scenarios 1, 2 and 3

S-1	r_1	r_2	r_3	r_4	S-2	r_1	r_2	r_3	r_4	S-3	r_1	r_2	r_3	r_4
m_1	0.25	0.003	0.17	0.001	m_1	0.15	0.18	0.003	0.005	m_1	0.3	0.22	0.17	0.002
m_2	0.16	0.002	0.26	0.0012	m_2	0.18	0.35	0.23	0.15	m_2	0.25	0.005	0.001	0.003
m_3	0.10	0.0015	0.17	0.22	m_3	0.001	0.45	0.001	0.25	m_3	0.002	0.15	0.006	0.001

Table 3 Supply disruption probabilities (β) in scenarios 4 and 5

S-4	r_1	r_2	r_3	r_4	S-5	r_1	r_2	r_3	r_4
m_1	0.3	0.25	0.1	0.17	m_1	0.16	0.003	0.15	0.17
m_2	0.002	0.16	0.12	0.26	m_2	0.23	0.16	0.22	0.26
m_3	0.15	0.001	0.2	0.5	m_3	0.18	0.1	0.35	0.005

Table 4 Quantity shipment decisions in scenarios 1, 2 and 3

S-1	r_1	r_2	r_3	r_4	S-2	r_1	r_2	r_3	r_4	S-3	r_1	r_2	r_3	r_4
m_1	0	0	2.8	7.2	m_1	0	7.8	0	2.1	m_1	0	0	2.8	0
m_2	13	2.1	0	0	m_2	0	15	0	0	m_2	0	15	0	0
m_3	0	14	8	0	m_3	10.8	0	9	6.7	m_3	10.8	1.5	8	7.2

Table 5 Quantity shipment decisions in scenarios 4 and 5

S-4	r_1	r_2	r_3	r_4	S-5	r_1	r_2	r_3	r_4
m_1	0	0	1.3	8.6	m_1	0	0	10	0
m_2	10.8	0	4.1	0	m_2	14	0	0	0
m_3	0	16.2	5.1	0	m_3	0	18	0	7.2

noted that whenever the routes are in State-1, the corresponding value of β is small, which indicates that reliable transportation routes have low disruption.

The total cost of the SC is INR 132,591.2. This cost considers disruption possibility in all the routes simultaneously, along with demand uncertainties. The quantity shipment decisions in all the scenarios are shown in Table 4 (scenarios 1, 2 and 3) and Table 5 (scenarios 4 and 5). We also observed that all the three manufacturers are required to be open in the resilient design.

We also analyse the disruption-free model, in which we consider the transportation routes as being disruption-free and demand, as being the only the source of uncertainty. Table 6 shows the design decisions for the disruption-free model. This model selects only two manufacturers out three and, hence, the disruption-free model yields a design that is different from that of the resilient model. The total cost of the disruption-free SC is INR 116,888.82. It can be observed that the total cost of the SC is higher in the resilient model.

Table 6 Design decisions for disruption-free model

Quantity shipment decisions					Location decisions	
	r_1	r_2	r_3	r_4	m_1	0
m_1	0	0	0	0	m_2	1
m_2	10.8	4.2	0	0	m_3	1
m_3	0	12	9	7.2	Decisions: 1—open 0—closed	

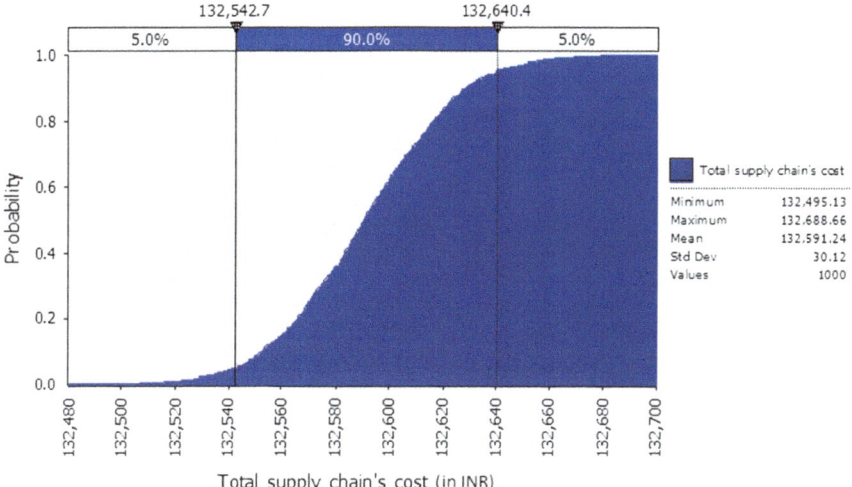

Fig. 1 Variations in SC's total cost due to uncertainty in transportation routes

3.1 Uncertainty Analysis

This section analyses the disruption effects that are present in the transportation routes. We have assumed that these disruptions are uncertain and that they follow a normal distribution with known mean and variance. The uncertainty effect is analysed through various graphs by executing a simulation of 1000 iterations through @Risk[2]. Through simulation, it is observed that the overall cost of the SC would lie between INR 132,542.7 and INR 132,640.4 with 90% confidence. The chance that the total cost exceeds INR 132,640.4 is only five percent. The mean value of the SC's cost is INR 132,591.2. Figure 1 statistically summarises the objective function.

We now analyse the effect of the disruption parameters, β on the SC. Figure 2 shows the tornado graph of the top ten significant β. We observe that the disruption in transportation route is most dominant between manufacturer, m_3, and retailer, r_3, in scenario 1. The R^2 value in this route is 0.883, which means that disruption in this route causes 88.3% variability in the total SC cost. Similarly, there is little variation

---

[2]http://www.palisade.com/risk/.

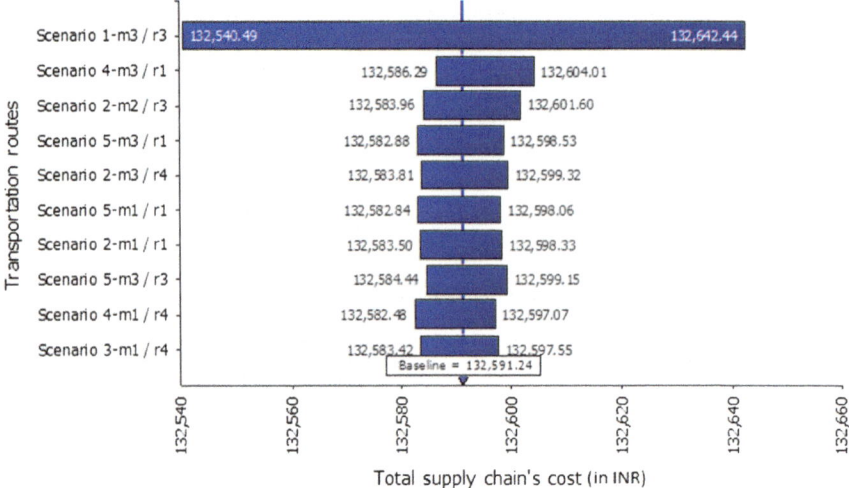

Fig. 2 Effect of disruption on SC's total cost

between $m_1 - r_4$ in scenario 5 and the variability due to this route is 0.06%. The first route is the most risky and is best for a risk-seeking decision maker, whilst a risk-averse decision maker could opt for the last route because the variation in the total cost is the least in this route.

3.2 Risk Averse Analysis

We now analyse our second model which is the risk-averse model. This would help us to compare the difference in the final decision when the decision maker is risk-neutral

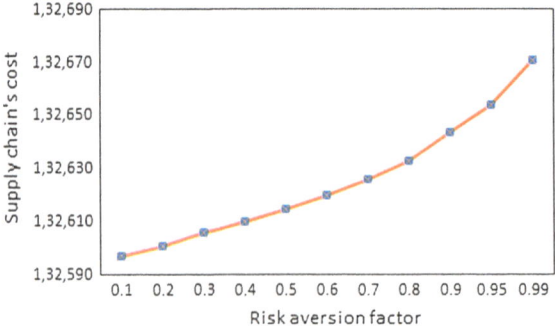

Fig. 3 Total supply chain's cost versus λ

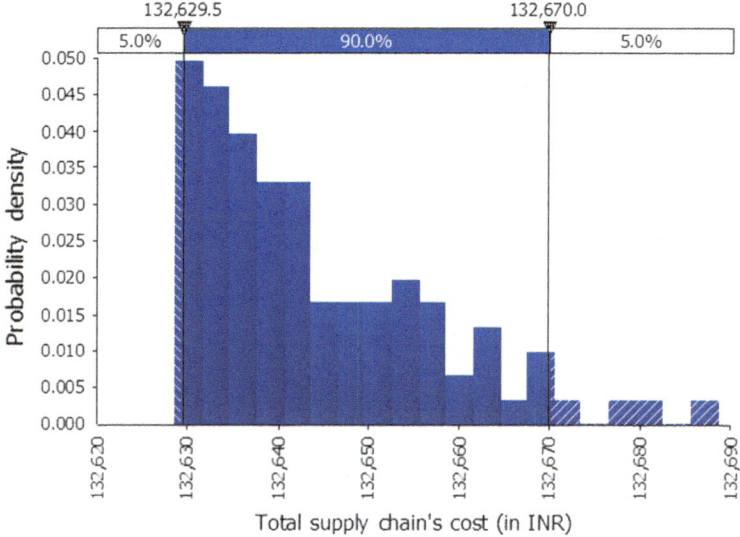

Fig. 4 Variation of CVaR with uncertainty at $\lambda = 0.90$

and risk averse. The risk-aversion is characterised by λ. As the value of λ increases, the risk aversion increases, and the range of worst-case scenarios becomes tighter. This is because the decision maker would be ready to spend more in order to avoid the occurrence of the worst-case scenarios. Figure 3 shows how CVaR varies as λ is varied from 0.1 to 0.99.

Further, Fig. 4 shows the uncertainty effect on CVaR. We observed that at $\lambda = 0.90$, there is a 90% chance that the worst-case lies between INR 132,629.5 and INR 132,670.0 with the mean value of INR 132,644.3.

4 Conclusion and Future Work

In this chapter, we have proposed two types of policies for decision-making under uncertainty and formulated the SC network design problem of a two-echelon SC system for a perishable product. The developed two-stage stochastic model addresses the decision variable, which corresponds to the location of the manufacturer and the quantity flow from the manufacturer to the retailer. The disruption in the transportation routes along with uncertainty in the demand are considered simultaneously. We observed that the SC decisions in the resilient model and the disruption-free model are not the same. It is also observed that as the risk aversion increases, the range of worst-case cost also increases. We have also carried out an extensive analysis of the uncertain disruptions that are present in the transportation link between the manufacturers and retailers and statistically studied the overall nature of the cost function.

This study can be extended by realising a more realistic SC that has a greater number of echelons. Also, in the present study it is assumed that the facilities are disruption-free. But, in practice, the facilities may be prone to disruption uncertainties. We have assumed a single-product and single-period model, however, the model can be extended for multiple products and multiple periods. This study can be further extended by considering other sources of uncertainties such as uncertainties in transportation cost, lead time and so on.

References

1. F. Aqlan, S. Lam Sarah, Supply chain risk modelling and mitigation. Int. J. Prod. Res. **53**(18), 5640–5656 (2015)
2. L.V. Snyder, Z. Atan, P. Peng, Y. Rong, A.J. Schmitt, B. Sinsoysal, OR/MS models for supply chain disruptions: a review. IIE Trans. **48**(2), 89–109 (2016)
3. G. Bhatia, C. Lane, A. Wain, Building resilience in supply chains, in *An Initiative of the Risk Response Network in collaboration with Accenture*. In World Economic Forum, Geneva, Switzerland (2013)
4. A. Baghalian, S. Rezapour, Z. Farahani Reza, Robust supply chain network design with service level against disruptions and demand uncertainties: a real-life case. Eur. J. Oper. Res. **227**(1), 199–215 (2013)
5. N. Azad, H. Davoudpour, K.D. Saharidis Georgios, M. Shiripour, A new model to mitigating random disruption risks of facility and transportation in supply chain network design. Int. J. Adv. Manuf. Technol. **70**(9–12), 1757–1774 (2014)
6. H. Shrivastava, P. Dutta, M. Krishnamoorthy, P. Suryawanshi, Designing a resilient supply chain network for perishable products with random disruptions, in *Proceedings of The International MultiConference of Engineers and Computer Scientists 2017*. Lecture Notes in Engineering and Computer Science. 15–17 March 2017, Hong Kong (2017), pp. 870–875
7. S. Negi, N. Anand, Issues and challenges in the supply chain of fruits & vegetables sector in India: a review. Int. J. Manag. Value Suppl. Chains **6**(2), 47–62 (2015)
8. V.R. Reddy, S.K. Singh, V. Anbumozhi, Food supply chain disruption due to natural disasters: entities, risks, and strategies for resilience. ERIA Discussion Paper 18 (2016)
9. Y. Merzifonluoglu, Risk averse supply portfolio selection with supply, demand and spot market volatility. Omega **57**, 40–53 (2015)
10. A. Madadi, M.E. Kurz, K.M. Taaffe, J.L. Sharp, S.J. Mason, Supply network design: risk-averse or risk-neutral? Comput. Ind. Eng. **78**, 55–65 (2014)
11. T. Sawik, Selection of resilient supply portfolio under disruption risks. Omega **41**(2), 259–269 (2013)
12. R.T. Rockafellar, S. Uryasev, Optimization of conditional value-at-risk. J. Risk **2**, 21–42 (2000)
13. M.C. Georgiadis, P. Tsiakis, P. Longinidis, M.K. Sofioglou, Optimal design of supply chain networks under uncertain transient demand variations. Omega **39**(3), 254–272 (2011)
14. M. Khouja, The single-period (news-vendor) problem: literature review and suggestions for future research. Omega **27**(5), 537–553 (1999)
15. P. Dutta, D. Chakraborty, Incorporating one-way substitution policy into the newsboy problem with imprecise customer demand. Eur. J. Oper. Res. **200**(1), 99–110 (2010)
16. R. Karuppiah, M. Martin, I.E. Grossmann, A simple heuristic for reducing the number of scenarios in two-stage stochastic programming. Comput. Chem. Eng. **34**(8), 1246–1255 (2010)
17. N.S. Sadghiani, S.A. Torabi, N. Sahebjamnia, Retail supply chain network design under operational and disruption risks. Trans. Res. Part E: Logistics Trans. Rev. **75**, 95–114 (2015)

The Relationships Between Education Level, Income Level and Prior Experience, and the Attitude of College Students About Online Shopping in Hong Kong Higher Education

Hon Keung Yau and Lai Ching Tsang

Abstract This study aims to examine the attitude of college students about online shopping in Hong Kong and how the factors education level, income level and prior experience affect the attitude of college students about online shopping. The survey was used in this study and 160 questionnaire had been collected. The results showed that College student in a higher level of education will not have a more positive attitude towards online shopping; (i) College student with a higher income level will have a more positive attitude towards online shopping, and (ii) College student with prior experience in online shopping will have a positive attitude towards online shopping.

Keywords Attitude · College students · Education level · Income level
Online shopping · Prior experience

1 Introduction

Different online shopping attitudes and patterns happen in different location around the world. Thus, researches on specific location for online shopping behavior for different types of consumers were studied. For example, in Malaysia, further investigation for online consumers' attitude towards online shopping as well as the influence of personal factors to their attitude is being studied [1] as online shopping is still considered as in the beginning stage of development.

H. K. Yau (✉) · L. C. Tsang
Department of Systems Engineering and Engineering Management,
City University of Hong Kong, Kowloon Tong, Kowloon, Hong Kong
e-mail: honkyau@cityu.edu.hk

L. C. Tsang
e-mail: lctsang5-c@my.cityu.edu.hk

© Springer Nature Singapore Pte Ltd. 2018
S.-I. Ao et al. (eds.), *Transactions on Engineering Technologies*,
https://doi.org/10.1007/978-981-10-7488-2_7

Shwu-Ing [2] showed the main factor affecting the likelihood of online shopping, Yau and Tsang [3] have studied the gender difference on the online shopping in higher education. In this study, we aim to study the attitude of college students about online shopping in Hong Kong and how the factors education level, income level and prior experience affect the attitude of college students about online shopping. The research question is "what are the relationships between education level, income level and prior experience, and the attitude of college students about online shopping in Hong Kong higher education"?

2   Literature Review

Level of education can add the motive to online shopping in term of capability of performing online shopping and awareness of consequences from online shopping [4]. And there is also research [5] revealing that prominent positive relationship exists between the education level of the consumer and his/her corresponding attitude towards online shopping behavior. Thus, hypothesis 1 is drawn as followed.

H1: College student in a higher level of education will have a more positive attitude towards online shopping.

Income level has a direct effect on the online shopping behavior. Study [6] shows that the income level is directly related to the frequency of online shopping, which means that a higher income level adds motivation for shopping online. In addition, a higher income also brings extra advantages in online shopping, like a higher credit ratings in credit cards payment and a higher change to afford expenses on high-speed internet service, implying a higher chance for more online purchases compared to others [7]. Thus, hypothesis 2 is drawn as followed.

H2: College student with a higher income level will have a more positive attitude towards online shopping.

Prior experience on online shopping could have influence on students' attitude towards online shopping behavior. Naseri and Elliot [7] state that the more products/services the consumers purchase in the internet, the more experienced they would be as an online consumers to give exclusive shopping experience to them. Thus, hypothesis 3 is drawn as followed.

H3: College student with prior experience in online shopping will have a positive attitude towards online shopping.

3   Research Methodology

Questionnaire survey has been chosen as the research method in this study to collect data for examining the gender difference of behavior intention towards online shopping. In the Technology Acceptance Model (TAM) [8], there are 4 questions

Table 1 Items of questionnaire (attitude)

Question	Items
1.	My general attitude towards online shopping is positive
2.	I prefer online shopping to traditional shopping
3.	The thought of buying a product/service online is appealing to me
4.	I recommend online shopping to my friends

concerning attitude (Table 1). These items were rated by a 5-point Likert type scale, ranging from 1 "strongly disagree" to 5 "strongly agree".

When the distributable version of questionnaire is completed, pilot test is conducted. The purpose of pilot test is to ensure that the subjects can understand the questions and also understand them in the same way. 10 questionnaires were distributed to my friends and classmates who are currently under higher education. Participants were asked to give feedback about the questionnaire individually. It was found that the questionnaire could be understood by all of the participants in this pilot study. The questionnaires are distributed after finalized. The target group of this survey is the students under higher education in Hong Kong.

Questionnaires were distributed during the short breaks of lectures, as this is the easiest way to distribute large number of questionnaires and the return rate is high. Also, questionnaire distribution in canteens during lunch time can be a possible way to boarder the variety of respondents. For the students in other universities, it may need the kind help from my friends for the distribution of questionnaires in their universities.

240 questionnaires distributed and 193 questionnaires were returned. There were 160 out of 240 questionnaires fully completed. Therefore, the response rate was:

Effective response rate = Total number of complete questionnaires

$$\text{returned/Total number of questionnaires distributed}$$
$$= 160/240 \times 100\%$$
$$= 66.67\%$$

The Cronbach's alpha value of "Attitude" was 0.795, the result was considered to be consistent and reliable. The factor loadings of 4 items of attitude ranged from 0.739 to 0.856 and all factor loadings were greater than 0.3. Therefore, four items were retained.

4 Results and Discussion

The summaries of percentage of respondents on education level, income level and prior experience are shown in Table 2, Table 3 and Table 4 respectively.

One-way ANOVA is used to find out if there is any significant difference of means between two or more independent groups, in which a post hoc test is used to

Table 2 Percentage of respondents' education levels

		Frequency	Percent	Valid percent	Cumulative percent
Valid	High school degree	21	13.1	13.1	13.1
	Associate degree	3	1.9	1.9	15.0
	Bachelor's degree	132	82.5	82.5	97.5
	Others	4	2.5	2.5	100.0
	Total	160	100.0	100.0	

Table 3 Percentage of respondents' average monthly income

		Frequency	Percent	Valid percent	Cumulative percent
Valid	Less than $4,000	113	70.6	70.6	70.6
	$4,000–$7,999	36	22.5	22.5	93.1
	$8,000–$11,999	8	5.0	5.0	98.1
	More than $20,000	3	1.9	1.9	100.0
	Total	160	100.0	100.0	

Table 4 Summary of prior experience statistics on attitude

	Prior experience	N	Mean	Std. deviation	Std. error mean
ATT	Less experience	80	3.1531	0.65179	0.07287
	More experience	80	3.5688	0.66081	0.07388

find out which specific groups are significantly different from each other. Meanwhile, multiple comparisons table presents the results of post hoc test in which the table would show which groups differed from others.

If the significance level under ANOVA is $=<0.05$, it indicates that there is a significant difference among the mean scores of different groups. Or otherwise, it shows there is no significant difference in the mean of different groups.

If the significance level under post hoc test is $=< 0.05$, it indicates that there is a significant difference between two different groups. Or otherwise, it shows there is no significant difference between the two groups (Table 5).

Comparison of ATT towards online shopping between Educational Levels Groups (Table 6).

Table 5 showed that the significance level in ANOVA under this category was 0.120 (F = 1.976, $p > 0.05$). It indicated that the education level groups had no

Table 5 ANOVA for education level groups on attitude

	Sum of squares	df	Mean square	F	Sig.
Between groups	2.744	3	0.915	1.976	0.120
Within groups	72.224	156	0.463		
Total	74.968	159			

significant difference in attitude towards online shopping. Table 6 shows the descriptive for education level groups on attitude. It shows that Bachelors's degree and high school students have more strong attitude. The post hoc test in Table 7 showed that all the significant levels between education level groups towards the attitude of online shopping are $p > 0.005$, meaning that there is no significant different in attitude towards online shopping between education level groups for college student. And in Table 3, it showed that the mean values for the 4 education level groups had no significant difference, which implied that students with different education level had no significant difference in attitude towards online shopping. Hence, the hypothesis, H1: College student in a higher level of education will have a more positive attitude towards online shopping, is rejected based on the above findings.

From the result above, it was found that educational levels was having no significant effect on attitude towards online shopping. Students with higher educational levels had similar attitude towards online shopping when they were compared to the students with lower educational levels. The reason of the failure of

Table 6 Descriptive for education level groups on attitude

	N	Mean	Std. deviation	Std. error	95% Confidence interval for mean	
					Lower bound	Upper bound
High school	21	3.2024	0.67832	0.14802	2.8936	3.5111
Associate degree	3	2.8333	0.72169	0.41667	1.0406	4.6261
Bachelor's degree	132	3.3807	0.67639	0.05887	3.2642	3.4971
Others	4	3.9375	0.82601	0.41300	2.6231	5.2519
Total	160	3.3609	0.68666	0.05429	3.2537	3.4682

Table 7 Multiple comparisons between education level groups on attitude

(I) 3. What is your educational level?	(J) 3. What is your educational level?	Mean difference (I–J)	Std. error	Sig.	95% Confidence interval	
					Lower bound	Upper bound
High school degree	Associate degree	0.36905	0.41997	0.816	−0.7216	1.4597
	Bachelor's degree	−0.17830	0.15986	0.681	−0.5934	0.2368
	Others	−0.73512	0.37120	0.200	−1.6991	0.2289
Associate degree	High school	−0.36905	0.41997	0.816	−1.4597	0.7216
	Bachelor's degree	−0.54735	0.39728	0.515	−1.5791	0.4844
	Others	−1.10417	0.51968	0.150	−2.4537	0.2454
Bachelor's degree	High school	0.17830	0.15986	0.681	−0.2368	0.5934
	Associate degree	0.54735	0.39728	0.515	−0.4844	1.5791
	Others	−0.55682	0.34533	0.375	−1.4536	0.3400
Others	High school	0.73512	0.37120	0.200	−0.2289	1.6991
	Associate degree	1.10417	0.51968	0.150	−0.2454	2.4537
	Bachelor's degree	0.55682	0.34533	0.375	−0.3400	1.4536

this hypothesis could be explained by the irrelevance of one's academic knowledge and its capability of using the internet for the purpose of shopping in college, like cases for the study aiming to investigate the online shopping behavior of international students by Aiyami and Spiteri [9]. Therefore, hypothesis H1 is rejected in this study.

Comparison of ATT towards online shopping between Income Groups.

Table 8 showed.that the significance level in ANOVA under this category was 0.001 (F = 5.556, $p < 0.05$). It indicated that the income groups had significant difference in attitude towards online shopping (ATT). In Table 9, it showed that the mean for the 3 income groups are: 3.3540 for income less than $4,000; 3.3958 for income ranged within $4,000–$7,999; 3.8125 for income ranged within $8,000–$11,999. The mean increased as the income increased, which implied that students with higher income were having a more positive attitude towards online shopping in the result. The post hoc test in Table 10 showed that significant difference between income groups towards the attitude of online shopping in which $p = 0.003$ ($p < 0.005$) for "Less than $4,000" and "More than $20,000"; $p = 0.003$ ($p < 0.005$) for "$4,000–$7,999" and "More than $20,000"; $p = 0.000$ ($p < 0.005$) for "$8,000–$11,999" and "More than $20,000" respectively. Hence, the hypothesis H2, College student with a higher income level will have a more positive attitude towards online shopping, is supported based on the above findings.

Based on the result above, it was found that income level was having a significant effect on attitude towards online shopping. Students with higher monthly income perceived a more positive attitude towards online shopping when they were compared to the students with less monthly income.

Research showing that teacher students with more average monthly income were having a more positive attitude and behavior intention towards online shopping [10]. It could be explained that as the students with higher income not only possessed a higher economic ability in consuming goods in the internet, but also have a better credit rating for using the credit card in which it facilitated the online shopping experience [7]. Thus, the students with higher income could enjoy a better shopping experience and had a more positive attitude towards online shopping. Therefore, hypothesis H2 is supported.

Comparison of Attitude towards online shopping between Prior Experience

The significance (2-tailed) under t-test is $p = 0.000$ ($p < 0.05$) which indicate there is a significance difference between two groups with different prior experience in online shopping.

Table 8 ANOVA for income groups on attitude

	Sum of squares	df	Mean square	F	Sig.
Between groups	7.237	3	2.412	5.556	0.001
Within groups	67.731	156	0.434		
Total	74.968	159			

Table 9 Descriptive for income groups on attitude

	N	Mean	Std. deviation	Std. error	95% Confidence interval for mean	
					Lower bound	Upper bound
Less than $4,000	113	3.3540	0.61666	0.05801	3.2390	3.4689
$4,000–$7,999	36	3.3958	0.71807	0.11968	3.1529	3.6388
$8,000–$11,999	8	3.8125	0.85304	0.30160	3.0993	4.5257
More than $20,000	3	2.0000	1.0000	0.57735	−0.4841	4.4841
Total	160	3.3609	0.68666	0.05429	3.2537	3.4682

Table 10 Multiple comparisons between income groups on attitude

(I) 7. What is your monthly income on average?	(J) 7. What is your monthly income on average?	Mean difference (I–J)	Std. error	Sig.	95% Confidence interval	
					Lower bound	Upper bound
Less than $4,000	$4,000–$7,999	−0.04185	0.12611	0.987	−0.3693	0.2856
	$8,000–$11,999	−0.45852	0.24107	0.231	−1.0846	0.1675
	More than $20,000	1.35398[a]	0.38544	0.003	0.3530	2.3550
$4,000–$7,999	Less than $4,000	0.04185	0.12611	0.987	−0.2856	0.3693
	$8,000–$11,999	−0.41667	0.25755	0.372	−1.0855	0.2522
	More than $20,000	1.39583[a]	0.39596	0.003	0.3675	2.4241
$8,000–$11,999	Less than $4,000	0.45852	0.24107	0.231	−0.1675	1.0846
	$4,000–$7,999	0.41667	0.25755	0.372	−0.2522	1.0855
	More than $20,000	1.81250[a]	0.44609	0.000	0.6540	2.9710
More than $20,000	Less than $4,000	−1.35398[a]	0.38544	0.003	−2.3550	−0.3530
	$4,000–$7,999	−1.39583[a]	0.39596	0.003	−2.4241	−0.3675
	$8,000–$11,999	−1.81250[a]	0.44609	0.000	−2.9710	−0.6540

[a]The mean difference is significant at the 0.05 level

Hence, the hypothesis, H3: College student with prior experience in online shopping will have a positive attitude towards online shopping, is proved to be supported based on the above findings.

From the result above, it was found that there was a significant difference in attitude towards online shopping between the more experienced students and less experience students. More experienced students perceived a more positive attitude towards online shopping when they were compared to the less experienced students.

Studies showed that experience in using internet will moderate the customer's attitudes to shop online [11, 12]. The students can use the prior experience to assist their new shopping behavior with reduced risk and thus having a more positive attitude towards online shopping as well as behavior intention to shop online [13]. And it can also be used to explain the result in this study for the case of online shopping behavior in college in Hong Kong. Therefore, hypothesis H3 is supported in such case.

5 Conclusion

It can be concluded that (i) College student in a higher level of education will not have a more positive attitude towards online shopping; (ii) College student with a higher income level will have a more positive attitude towards online shopping, and (iii) College student with prior experience in online shopping will have a positive attitude towards online shopping. Since the target participants were students in higher education, the findings of this study contributed to those educators who are teaching in higher education. Based on the findings, the income level and prior experience were proved to be the significant factors in current circumstance in Hong Kong.

The limitations of this study are small sample size and uneven distribution of education level. If sufficient resource is provided, the sample size is enlarged which the education level will be more evenly distributed.

As this study revealed that the income level and prior experience are the major factor, future studies can be done to investigate other factors, such as age, study mode and grade level.

References

1. A. Haque, J. Sadeghzadeh, A. Khatibi, Identifing potentiality online sales in malaysia: a study on customer relationships online shopping. J. Appl. Bus. Res. **22**(4), 119–130 (2006)
2. W. Shwu-Ing, The relationship between consumer characteristics and attitude toward online shopping. Manage. Intelligen. Plan. **21**(1), 37–44 (2003)
3. H.K. Yau, L.C. Tsang, Gender difference of behavior intention on online shopping: an empirical study in Hong Kong higher education, in Lecture Notes in Engineering and Computer Science Proceedings of the International MultiConference of Engineers and Computer Scientists, 15–17 March, Hong Kong, 901–902 (2017)
4. R.E. Burroughs, R. Sabherwal, R, Determinants of retail electronic purchasing: a multi-period investigation, Informat. Syst. Operat. Res. **40**(1), pp. 35–56, 2002, viewed 24 December 2015, http://www.proquest.com/
5. B.C. Yin-fah, Undergraduates and online purchasing behavior. Asian Soc. Sci. **6**(10), 133–147 (2010)
6. Y. Wan, M. Nakayama, N. Sutcliffe, The impact of age and shopping experiences on the classification of search, experience, and credence goods in online shopping'. IseB **10**(1), 135–148 (2012)
7. M.B. Naseri, G. Elliott, Role of demographics, social connectedness and prior internet experience in adoption of online shopping: applications for direct marketing. J. Target. Measure. Anal. Market. **19**(2), 69–84 (2011)

8. V. Venkatesh, D.F. Davis, A model of the antecedents of perceived ease of use: development and test, Decision Sci. **27**(3), 451–481 (1996)
9. E. Alyami, L. Spiteri, International University Students' Online Shopping Behaviour. World J. Soc. Sci. **5**(3), 227–243 (2015)
10. M. Kiyici, Internet shopping behavior of college of education students. TOJET Turkish Online J. Education. Technol. **11**(3), 202–214 (2012)
11. S.L. Jarvenpaa, N, Tractinsky et al. Consumer trust in an internet store. Informat. Technol. Manage. **1**(1), 45–71 (2000)
12. J.A. Castaneda, F. Munoz-Levia, T. Lupue, Web Acceptance Model (WAM): Moderating effects of user experience. Informat. Mange. **44**(4), 384–396 (2007)
13. S. Shim, M.F. Drake, Consumer intention to utilize electronic shopping: The Fishbein behavioral intention model. J. Direct Market. **4**(3), 22–33 (1990)

Risk Mitigation Strategy for Public Private Partnership (PPP) of Airport Infrastructure Development in Indonesia

Rusdi Usman Latief

Abstract The capacity of Indonesian Government to grow the country to become developed country in 2025 as mentioned in The Master Plan for Acceleration and Expansion of Indonesia's Economic Development (abbreviated MP3EI) faces many challenges, one of them is financial capacity. Huge archipelago like Indonesia needs to be connected by good infrastructure, especially an airport, and it is needed enough funding for the government to make an equitable development. The government realizes the limitation or gap in funding the infrastructure needs. The government comes with the concept to form the cooperation with private sectors as known as a Public Private Partnership (PPP). The PPP schemes are expected to fulfill the financial problems in infrastructure service providing. Therefore, the aim of the study is to develop risk mitigation strategy for airport infrastructure and expected to solve the lack of succession of PPP. The research study was conducted by collecting data from Indonesian airports. The data used were primary and secondary data with validity test and reliability also descriptive analysis method. The primary based on field survey, while secondary according to the study on various literatures of the success with the implementation of PPP in Indonesia and abroad. The findings and recommendations are risk response and risk strategy in order to solve of each risk that already identified. Risk response can be: Retention, Avoidance, Reduction, or Transfer.

Keywords Airport · Mitigation · Infrastructure · Public private partnership
Risk · Risk response

R. U. Latief (✉)
Civil Engineering Hasanuddin University, Jl. Poros Malino Km. 6. Gowa,
92171 Makassar, Indonesia
e-mail: rusdiul@gmail.com

© Springer Nature Singapore Pte Ltd. 2018
S.-I. Ao et al. (eds.), *Transactions on Engineering Technologies*,
https://doi.org/10.1007/978-981-10-7488-2_8

1 Introduction

Master plan for acceleration and expansion of Indonesia Economic Development (MP3EI) is an ambitious plan by the government of Indonesia to be a developed country in 2025 [1]. To obtain this plan, it is required for the real economic growth 6.4–7.5% during the period 2011–2014 and about 8.0 to 9.0 for the period 2015–2025. However, the inflation shall be declined from 6.5% in the period 2011–2014 to 3.0% for the period 2025, which shows the characteristic of developed countries. The demographic potential, the wealth of natural resources and geographical position will support the acceleration and expansion of Indonesia's economic development. Although Indonesia has good strategic geographic position, but a number of challenges are needed to be solved. One of them is the lack of quality and quantity of infrastructures so that becomes the reducing of competitiveness and attractiveness of climate investment. Another challenge is that the main investment problem of the Government of Indonesia in infrastructures is less of financial resources. Therefore, the participation of the private sector and both foreign and domestic sectors is desired. The presence of Public Private Partnership (PPP) has been participating as an alternative and effective method to mobilize additional financial resources to invest in infrastructures. PPP is not only for financial mechanism, but can also be appropriate to finance resource for infrastructure projects. Currently, PPP is widely taken a part in the sector of water treatment and highways due to these sectors guarantees the financial benefits. Moreover, PPP has been participating in other sectors such as railway/railroad, housing, and airports.

The need of air transportation in Indonesia grows rapidly, the increase of passenger and cargo reflects the circumstance each year. Surely it should be countered by the enlargement of airport infrastructure [2]. However, Fiscal limitation which is from the national budget causes the expansion of infrastructure capacity in Indonesia hampered. Amongst of 2010–2014, it is estimated the need of investment about an IDR1.450 quintillion [3]. One step the government has taken to cope the infrastructure deficit is to encourage active participation of the private sector, where private is allowed to join in developing infrastructure through *Public—Private Partnership* (PPP) scheme including risks share to each other. Until now there has been no implementation of a successful PPP airport in Indonesia due to the high risk for private parties. Risk mitigation must be done to reduce the chances of occurrence of the risk or impact of a risk. Therefore, this study aims to develop risk mitigation strategy in airport infrastructure development and expected to be problem solving for the lack of succession of PPP in Indonesia and as reference to all participant to arrange the strategy for collaboration on PPP schemes [4].

2 Literature Reviews

A. PPP Airport Infrastructure Project in Indonesia

The airport can be defined as one or more runways and facilities to complement the aircraft (taxiway, apron area) along with Union terminal and facilities to lower passenger and cargo [5]. Airport operators are responsible for the provision and maintenance of airport infrastructure, and on conditions of service, including the main searches of passengers, and security, fire, hygiene and maintenance areas of the passenger terminal. The operational of the airport receives income from aircraft service and marketing.

Major infrastructure airport terminal operations, consisting of runway operations, and taxiway facilities, are such engineering facilities. The cargo, plane maintenance facility, ARRF (Airport Rescue and Fire Fighting) fuel, logistics facilities, administration, service aircraft, traffic and utilities are also the main of airport infrastructures [6]. Airport operation such as service aircraft, service fleets, and marketing activities is a part of the main airport of service provision. This infrastructure becomes a reference for the national development and planning agency of the Republic of Indonesia (BAPPENAS) to develop the airport in the form of cooperation PPP. The type of project that exists in PPP book 2010–2014 BAPPENAS diverse. The project will be soon in tender's supports airport infrastructure in Indonesia, especially in the form of PPP cooperation entered into PPP Book list (2010–2014) BAPPENAS.

B. Risk of PPP Airport Infrastructure

The main risk in investment related to the Airport directly with the basic parameters of the PPP infrastructure investment in the airport that is the decisive variable the magnitude as the cost of the investment. PPP risks are divided the airport into Air Traffic Forecast, Airport Development Proposal, Airport Transport Risk, Revenue Estimation, Capital Cost Estimates, Concessionaire Competition & Culture, Institutional Influence, Effect of Term of Reference for Privatization [7]. PPP risks are among other airport Revenue Risk, Operating Risk, Regulatory Risk, and Review of Policy on Water Infrastructure [8]. In addition, BAPPENAS [9], the risk of PPP is the airport land acquisition, tariffs, demand, political risk and country risk, as well as the main buyer of creditworthiness (off-taker). Risk response preference has been investigated on how to assess the risk of PPP airport infrastructures development project in Indonesia [10]. This will give the description to the Government of Indonesia that risk mitigation is significant issue for developing the airport infrastructures in Indonesia.

C. Risk Mitigation

Risk response is the response or reaction to the risks undertaken by any person or company in making decisions, which are influenced by the risk attitude of decision makers. Action taken to reduce this risk is called mitigation/risk management (risk mitigation). Occasionally, risks can be not eliminated completely but can only be reduced so that there will be the residual risk. Thus, this can be done by risk retention, risk reduction, risk transfer and risk avoidance.

3   Research Methodology

A. Research Frameworks
This study has a framework with five main steps to achieve the purpose. Schematically, the research frameworks following the steps as below.

(1) Literature Study

This research was conducted by an observation for the recent condition of airport infrastructures of Indonesia and research related in the model. The references are collected to identify the risks then determine all of variable risks for research. Moreover, conducted the preliminary study was not only for the risk respond but also the risk strategy model as reference.

(2) Preparing Questionnaires

Preparing a questionnaire with setting the total respondents, draft, and the conceptual assessment model risk respond and mitigation for PPP airport in Indonesia. The pilot survey is conducted to verify the feasibility of the questionnaire through discuss with professionals in airport infrastructures before doing the main survey with a fix draft.

(3) Collecting Data

The questionnaires were distributed to all primary stakeholders as respondents to get the model of risk mitigation by e-mail, post mail, or direct interview.

(4) Analysis and Discussion

The feedback of the questionnaire from respondents is a primary data, then analysis to get the interpretation and further information.

(5) Result and Conclusions

The result describes the interpretation of analysis data and new finding of this research.

B. Method of Data Collection

To collect the data, this study has adopted methods from literature studies. Then, primary data were collected by spreading the questionnaires to public and private

Table 1 Primary respondent of PPP airport projects

Primary stakeholder		Sum of respondents
Public	The Ministry of Aviation of Indonesia	12
	BAPPENAS	2
	The Ministry of Finance of Indonesia	2
	Total Public	16
	Kertajati Int. Airport, West Java	1
	South Banten Airport, Pandeglang	1
	Development of Singlawang, West Borneo	1
	Expansion of Dewandaru Karimun Jawa, Java	1
	Expansion of Tjililriwut, Central Kalimantan	1
	Development of New Samarinda, East Borneo	1
	Development of New Bali Airport	1
	Radin Inten II Airport (Bandar Lampung)	1
	Mutiara Airport (Palu)	1
Private	Haluoleo Airport (Kendari)	1
	Komodo Airport (Labuan Bajo)	1
	Sentani Airport (jayapura)	1
	Juwata Airport (Tarakan)	1
	Tjilik Riwut Airport (Palangkaraya)	1
	Fatmawati Airport (Bengkulu)	1
	Hananjoeding Airport (Tj. Pandan)	1
	Total Private	16
Total respondents		32

related to variables study. Table 1 shows the respondents of primary stakeholders on PPP airports. The respondents consist of public and private with differently number. For public in the table, total respondents are 16 from three agencies. The private also have 16 respondents from sixteen ongoing project of airport. Moreover, collecting secondary data also conducted by studies journal related to PPP airport projects.

C. Method of Analysis

To analysis the data, this study used test validity and reliability methods and descriptive analysis method. Both of these methods were used to analysis the data respondents from public and private sectors. Each respondent provided response options to each risk variable of risk reduction, risk retention, risk transfer, and risk avoidance. Further, the analysis shows the response from high and extreme categories level of risks.

4   Result and Discussions

A. Data Analysis

The method of study was carried out by distributing the questionnaires to some respondents who related in this study. The following will be explained about the profile of the respondents based at the level of education, positions at the agency/institution, type of institution/agencies, and work experience in their institutions. The following results as below.

(1) Business Entity

Table 2 shows that the majority of respondents worked in business government with the percentage reached 54.2 and 45.8% working in private business enterprises.

(2) Position in Institution

Based on data in Table 3, it can be seen that the first position of respondent is senior manager with percentage of 50% or as much as twelve respondents. The second position is as section chief with the percentage of 16.7%. The managing director and senior administrator are in third position with same percentage of 12.5%. The last position is airport project advisor with percentage of 8.3% or as much as two respondents.

(3) Level of Education

Table 4 shows the majority of respondent has level of education in master degree with percentage of 54.2% or 13 respondents. Ten respondents have level of

Table 2 Type of business

Type of business	Entity frequency	Percent (%)
Government	13	54,2
Private	11	45,8
Total	24	100

Table 3 Positions

Positions	Frequency	Percent (%)
Managing director	3	12,5
Section chief	4	16,7
Senior manager	12	50,0
Airport project advisor	2	8,3
Senior administrator	3	12,5
Total	24	100

Table 4 Level of education

Level of education	Frequency	Percent (%)
Diploma	1	4,2
Bachelor	10	41,7
Master	13	54,2
Total	24	100

education in Bachelor degree with percentage of 41.7%. The last is Diploma degree with percentage with percentage 4.2% or only one respondent.

(4) Work Experience

On below Table 5 is seen that the respondents have working experience less than 5 years until 20 years. Majority respondents has working experience 11–20 years with percentage of 29.2% or has 7 respondents. Respondents have percentage of 25% with work experience less 5 years and 5–10 years. The last is minority respondents have work experience more than 20 years with percentage of 20.8% or 5 respondents.

B. Risk Mitigation and Recommendation Strategy

The existence of the risks will give effect to the development of airport infrastructure of PPP projects that will require the measuring mitigation to reduce its impact. The important risks were already knew the need to follow up with responds and mitigates. Risk mitigation can be done by risk reduction, risk retention, risk transfer and risk avoidance. Mitigation action undertaken for this study was obtained from the results on the analysis, interviews with experts and reference supported from previous studies.

The preference of risk response presents the data of percentage value for each variable risk from public and private sector. The management of risk mitigation as one of the parts in cooperation schemes. Most of the guidance in PPP risk describes clearly about the risk allocation and action will be charged to all parties who are involved in the airport project. The alternative approach also suggested to reduce PPP risk that comes from any source.

Table 5 Work experience

Work experience	Frequency	Percent (%)
<5 years	6	25,0
5–10 years	6	25,0
11–10 years	7	29,2
>20 years	5	20,8
<5 years	6	25,0
Total	24	100

Table 6 shows the results analysis with a survey of risk responses according to respondents who works in public and private sectors. Risk variable of land acquisition in response to the preferences of 70.83% retention risk responses. The risk avoidance response preference as much 13%, for 8% of a preference response risk reduction, and amount to 8% of transfer of risk response preferences. Based on the results of survey variables, land acquisition has mainly retention of preference of risk response due to the number of respondent.

The airline and terminal design and capacity and site expandability also has same the preference in retention risk responses but different in risk avoidance and risk reduction. Totally the preference of risk response describes the value of all variable risk and it has important role as reference to arrange the risk mitigation.

Moreover, the preferences of risk responses mitigation if categorizes according to implementing risk assessment divide into extreme and high [2]. The majorities

Table 6 Result analysis of a survey of risk responses

No	Risk variables	Preference of risk response (%)			
		Retention	Avoidance	Reduction	Transfer
1	Land acquisition	**70.83**	13	8	8
2	Airline and terminal design	**70.83**	4	17	8
3	Capacity and site expandability	**70.83**	4	4	21
4	Changes in aircraft mix	37.5	17	**42**	4
5	Competing airport	**54.17**	33	13	0
6	Airline alliance	**58.33**	17	21	4
7	Capital cost estimates	**58.33**	8	33	0
8	Concessionaire composition and culture	**41.67**	13	21	25
9	Institutional influences	25	21	**50**	4
10	Effect of terms of reference (TOR)	29.17	13	**58**	0
11	Corporate governance	**37.5**	33	21	8
12	Center state relations	**54.17**	33	13	0
13	Continuity of political leadership	37.5	**42**	21	0
14	Activity of local politics	25	**54**	13	8
15	Demand	**54.17**	25	21	0
16	Price	**58.33**	8	33	0
17	Cost escalation	**54.17**	4	29	13
18	Staffing	**58.33**	4	21	17
19	Labor unions	**45.83**	8	29	17
20	Coordination with governmental agencies	**75**	17	4	4
21	Classification and licensing	**58.33**	8	25	8
22	Revenue sharing	**79.17**	4	13	4
23	Risk country and risk politics	**41.67**	29	13	17
24	Risk enclave (civil and military)	**45.83**	33	17	4

Fig. 1 Preference risk response mitigation based on risk level categories

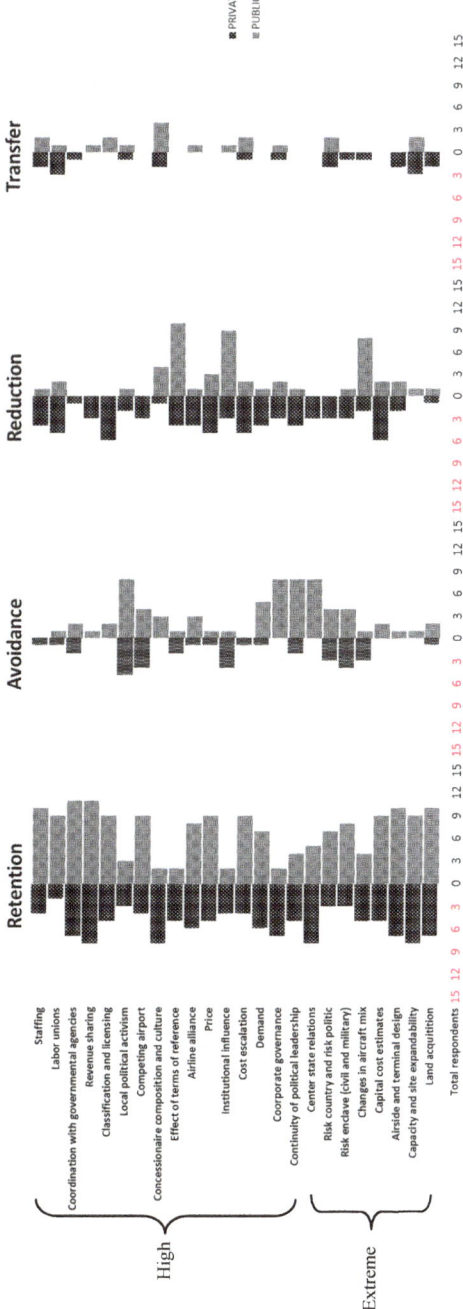

Table 7 Risk mitigation as recommendation strategy

No	Risk variables	Risk mitigation	
		Risk responds	Recommendation strategy
1	Land acquisition	Retention	Socialize the rules. Invite land owners to invest
			Replacement land into noncash payment
2	Airline and terminal design	Retention	Private use experts to forecast traffic passenger to determine the capacity of airports
			The government make master plan and spatial planning airport development and disseminating the tender process of PPP project
3	Capacity and site expandability	Retention	Government provides airport development concept clearly to the private sector
			Private use professional engineers and experienced in the process of making design
4	Changes in aircraft mix	Retention	Use experts to develop scenarios to forecast the growth of passengers and travel route
5	Competing airport	Retention	Private improves services and promotions to airports user
6	Airline alliance	Retention	Governments makes regulations regarding operational technical airline alliance airport
7	Capital cost estimates	Retention	Using professionals in estimating costs
8	Concessionaire composition and culture	Retention	Government as regulator sets minimum standards of expertise in concessionaire competition
9	Institutional influences	Retention	Government provides compensation to the private sector if there are institutions disrupt the project
10	Effect of terms of reference (TOR)	Retention	Provide a complete and clear TOR
11	Corporate governance	Retention	Set up management infrastructure such as SOP as a reference work for business entities
12	Center state relations	Retention	Synchronization rules against institutions associated with the course of the project
13	Continuity of political leadership	Avoidance	Build emotional attachment to all political stakeholders
14	Activity of local politics	Avoidance	Place a project manager who understand the dynamics of local politics
15	Demand	Retention	Make a good quality of service to passenger and airlines
16	Price	Retention	The government provides a guarantee in the even of a decrease in revenue due to level of demand that below the agreed level
17	Cost escalation	Retention	Using a competent estimators in calculating the price adjustment at the time of initiation and operation the concession (multi-year)
18	Staffing	Retention	Provide regular training of staff

(continued)

Table 7 (continued)

No	Risk variables	Risk mitigation	
		Risk responds	Recommendation strategy
19	Labor unions	Retention	Regular coordination with union leaders
20	Coordination with government agencies	Retention	The private sector/business entities to coordinate with the various parties involved in the airport such as the police, immigration, and customs
21	Classification and licensing	Retention	The government will accomodate and foster business entity in order to obtain licenses and international classification
22	Revenue sharing	Retention	Creating and selecting appropriate forms of cooperation on revenue sharing and allocation risk
23	Risk country and risk politics	Retention	Government guarantees to investors in case of disruption due to the political situation in the country
24	Risk enclave (civil and military)	Retention	Government makes regulations to control the airport by the military

respondents' retention for the high risks as much as 36 and 16% for extreme, it has shown on Fig. 1 on the preference of risk response mitigation based on risk level.

The figures show bar with the total of respondents from public and private prefer to retention, as a sequence followed by reduction, avoidance, and transfer. This show all risk prefer to retention and action responses to the all risks variable from public and private respondents.

Therefore, on below Table 7 shows the risk mitigation model in sequenced according to risk level as description of recommendation strategies to the airport infrastructures in Indonesia. The strategy is created with study literature and interview an expert as the suggestion to formulate the strategy for each variable on PPP airport of Indonesia. Recommended strategy is conducted according responds for all variable risks. For 24 risks variable, each of them could have two until three recommendation strategies in accordance with situation, suggestion, and literature review. These strategies were expected as approaching reference to who are interest to PPP schemes especially for airport infrastructure projects in Indonesia. The recommendation probably has a difference condition to another country.

5　Conclusion

Risk Mitigation on Public Private Partnership (PPP) to the airport infrastructures development project in Indonesia has been analyzed and given the recommendation strategies in order to solve the problem of airport infrastructure project. Therefore, this study can briefly give the conclusion as follows:

(1) Risk response based on public private partnership conformed that risk retention reaches about 52%. Other respondents showed that risk reduction and risk avoidance are 22% and 19% respectively. The minority respondents choose risk transfer by 7% response. This shows that majority of participants on PPP schemes are ready to retention the risk.

(2) According to risk level high and extreme, the majorities respondent retention for the high risks as much as 36 and 16% for extreme.

(3) Recommendation of risk mitigation strategies is contrary to the four highest risks which is described as follows:

   a. Land Acquisition: socialize the rules, invite the land owners to invest their land, replace the land to the non-cash.
   b. Capacity and Expansion of the airport: use experts for forecasting passenger traffic to determine the capacity of airports, create a master plan and spatial planning of airport development and socialize the tender process of PPP projects.
   c. Airside and terminal design: the government provides the airport development concepts clear to the private sector, the private sector using professional engineers and experienced in the process of making design.

Estimation on capital cost: use professional engineers and experienced in the process of making design.

Acknowledgements The authors thank to Budi Prasetyo, the Airport Ministry of Indonesia for advice and suggestions in pilot survey. Special thanks to family for cheering and all lecturer of civil engineering support in writing of this manuscript.

References

1. Rusdi Usman Latief, Risk Mitigation Strategy for Public Private Partnership (PPP) of Airport Infrastructure Development Projects in Indonesia, in *Lecture Notes in Engineering and Computer Science: Proceedings of The International MultiConference of Engineers and Computer Scientists*, pp. 1068–1073 (15–17 March 2017, Hong Kong)
2. S. Adjisasmita, *PPP Scheme in Airport*. Jakarta (2010)
3. R. Magagi, *Increasing Trend Towards Airport PPPs in Emerging Markets* (IFC's Global Airport PPP Seminar, Dubai, 2011)
4. International Association of Engineers. Available: http://www.iaeng.org
5. Asian Development Bank, *Airport and Air Traffic Control* (ADB, Philippines, 2000)
6. Dewey and Lebouf, *PPP in Airport Infrastructure* (University Press, Pulkovo, 2006)
7. V. Craig, *Risk and Due Diligence in Airport Privatization* (Air Transport, Malaysia, 2010)
8. B. Varkey, *Public Private Partnership in Airport Development* (Oxford University Press, New Delhi, 2002)
9. Pemerintah Republik Indonesia, *KPS dan Panduan Bagi Investor untuk Investasi* (Bappenas, Jakarta, 2010)
10. U.L. Rusdi, S. Pallu, S. Adisasmita, S.A. Aly, SH. Suyuti, A. risk response preference on public private partnership (PPP) in Indonesia airport infrastructure development. Int. J. Appl. Innov. Eng. Manag. (IJAIEM) **3**, 120–124. Japan (2014)

An Experimental Study of Adhesion Between Aggregate Minerals and Asphalt Binders Using Particle Probe Scanning Force Microscopy

T. Tan, Yujie Li and Jie Yang

Abstract Accurate determination of adhesion between asphalt binders and aggregates is essential to the performance of pavement structures. In this study, particle probe scanning force microscopy was used to measure the adhesion between mineral microspheres representing the primary aggregate constituents and various control and modified binders. Average unit surface energies were applied to distinguish adhesion between various aggregate-binder pairs. Results showed that the alumina-binder pair exhibited higher adhesion than those of silica-binder and calcium carbonate-binder pairs. Microstructure variations were also detected for different modified binders that could lead to the adhesion differences.

Keywords Adhesion · Aggregates · Asphalt binder · Modifier
Pavement · Particle probe scanning force microscopy

1 Introduction

Asphalt materials are very valuable for constructing pavement around the world. One of the most important pavement materials is hot mixed asphalt (HMA), i.e., a composite produced by mixing asphalt binders and aggregates at elevated temperatures. HMA has been used for roads in the United States since 1860s [1]. The performance of HMA is highly dependent on the mechanical properties of asphalt binders, aggregates and the adhesive properties between them. After refining crude

T. Tan (✉) · Y. Li
Civil and Environmental Engineering, The University of Vermont, Burlington,
VT 05405, USA
e-mail: ting.tan@uvm.edu

Y. Li
e-mail: Yujie.Li@uvm.edu

J. Yang
Physics Department, The University of Vermont, Burlington, VT, USA
e-mail: Jie.Yang@uvm.edu

© Springer Nature Singapore Pte Ltd. 2018
S.-I. Ao et al. (eds.), *Transactions on Engineering Technologies*,
https://doi.org/10.1007/978-981-10-7488-2_9

oil to obtain various petroleum products, asphalt binders are produced from the thick, heavy residue when the specifications are satisfied. Thus the chemical compositions of asphalt binders are very complex. Based on the SARA fractions in ASTM, binders can be roughly classified into four groups based on their polarity and solubility, i.e., asphaltenes, resins, aromatics and saturates [2]. On the other hand, aggregates include fine and coarse solids, such as sand, gravel and crushed stone, which can be obtained either through nature or manufacturing. Natural aggregates are extracted from rock in an open quarry or mine. The mineral compositions largely determines the mechanical properties of aggregates. For example, limestones is primarily comprised of calcium carbonate ($CaCO_3$). Granite is comprised of ~72.04 wt% of silica (SiO_2) and ~14.42 wt% of alumina (Al_2O_3). Sand is primarily comprised of silica (SiO_2) [1]. Thus, the major chemical compositions of aggregates are silica (SiO_2), calcium carbonate ($CaCO_3$), and alumina (Al_2O_3).

The adhesive behavior between aggregates and binders is essential to the performance of asphalt admixtures. Various testing methods have been proposed to study the interfacial fracture of asphalt mixtures at macroscale [3]. Recently, Atomic Force Microscopy (AFM) has been used to study the adhesive and cohesive properties of asphalt binders. For example, Tarefder and Zaman [4] measured adhesion between bare silicon nitride AFM probes and binders, as well as the cohesion between functionalized AFM probes and asphalt binders. By using the same type of probes, Lyne et al. [5] reported higher adhesive forces and elastic moduli in the catana and peri phases than those in the para phase. Yu et al. [6] investigated the effects of compressive forces and scan speed in measuring adhesive forces between silicon tips and binders. Al-Rawashdeh and Sargand [7] performed tests to measure the adhesion between silica/calcium carbonate particles and three asphalt binders, and cohesion between carboxylic acid functionalized silica particles and plain or modified PG 70-22 binders. Despite these studies, two essential issues remain to be addressed. The tip radii of conventional AFM probes are on the order of nanometers, making the adhesion characterization local to asphalt binders. Furthermore, the adhesive measurements between silicon or silicon nitride and asphalt binders do not present the adhesion between primary aggregate minerals, i.e., calcium carbonate, silica and alumina, and the corresponding binders.

In this work, we used particle probe scanning force microscopy to study adhesion between primary aggregate minerals and various plain and modified asphalt binders. First, we created particle probes using microspheres made from calcium carbonate, silica and alumina to represent the primary minerals of aggregates. Then, adhesive forces were collected between microspherical probes and various binder substrates at ambient conditions, which are normalized to average surface energy per unit area to differentiate various binders. Morphology characterization of different plain and modified binders illustrates the relationship between microstructural variations of binders and their adhesive properties to aggregates. Particle probe scanning force microscopy directly quantifies the adhesion induced by intermolecular forces between aggregate minerals and binders [8, 9], but the interlocking mechanisms of adhesion [10, 11] are not included.

2   Materials and Methods

2.1   Scanning Force Microscopy

Depending on the forces measured, Scanning Force Microscopy (SFM) can be used to measure mechanical, electrical and magnetic properties of materials [12]. SFM has been used to study adhesion between the probe and different materials [12], such as between glass and silicon [13, 14], between stainless steel and polymers [15, 16], between ligands and receptors [17, 18]. If conventional AFM tips with nanoscale radii are replaced by mineral microspheres, particle probes could be used to measure adhesion between primary aggregate minerals and binder substrates at the microscale. In Fig. 1, the working principle of SFM is illustrated. The reflected laser light on the back of the deflected cantilever is collected by a split photodiode, which is related to surface properties of the substrate. In approaching, the cantilever is initially away from the substrate. Once the microsphere contacts the substrate, the cantilever bends as driven forward resulting the linear, increasing signal. In retracting, due to adhesion between the microsphere and substrate, they stay in contact until the restoring force of the cantilever overcomes the adhesive force. Since the cantilever is strong enough, its deflection is linear. The snap-off force between the microsphere and substrate is measured using Hooke's law.

$$F = k\Delta x \tag{1}$$

where k is the spring constant of the cantilever, and Δx is the distance between the tip-sample contact and snap-off points.

2.2   Particle Probes

Three types of particle probes were created using calcium carbonate, silica, and alumina microspheres. The fabrication was achieved by attaching individual mineral microspheres to tipless cantilevers. A cantilever was driven to touch the epoxy, and

Fig. 1 Schematic of the working principle of particle probe scanning force microscopy

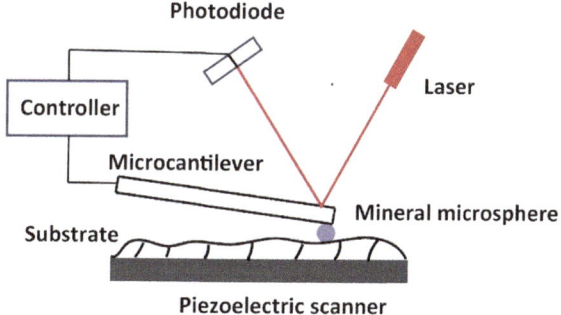

then moved to the target particle using a high resolution 3D translation stage (Newport, Irvine, CA) with the help of the optical microscope. When the particle was successfully attached to the free end of the microcantilever, the probe was cured in air for 24 h before adhesion measurements. The diameters of the mineral microspheres are ~6 μm. The plain cantilevers were of straight and rectangular, whose spring constants ranged from 2.5 to 10 N/m (NSG11, NT-MDT Inc., Ireland).

2.3 Binder Substrates

Substrates were prepared using various plain and modified binders. PG 64-22, PG 58-22 and PG 70-22 obtained from the SHRP's Materials Reference Library (MRL, Austin, TX) were used as control specimens. Three types of modified binders were prepared by adding styrene-butadiene-styrene triblock rubber (SBR), styrene butadiene styrene (SBS), and polyphosphoric acid (PPA) at the weight percentages of 2 and 4%, respectively. SBR modifications were prepared by adding the modifier (Ultrapave, Dalton, GA) into control binders at ~165 °C and mixed for 2 h. PPA modifications were prepared adding PPA (ICL Performance, St. Lousi, MO) into the control binder at 120 °C and mixed for ~30–45 min. SBS modifications were prepared by adding SBS (Bitumar, QC, Canada) to control binders at ~175 °C and mixed for 2 h. After that, binders were stirred using a rod for 5 min and then transferred onto a 5 × 5 mm wafer plate. The binder drop was gradually heated to 163 °C (325 °F) using a ceramic hotplate (Model 11-300-49SHP, Thermo Fisher Scientific Inc., MA), and spread over the wafer surface using a stainless steel spatula within ~5 min to ensure smooth binder coverage over the surface. All binder specimens were stored in oven at 30 °C overnight before adhesion and morphologies measurements were performed. The root mean square (RMS) of the substrate surface roughness was ~30 nm, which is smooth enough for microscale adhesion measurement.

2.4 Experimental Plan

In this study, adhesion measurements were collected using nanoscope E (Bruker Inc., Santa Barbara, CA). For each substrate, three spots were chosen to collect the adhesive forces, and ten curves were collected at each spot. Thus, thirty measurements were collected between one particle probe and one substrate. For each mineral-binder pair, data replicas were collected between two particle probes and three substrates, i.e., 180 measurements in total. All data were performed at ~25.7 °C and ~30% relative humidity. To characterize the binder microstructures, an Asylum MFP-3D-BIO atomic force microscope was used with conventional AFM probes (NanoAndMore USA Corp., Watsonville, CA) in AC mode at the frequency of ~300 kHz.

3 Results and Discussions

3.1 Mineral Particle Probes

All fabricated particle probes were examined using the scanning electronic microscope (JEOL JSM 6060, Peabody, MA). Figures 2a, b showed two representative mineral particle probes with silica and alumina microspheres. The mineral particles are uniform microspheres with consistent surface textures, which could ensure ideal contact to the substrates. Spring constants of particle-modified cantilevers were measured using the thermal tuning method [23, 24], which ranged from 1.0 to 5.7 N/m. Diameters of mineral spheres used ranged from 5.9 to 13.0 μm [20, 22].

Fig. 2 SEM images of mineral particle probes for scanning force microscopy. **a** Silica. **b** Alumina. Adapted from Ref. [19] with permission

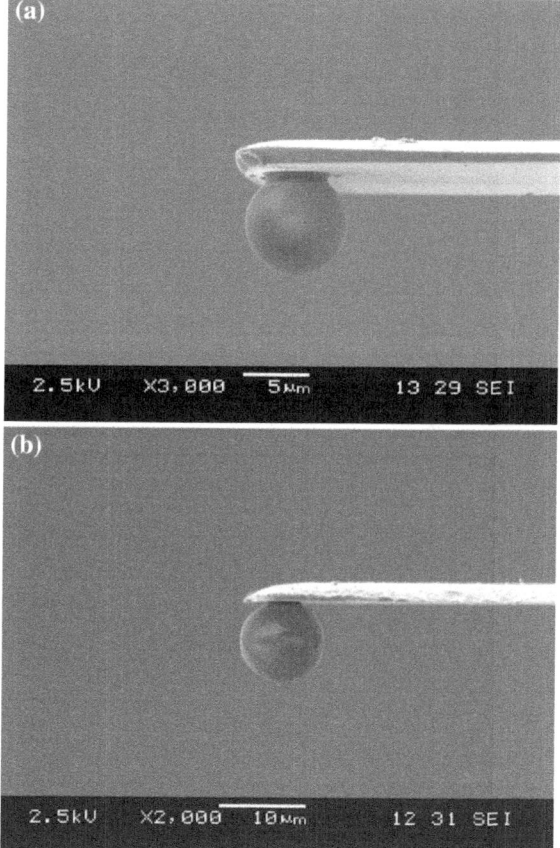

3.2 Force Curves

Two representative force curves collected by the particle probe scanning force microscopy were shown in Fig. 3 with illustrations of Δx, the contact and disengage points. The vertical axis is a measure of the signal in the arbitrary unit, and the horizontal axis is the z-travel of the tip in nm. In approaching (orange curves), the detected signal was constant when the particle microsphere did not contact the substrate. After the contact, the defection signal increased linearly since the cantilever was continuously driven forward. In retracting (blue curves), the cantilever bent reversely so the signal decreased with the same slope as expected. The spherical probe did not disengage from substrate until the restoring force of the cantilever overcame the adhesion between the microsphere and substrate. After that, the deflection signal became flat again since the microsphere disengaged from the substrate. For large Δx, the signal was flat at the bottom of the retracting curve since

Fig. 3 Two representative force curves for **a** the calcium carbonate and the binder; **b** the silica and the binder. Adapted from Ref. [19] with permission

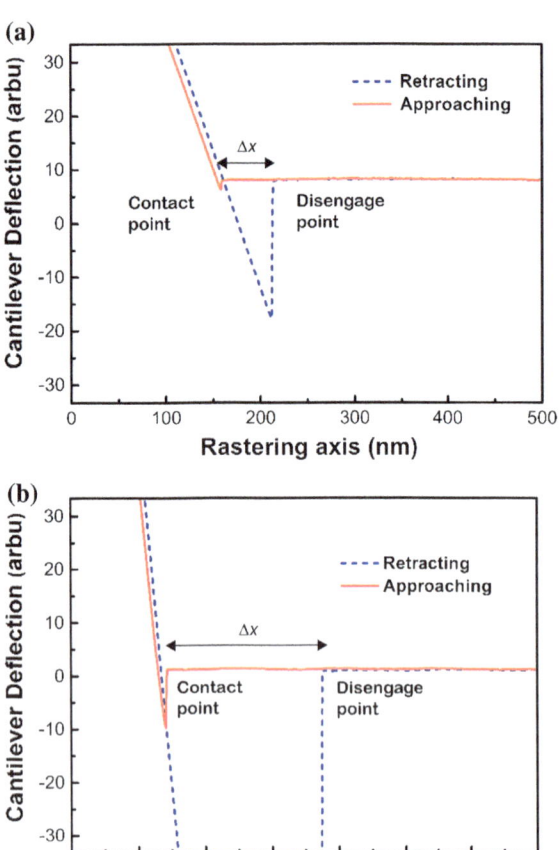

the deflection signals were beyond the instrumental range (Fig. 3b). But the measurement of Δx was not affected because the horizontal signal was within the range of the piezo scanner.

3.3 Adhesion Between Binders and Mineral Microspheres

Surface energy per unit area is used to quantify the adhesion between mineral microspheres and binder substrates based on the Derjaguin-Muller-Toporov (DMT) model [25], DMT model provided more realistic adhesion predictions between a hard sphere and a soft surface. Since two surfaces are created during fracture, the average surface energy per unit area, γ_{avg}, is given by

$$\gamma_{avg} = \frac{1}{2}(W_{ad}) = \frac{1}{2}\left(\frac{F_{ad}}{2\pi R}\right) = \frac{F_{ad}}{4\pi R} \qquad (2)$$

where F_{ad} is the adhesive force between the microsphere and the substrate, W_{ad} is the work of adhesion per unit area, and R is the radius of the microsphere.

Average unit surface energies between mineral microspheres and different binders are shown in Fig. 4. Among three types of mineral-binder pairs, i.e. calcium carbonate-, silica-, and alumina-binder pairs, surface energies of alumina-binder pairs are the largest. For the control specimens, i.e., PG 58-22 and PG 64-22 binders, the alumina-binder adhesion slightly increased as the binder softened but more variations in measurements occurred. In contrast, no monotonic trends were observed for adhesion between silica and calcium carbonate microspheres and control binders. For modifications using the same control binder, the alumina-binder adhesion increased as weight percentages of polymer modifiers (PPA, SBS, and SBR) increased (Figs. 4a–c). In the PG 58-22 series, the silica-binder adhesion increased as weight percentages of SBR increased (Fig. 4b). However, in PG 64-22 series, the silica-binder adhesion decreased as the weight percentage of PPA increased (Fig. 4c). Generally, no monotonic trends were observed for adhesion between silica and calcium microspheres and modified binders. These results showed that SBS, SBR, and PPA modifiers did affect the adhesion between binders and aggregate minerals.

The differences of average unit surface energies between various mineral-binder pairs are attributed to the chemical constituents of alumina, silica, and calcium carbonate particles. Among these particles, alumina spheres are the most polar, while calcium carbonate and silica are less polar [26]. According to the SARA classification in ASTM [2], a great portion of binder is resin, i.e., polar aromatic substances. Therefore, larger adhesion is achieved via stronger interactions between alumina microspheres and binders. More work is needed in the future to fully understand these differences.

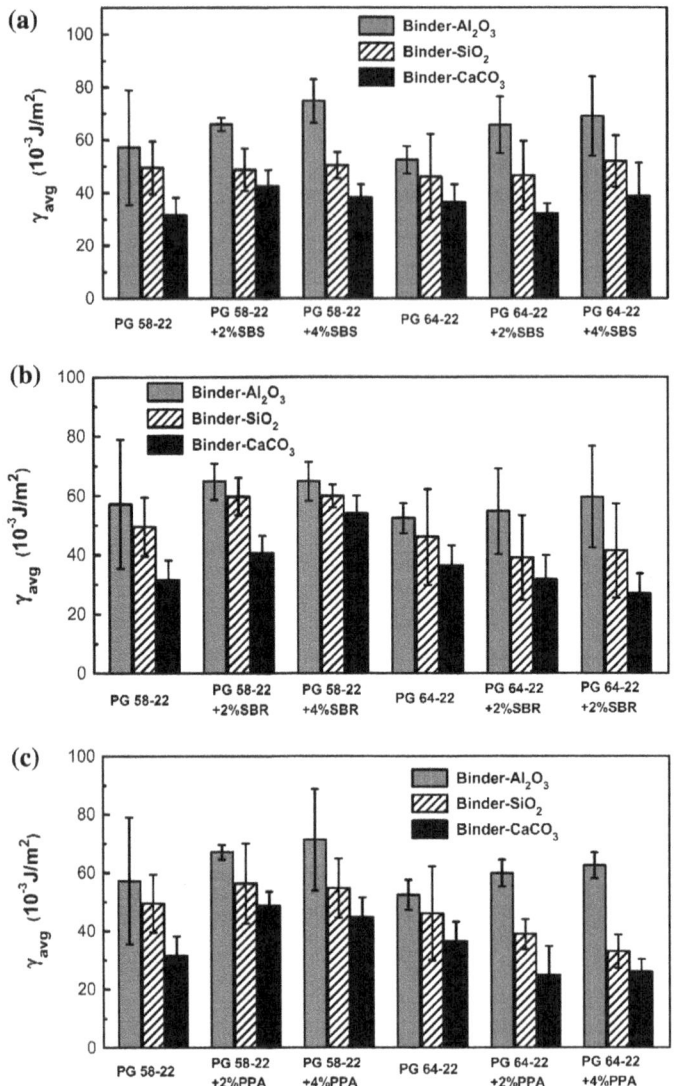

Fig. 4 Average unit surface energies of PG 58-22 and PG 64-22 binder series: **a** 2 and 4 wt% SBS modifications. **b** 2 and 4 wt% SBR modifications. **c** 2 and 4 wt% PPA modifications. Adapted from Ref. [21] with permission from ASCE

3.4 Statistical Analysis of the Adhesion Results

To evaluate the performance of particle probe scanning force microscopy, T-tests were used to analyze the adhesion results (Tables 1 and 2) based on the 95% confidence interval for various binder-aggregate pairs. For the group of control

Table 1 T-test results of adhesion in control binders

Particle	PG XX-22 series			PG 64-XX series		
	$P(22_{ab})$	$P(22_{ac})$	$P(22_{bc})$	$P(64_{ab})$	$P(64_{ac})$	$P(64_{bc})$
Al_2O_3	9.3E-3	1.2E-3	2.0E-4	2.5E-49	5.1E-42	5.2E-8
SiO_2	3.8E-3	3.7E-21	7.9E-4	4.3E-14	7.9E-2	1.6E-22
$CaCO_3$	9.9E-7	4.6E-7	4.2E-2	6.7E-27	4.9E-3	4.3E-9

Table 2 T-test results of adhesion in PG 58-XX binder series

Particle	SBR modified PG 58 series			PPA modified PG 58 series		
	$P(SBR_{02})$	$P(SBR_{04})$	$P(SBR_{24})$	$P(PPA_{02})$	$P(PPA_{04})$	$P(PPA_{24})$
Al_2O_3	4.1E-5	2.4E-7	9.7E-1	5.8E-9	5.4E-7	1.3E-3
SiO_2	1.7E-21	1.5E-19	8.5E-1	2.2E-8	8.7E-9	2.5E-1
$CaCO_3$	5.9E-26	8.2E-62	9.0E-32	1.0E-42	4.6E-24	3.0E-4

Notes PG XX-22 series: a. PG 58-22; b. PG 64-22; c. PG 70-22
PG 64-XX series: a. PG 64-16; b. PG 64-22; c. PG 64-28
SBR modified PG 58-22 series: (0) PG 58-22; (2) PG 58-22 + 2% SBR; (4) PG 58-22 + 4% SBR
PPA modified PG 58-22 series: (0) PG 58-22; (2) PG 58-22 + 2% PPA; (4) PG 58-22 + 4% PPA
The confidence interval is 95%. Adapted from Refs. [19, 21] with permission

binders (Table 1), the adhesion levels between alumina microspheres and control binders were significantly different from each other. The adhesion measurements using silica and calcium carbonate microspheres are different among most cases except for a few similar pairs, such as PG 64-16 and PG 64-28 when using silica particles. In the SBR modified PG 58-22 series (Table 2), adhesion measurements using mineral microspheres were different for most aggregate-binder pairs except for a few combinations. For example, similar adhesion were measured between alumina and silica microspheres and 2 and 4 wt% SBR modified PG 58-22 binders. Similar trends were obtained for adhesion results between mineral microspheres and other modifications [21]. Generally, particle probe scanning force microscopy is able to detect different adhesion among mineral-binder pairs successfully.

3.5 Morphologies of Modified Binders

Surface morphologies of control and modified binders were characterized using conventional AFM probes (Fig. 5). Height scans were shown for the PG 58-22 control binder, 2 and 4 wt% PPA modifications. Bee structures in the control PG 58-22 binder were clearly illustrated (Fig. 5a), where alternating ripples were orthogonal to the longitudinal axis. In PPA modified PG 58-22 binders, bee structures were surrounded by flocculent wing structures (Fig. 5b) that are different from those in the control specimens. As PPA increased from 2 to 4 wt%, the

Fig. 5 Height scans of
a control PG 58-22 binder.
b 2 wt% PPA modifications.
c 4 wt% PPA modifications
(40 × 40 μm). Adapted from
Ref. [21] with permission
from ASCE

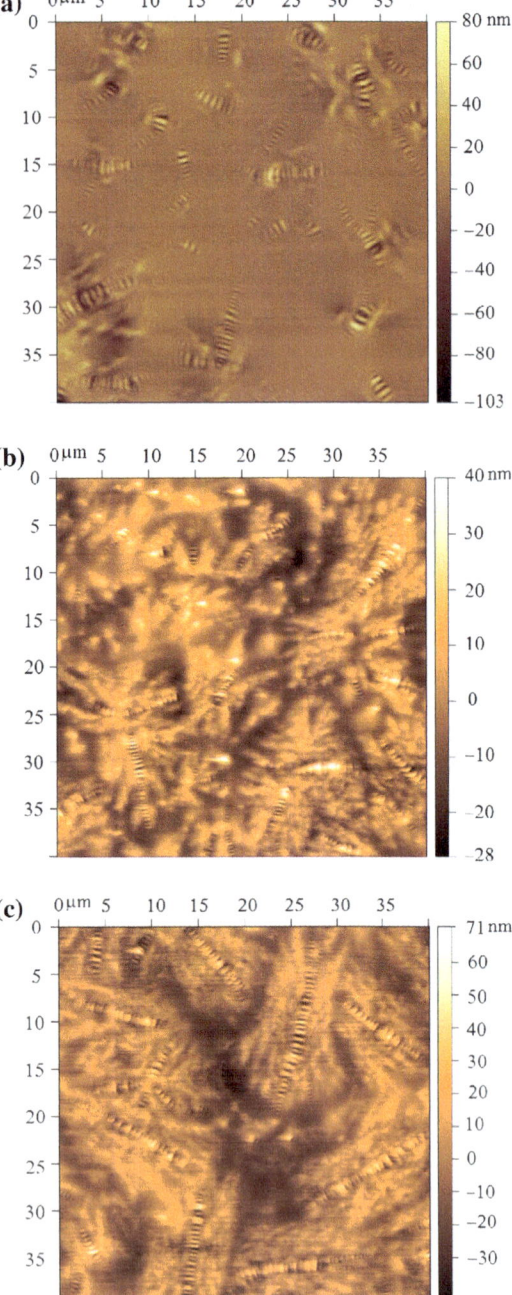

flocculent wings became more extent (Fig. 5c) compared to those in Fig. 5b. Variations of morphologies were also observed in other modified series, which are not shown here due to the page limit. These results show that binder morphologies are affected by modifier types and weight percentages.

4 Conclusions

Particle probe scanning force microscopy was used to measure the adhesion between aggregate minerals and plain or modified binders. Three types of particle probes were created using microspheres, i.e., silica, calcium carbonate and alumina, to represent the primary aggregate constituents. Control binders PG xx-22 and PG 64-xx series were used to prepare SBS, SBR, and PPA modifications. The average unit surface energies measured between microspherical probes and binder substrates were able to differentiate various binders, which are further proved by the statistical analyses. The results showed that alumina-binder pairs exhibit higher adhesion than silica- and calcium carbonate-binder pairs. The alumina-binder adhesion values increase as weight percentages of modifiers (PPA, SBS, and SBR) used in this study increase. Since alumina is the most polar among the three types of minerals, stronger interactions were induced in alumina-binder pairs than those in silica- and calcium carbonate-binder pairs. Microstructural variations were clearly demonstrated for different modified binders. These studies showed that particle probe scanning force microscopy is an effective tool to distinguish different asphalt binders and aggregates. The findings could lead to the development of high-performance asphalt binders and mixtures in the future.

Acknowledgements This study was supported by the College of Engineering and Mathematical Sciences at the University of Vermont (UVM). The authors would like to sincerely thank the Agency of Vermont Transportation for their support. We sincerely thank ICL Performance Company for providing PPA specimens, Bitumar Company for providing the SBS specimens, and Ultrapave Company for providing the SBR specimens. Appreciation is also extended to the UVM microscopy imaging center for the assistance of image characterization.

References

1. W.M. Kelly, Mineral industry of the state of New York (2011), http://www.nysm.nysed.gov/staffpubs/docs/20367.pdf
2. American Society for Testing and Materials, *Standard test method for separation of asphalt into four fractions, ASTM D4124* (West Conshohocken, PA, 2009)
3. A. Hanz, H. Bahia, K. Kanitpong, H.F. Wen, *Test method to determine aggregate/asphalt adhesion properties and potential moisture damage SPR 0092-05-12*, (Madison, WI, 2007)
4. R.A. Tarefder, A.M. Zaman, Nanoscale evaluation of moisture damage in polymer modified asphalts. J. Mater. Civ. Eng. **22**(7), 714–725 (2009)

5. Å.L. Lyne, V. Wallqvist, B. Birgisson, Adhesive surface characteristics of bitumen binders investigated by atomic force microscopy. Fuel **113**, 248–256 (2013)
6. X. Yu, N.A. Burnham, R.B. Mallick, M. Tao, A systematic AFM-based method to measure adhesion differences between micron-sized domains in asphalt binders. Fuel **113**, 443–447 (2013)
7. A.S. Al-Rawashdeh, S. Sargand, Performance assessment of warm asphalt binder in presence of water by using surface free energy concepts and nano-scale techniques. J. Mater. Civ. Eng. **26**(5), 803–811 (2013)
8. D. Cheng, D. Little, R. Lytton, J. Holste, Moisture damage evaluation of asphalt mixtures by considering both moisture diffusion and repeated-load conditions. Trans. Res. Rec. J. Trans. Res. Board **1832**, 42–49 (2003)
9. J.N. Israelachvili, *Intermolecular and surface forces: revised*, 3rd edn. (Academic press, Oxford, UK, 2011)
10. K. Kendall, Adhesion: molecules and mechanics. Science **263**(5154), 1720–1725 (1994)
11. A. Kinloch, *Adhesion and adhesives: science and technology* (Chamman and Hall, NY, 2012)
12. J. Drelich, K.L. Mittal, *Atomic force microscopy in adhesion studies* (CRC Press, Boca Raton, FL, 2005)
13. R. Jones, H.M. Pollock, J.A.S. Cleaver, C.S. Hodges, Adhesion forces between glass and silicon surfaces in air studied by AFM: effects of relative humidity, particle size, roughness and surface treatment. Langmuir **18**, 8045–8055 (2002)
14. T. Eastman, D.M. Zhu, Adhesion forces between surface-modified AFM tips and a mica surface. Langmuir **12**(11), 2859–2862 (1996)
15. J. Meng, A. Orana, T. Tan, K. Wolf, N. Rahbar, H. Li, G. Papandreou, C. Maryanoff, W.O. Soboyejo, Adhesion and interfacial fracture in drug-eluting stents. J. Mater. Res. **25**(4), 641–647 (2010)
16. T. Tan, J. Meng, N. Rahbar, H. Li, G. Papandreou, C.A. Maryanoff, W.O. Soboyejo, Effects of silane on the interfacial fracture of a parylene film over a stainless steel substrate, Mater. Sci. Eng. C **32**(3), 550–557
17. V.T. Moy, E.L. Florin, H.E. Gaub, Intermolecular forces and energies between ligands and receptors. Science **266**(5183), 257–259 (1994)
18. J. Yang, AFM as a high-resolution imaging tool and a molecular bond force probe. Cell Biochem. Biophys. **41**(3), 435–449 (2004)
19. Y. Li, J. Yang, T. Tan, Study on adhesion between asphalt binders and aggregate minerals under ambient conditions using particle-modified atomic force microscope probes. Constr. Build. Mater. **101**, 159–165 (2015)
20. Y. Li, J. Yang, T. Tan, Measuring adhesion between modified asphalt binders and aggregate minerals: use of particle probe scanning force microscopes, Transp. Res. Rec. J. Transp. Res. Board, 117–123 (2016)
21. Y. Li, J. Yang, T. Tan, Adhesion between modified binders and aggregate minerals at ambient conditions measured with particle-probe scanning force microscopes. J. Mater. Civ. Eng. **29**(8), 4017068 (2017)
22. T. Tan, Y. Li, J. Yang, Quantification of aggregate-binder adhesion in asphalt mixtures using particle probe scanning force microscopes, in *Proceedings of the International MultiConference of Engineers and Computer Scientists 2017*, 15–17 March, 2017, Hong Kong, pp. 1063–1067
23. J.L. Hutter, J. Bechhoefer, Calibration of atomic-force microscope tips. Rev. Sci. Instrum. **64**(7), 1868–1873 (1993)
24. J.L. Hutter, M. Jaschke, Calculation of thermal noise in atomic force microscopy. Nanotechnology **6**(1), 1 (1995)
25. B.V. Derjaguin, V.M. Muller, Y.P. Toporov, Effect of contact deformations on the adhesion of particles. J. Colloid Interface Sci. **53**(2), 314–326 (1975)
26. C.T. Joseph, *Practice of thin layer chromatography* (Wiley, NY, 1992)

Approaches to a Descent Underwater Vehicle Stabilization in Conditions of a Sea Disturbance

Sergey An. Gayvoronskiy, Tatiana Ezangina and Ivan Khozhaev

Abstract The article deals with the position stabilization mode of a descent underwater vehicle under the conditions of sea disturbance. This descent underwater vehicle is connected with a carrier—ship by an elastic rope. A shock-absorbing hoist installed on the descent underwater vehicle is used to damp its oscillation. The analysis of the automatic positioning of the descent underwater vehicle at a selected depth using various stabilization systems is carried out. The systems difference is in the use of various sensors (measuring converter of the rope length, tension deviation sensor and their combination), measuring external disturbance from sea disturbance. The considered systems were simulated and the conclusions were made.

Keywords Descent underwater vehicle · Measuring converter of the rope length · Robust control · Sea disturbance · Stabilization systems Tension deviation sensor

1 Introduction

At present different practical tasks of the World ocean exploration are solved using tethered underwater vehicle tied up with a carrier-ship by means of a cable-rope [1–12]. The activities of the World ocean exploration include geological

S. A. Gayvoronskiy (✉) · T. Ezangina · I. Khozhaev
National Research Tomsk Polytechnic University,
Lenina avenue 30, Tomsk, Russia
e-mail: saga@tpu.ru

T. Ezangina
e-mail: eza-tanya@tpu.ru

I. Khozhaev
e-mail: ivh1@tpu.ru

© Springer Nature Singapore Pte Ltd. 2018
S.-I. Ao et al. (eds.), *Transactions on Engineering Technologies*,
https://doi.org/10.1007/978-981-10-7488-2_10

Fig. 1 Allocation of a carrier
vessel and a DUV

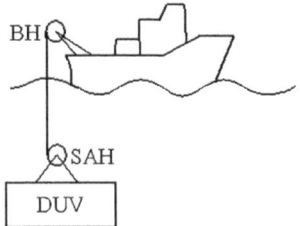

prospecting, oceanographic and other kinds of works. The use of tethered underwater vehicle is very promising since they can increase the operation life of
autonomous underwater vehicle. For instance, the descent battery-charging stations
are able to charge the storage batteries of autonomous underwater vehicle without
taking them aboard. Using descent containers autonomous underwater vehicle can
be submerged to the predetermined depth and taken again aboard. An example of
the descent underwater vehicle and carrier vessel allocation is shown in the Fig. 1.

The following symbols are introduced in Fig. 1: SAH—is shock-absorbing
hoist; BH—is boat hoist; DUV—is descent underwater vehicle.

The main problem by using different descent underwater vehicle is to ensure the
precise stabilization to the predetermined depth in the sea disturbance environment.
This problem is associated with vertical oscillations of the tethered descent
underwater vehicle affected by sea disturbance. These oscillations can transform
into resonance ones and cause the rope break and the tethered descent underwater
vehicle hitting the ground. Therefore, the damping of the descent underwater
vehicle should be ensured by the stabilization system. To design such systems their
mathematical models are required. These models take into consideration the abil-
ities of the elastic "rope-descent underwater vehicle" link (the interval uncertainty
of their parameters, its friction on water, etc.). A range of various control modes
based on the measurement of their defined coordinates can be used. There are
several ways to stabilize a descent underwater vehicle with the help of different
control signals. It is proposed to implement three ways of a descent underwater
vehicle stabilization, based on such control signals:

- a signal from the rope length measuring converter (MC);
- a signal from the rope tension deviation sensor (SRTD);
- a combination of two previous signals.

The main aim of the research is to analyze accuracy of these three ways of a
DUV stabilization.

2 Block Diagrams of Position Stabilization Systems of a Descent Underwater Vehicle

2.1 Structure of a Descent Underwater Vehicle Stabilization System on the Base of Rope Length Measuring Converter Signal

Let us develop a structure of the descent underwater vehicle stabilization system, based on a rope length measuring converter signal (Fig. 2).

The following symbols are introduced in Fig. 2: SAH—is shock-absorbing hoist; MC—is the measuring converter of rope length using the reducer on the shaft of the SAH; ED2—is electric drive 2, CS—is comparison-summator; CB2—the second control block; 1—cable–rope, 2—a lock joint, 3—fastening, 4—metal rod, 5—rope, 6—the drum of a SAH, FD—is force of disturbance.

In this case the shock-absorbing hoist is controlled on the base of comparison of the disturbance ordinate Δl_{dis} and the rope length change Δl_r between the lock joint and descent underwater vehicle. The difference signal $(\Delta l_{dis} - \Delta l_r)$ is the control error. The signal is transmitted via the control block 2 to the shock-absorbing hoist,

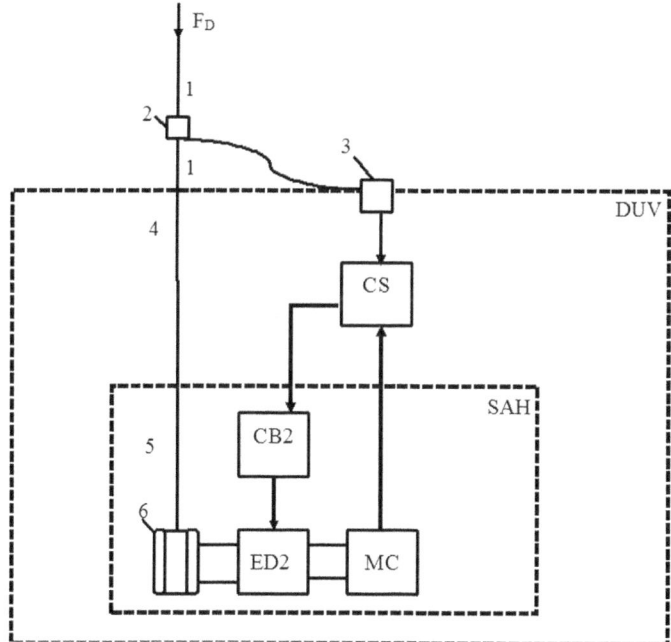

Fig. 2 Descent underwater vehicle stabilization system on the base of rope length measuring converter signal

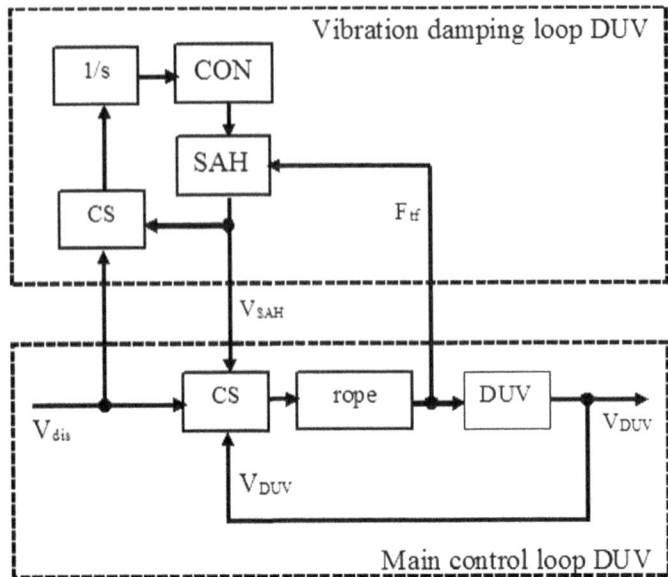

Fig. 3 Functional diagram of a descent underwater vehicle stabilization system on the base of rope length measuring converter signal

which damps the oscillation of the descent underwater vehicle. Functional diagram of the system is shown in the Fig. 3.

The following symbols are introduced in Fig. 3: CON—is a shock-absorbing hoist regulator; V_{dis}—the speed of sea disturbance; V_{DUV}—is speed of descent underwater vehicle, V_{SAH}—linear speed of shock-absorbing hoist; F_{tf}—is tension force in the cable-rope, CS—is comparison-summator.

To develop a structural diagram of such system, let us derive equations for system elements separately. The equation of vertical motion has the following form $m\frac{dV_{DUV}}{dt} = F_{tf}$, where $m = (m_{DUV} + \mu)$—is the mass of a descent underwater vehicle consideration associated water mass, μ—is associated water mass, m_{DUV}—is the mass of a descent underwater vehicle in water. Inertia of a DUV is defined by the transfer function $W_{DUV}(s) = 1/ms$.

The electric drive of shock-absorbing hoist is described by the following equation $J_2 \frac{d\omega_{SAH}}{dt} = M_d + M_{tf}$, where ω_{SAH}—is angular velocity of the drum rotation of shock-absorbing hoist; J_2—is moment of inertia of shock-absorbing hoist. $M_{tf} = F_{tf}R_2$—is torque produced on shock-absorbing hoist by rope tension force, R_2—is drum radius of shock-absorbing hoist. $M_d = k_{m2}(U_c - U_e)$—is controlling torque of the drive of shock-absorbing hoist, U_c—is output voltage of the SAH controller, k_{m2}—is the transfer constant of SAH drive on torque. $U_e = k_{e2}\omega$—is back EMF voltage of the motor of shock-absorbing hoist, k_{e2}—is back EMF coefficient of the motor of shock-absorbing hoist. The transfer function of an electric drive of a SUH is as follows: $W_{SAH}(s) = V_{SAH}(s)/U_c(s) = k_{m2}R_2/(J_2s + k_{e2}k_{m2})$.

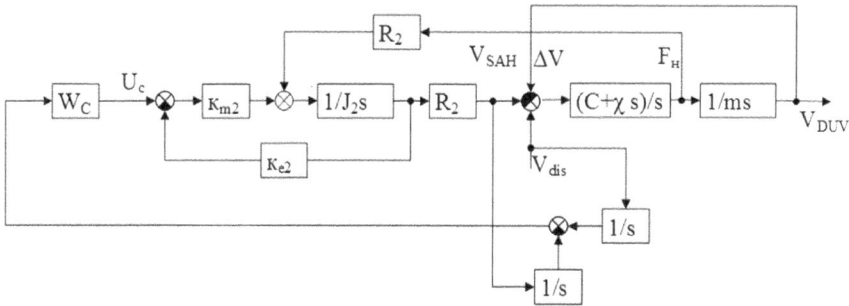

Fig. 4 Blok diagram of a descent underwater vehicle stabilization system on the base of rope length measuring converter signal

On the basis of the Hooke's law we will get the equation connecting the rope tension force and displacement force of its end $F_{tf} = \frac{C}{l_r}(x_{dis}) - (x_{SAH} + x_{DUV}) + \frac{\chi}{l_r}\frac{d(x_{dis}) - (x_{SAH} + x_{DUV})}{dt}$, where l_r—is the lengthening of the rope, x_{dis}— is the vertical movement of the ship x_{SAH}—is the movement of the rope on the dram of SAH, $C = C_{ssr}/l_r$—is stiffness coefficient of a rope, C_{ssC}—is specific stiffness of a rope, $\chi = \chi_{idr}/l_r$—is the damping coefficient of a rope, x_{tr}—is the displacement of the rope top end, χ_{idr}—is internal damping of a rope. Following interval parameters are used to describe a system: $[\chi]$, $[\chi_{idr}]$, $[l_r]$, $[C]$, $[C_{ssr}]$, $[m]$. It is to be noticed, that the reduction unit on which the measuring converter is based is an integral element which increases the system astaticism. Therefore, we suggest using P-controller instead of PI-controller. Its mathematical model is illustrated in Fig. 4.

2.2 Structure of Descent Underwater Vehicle Stabilization System on the Base of a Rope Tension Deviation Sensor Signal

In such system, a SRTD only is used to stabilize a descent underwater vehicle. Let us develop a structure of such stabilization system (see Fig. 5). The following symbols are introduced in Fig. 5: SRTD—is the sensor of rope tension deviation.

In this case the shock-absorbing hoist is controlled on the base of the rope tension deflection $\Delta F_{tf} = C(\Delta l_{dis} - \Delta l_r)$, where Δl_{dis}—is corresponds to the movement of the upper end of the rope, and Δl_r—is corresponds to the lower end of the rope. This expression shows that the sensor of rope tension deviation is able to measure the same control error as in the system with the measuring converter of rope length. Functional diagram of the system is shown in the Fig. 6.

A mathematical description of descent underwater vehicle motion and an electric drive of a shock absorbing windlass is the same as for the system with measuring converter. Input voltage of a controller of a shock-absorbing hoist is the sensor

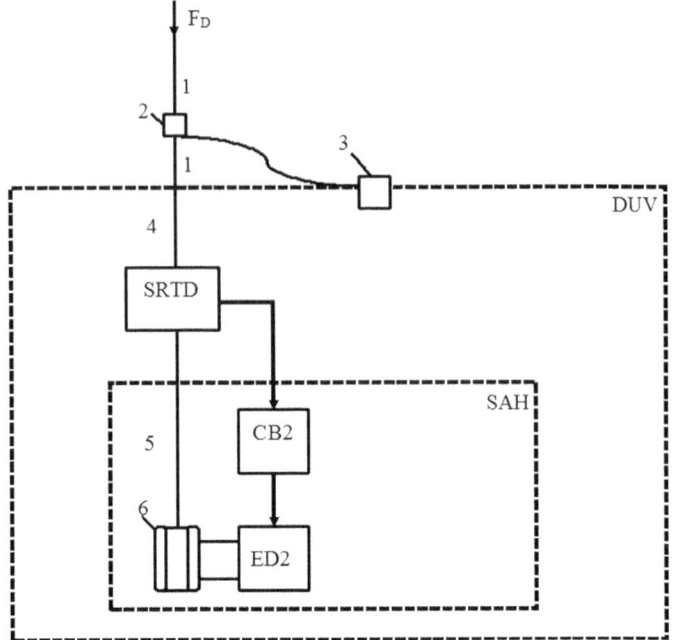

Fig. 5 Descent underwater vehicle stabilization system on the base of SRTD signal

signal of the rope deviation tension. Assuming that its transfer function has the following image: $W_{SCTD}(s) = k_{SRTD}$ we can conclude that the output sensor signal varies proportionally to the law $U_{SRTD} = F_{tf}k_{SRTD}$. To increase the rate of astaticism and minimize the errors in the control system we suggest using a PI-controller $(W_C(s) = (k_1 + k_2 s)/s)$ as a controller. Its mathematical model is illustrated in Fig. 7.

2.3 Structure of a Descent Underwater Vehicle Stabilization System on the Base of Rope Tension Deviation Sensor and Rope Length Measuring Converter Combined Signal

Let us assume that the signals from the sensor of rope tension deviation and measuring converter of rope length are transmitted to the control block illustrated in Fig. 8. The control is carried out on the base of the observed coordinates of the rope length change and its tension deflection. A functional diagram of such system can be developed on the base of the structure (Fig. 9).

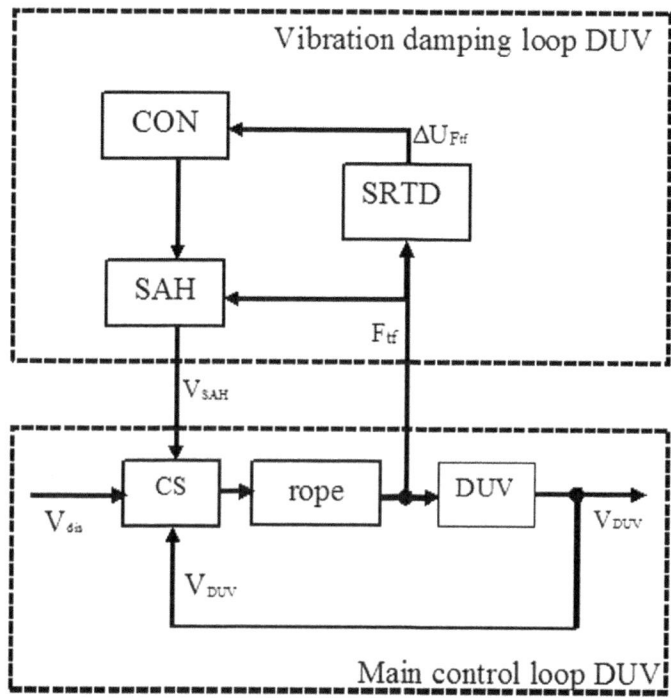

Fig. 6 Functional diagram of a descent underwater vehicle stabilization system on the base of SRTD signal

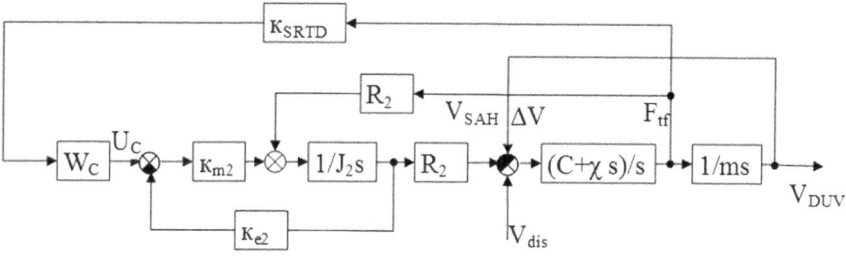

Fig. 7 Blok diagram of a descent underwater vehicle stabilization system on the base of SRTD signal

It should be noticed that the simultaneous use of the reduction unit for the measuring converter and PI-controller for the sensor of rope tension deviation can result in system instability. Thus, we can use the sensor of the rope tension deviation simultaneously with the measuring converter but with a P-controller. Block diagram of this system is presented in Fig. 10.

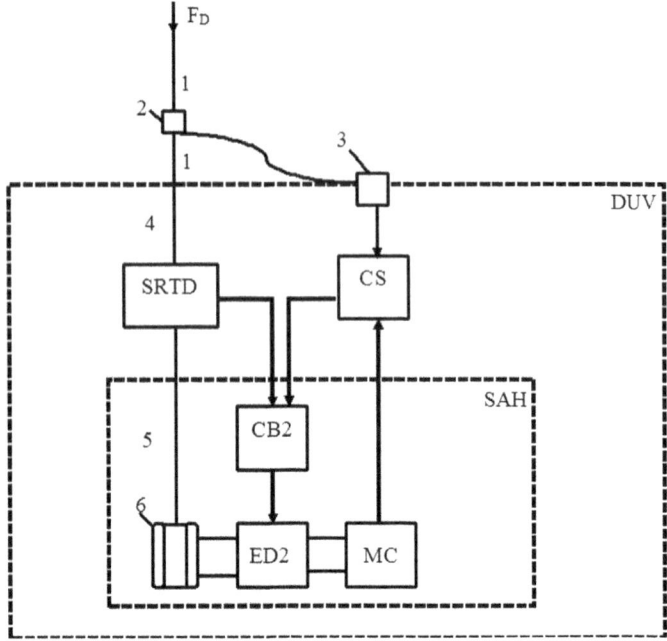

Fig. 8 Descent underwater vehicle stabilization system on the base of rope tension deviation sensor and rope length measuring converter combined signal

3 The Parametric Synthesis of the Robust Controllers

Since there are interval parameters in systems it is necessary to provide them with robust characteristics ensuring the permissible performance quality at any possible variations of unstable parameters [9–19]. It is suggested using a robust approach in the control loop of a shock absorbing hoist by the parametric synthesis of a controller. The interval expansion of the mathematical programming technique can be used as the basis of such approach [19]. To apply this approach we suggest combining the procedures of the system analysis and synthesis by affine and interval types of coefficients uncertainty of the polynomial. The work [13] presents the algorithm of such synthesis. In accordance with the developed algorithm the controller synthesis ensuring quasi maximal degree of stability [14] is carried out at the first stage. At the second stage the found parameterizations of the controller are substituted into the polynomial with affine uncertainty $D(s) = \sum\limits_{i=1}^{m} [T_i] A_i(s) + B(s)$, where $[T_i] = [\underline{T_i}; \overline{T_i}]$). Then on its basis the boundary vertex-edge route [20] is constructed for the polyhedron of interval system parameters. After that the found route mapped onto the root plane and the vertex which is the closest to the

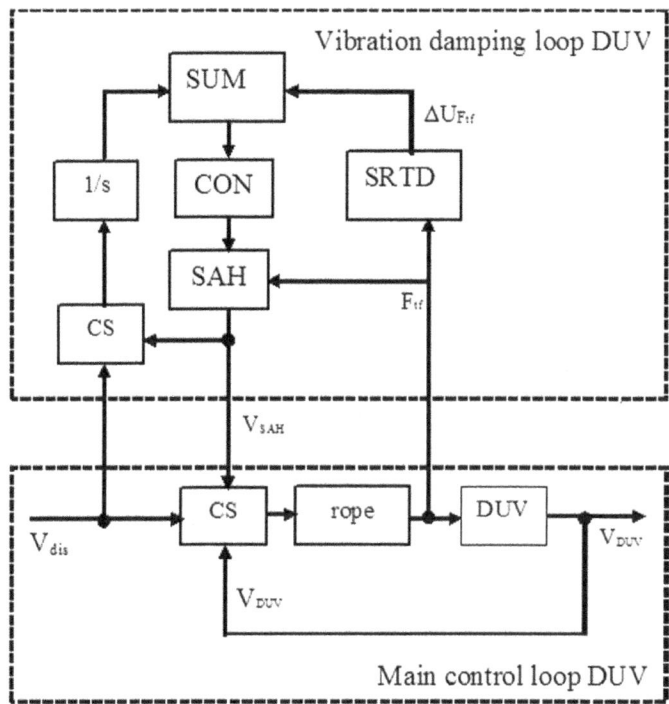

Fig. 9 Functional diagram of a descent underwater vehicle stabilization system on the base of rope tension deviation sensor and rope length measuring converter combined signal

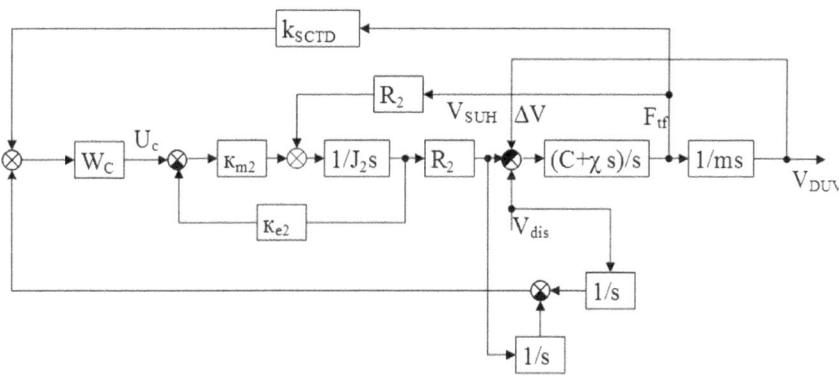

Fig. 10 Blok diagram of a descent underwater vehicle stabilization system on the base of rope tension deviation sensor and rope length measuring converter combined signal

imaginary axis is defined V_q(q—vertex number). Finally, previously calculated vertex coordinates are substituted into interval characteristic polynomial $D^q\left(\vec{k}, \alpha^*, \beta\right) = \mathrm{Re}D^q\left(\vec{k}, \alpha^*, \beta\right) + \mathrm{Im}D^q\left(\vec{k}, \alpha^*, \beta\right)$.

The following non-linear equation system is derived, based on the interval characteristic polynomial

$$\begin{cases} \operatorname{Re} D^q \left(\overrightarrow{k}, \alpha, \beta \right) = 0; \\ \operatorname{Im} D^q \left(\overrightarrow{k}, \alpha, \beta \right) = 0; \\ \partial \operatorname{Re} D^q \left(\overrightarrow{k}, \alpha, \beta \right) / \partial \alpha = 0; \\ \partial \operatorname{Im} D^q \left(\overrightarrow{k}, \alpha, \beta \right) / \partial \alpha = 0; \\ \dots \\ \partial^c \operatorname{Re} D^q \left(\overrightarrow{k}, \alpha, \beta \right) / \partial \alpha^c = 0; \\ \partial^c \operatorname{Im} D^q \left(\overrightarrow{k}, \alpha, \beta \right) / \partial \alpha^c = 0. \end{cases} \qquad (1)$$

Solving this problem we define the values of maximal degree of stability α and ensure its parameterizations of $\overrightarrow{k}$ controller.

Let us apply this algorithm for the combined synthesis of the second system. Upon the results of the first and second stages [13] the vertex has been found $V_{20}(\underline{T_1}; \overline{T_2}; \underline{T_3}; \overline{T_4}; \underline{T_5})$, where $[T_1] = [m]$, $[T_2] = [\chi]$, $[T_3] = [\chi][m]$, $[T_4] = [C][m]$, $[T_5] = [C]$. After that, the coordinates of the obtained vertex are substituted into the set of nonlinear Eqs. (1)

$$\begin{cases} \operatorname{Re} D^{20}(k_1, k_2, \alpha, \beta) = 0; \\ \operatorname{Im} D^{20}(k_1, k_2, \alpha, \beta) = 0; \\ \partial \operatorname{Re} D^{20}(k_1, k_2, \alpha, \beta) / \partial \alpha = 0; \\ \partial \operatorname{Im} D^{20}(k_1, k_2, \alpha, \beta) / \partial \alpha = 0. \end{cases}$$

Solving the set of nonlinear equations, we will find the desired controller parameters $k_1 = 0.77$, $k_2 = 2.2$ and maximal robust degree of stability $\alpha = 0.35$ [21].

As a result of applying of the combined synthesis algorithm the controller parameters for the first system $k_0 = 10^3$, and the third system is $k_0 = 5 \times 10^4$ was found [21].

4 Comparative Simulation of the Control Processes of the Position Stabilization Systems of the Descent Underwater Vehicle

To compare the stabilization accuracy of the descent underwater vehicle location at different modes with the sensor of rope tension deviation and the measuring converter the simulation of the corresponding systems at a depth of 6000 m was carried out. In the simulation was used the model of sea disturbance is maximally close to real conditions. Block diagram of this model is shown in Fig. 11.

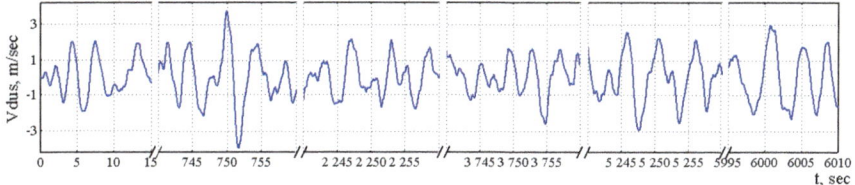

Fig. 11 The diagram of the ordinate variation of irregular sea disturbance

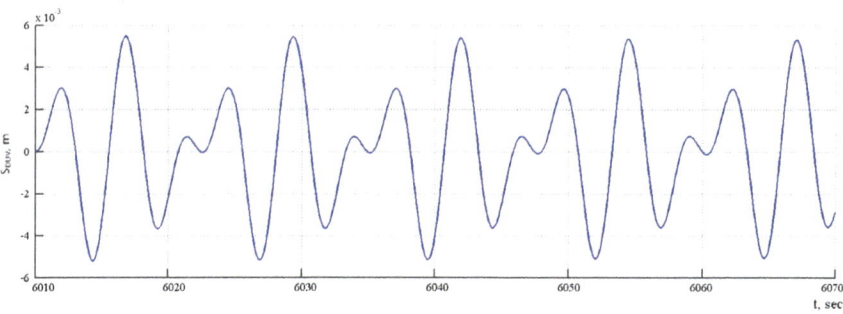

Fig. 12 Vertical movement variation curves of the descent underwater vehicle with the measuring converter

The simulation results are shown in Figs. 12, 13 and 14.

The system simulation allowed finding the mean square value of the vertical deviation of the descent underwater vehicle from the location stabilization $\sigma_1 = 2.7 \times 10^{-3}$ m (Fig. 12).

As a result of system simulation using the sensor of rope tension deviation we have found the mean square value of the vertical deviation of the descent underwater vehicle from the location stabilization $\sigma_2 = 0.11 \times 10^{-3}$ m (Fig. 13).

In the case of simultaneous use of the sensor of rope tension deviation and the measuring converter in the system, the mean square value of the vertical deviation of the descent underwater vehicle from the location stabilization makes up

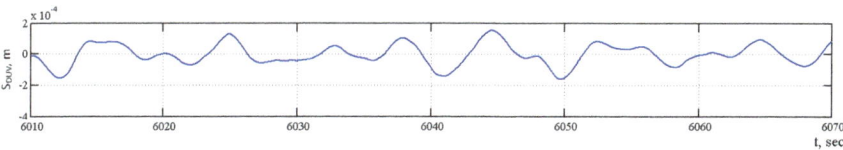

Fig. 13 Vertical movement variation curves of the descent underwater vehicle with the sensor of rope tension deviation

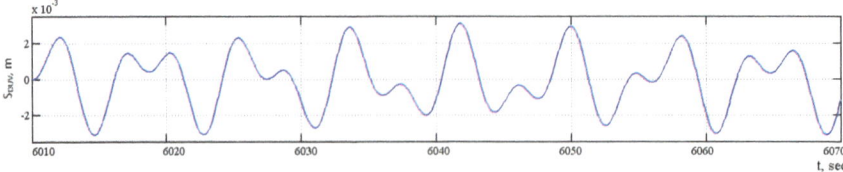

Fig. 14 Variation curves in the vertical movement stabilization mode of the descent underwater vehicle with the sensor of rope tension deviation and the measuring convert

$\sigma_3 = 1.6 \times 10^{-3}$ m. Upon the simulation results, we can conclude that the more accurate stabilization of the descent underwater vehicle is possible when using the sensor of rope tension deviation.

5   Conclusion

Upon the comparison of the mean square deviation of the vertical movement of the descent underwater vehicle for different systems it was determined, that $\sigma_2 < \sigma_3 < \sigma_1$. The inequality $\sigma_3 < \sigma_1$ is correct, since a correcting element in the form of the sensor of rope tension deviation is introduced in the system with the measuring converter. This inequality, where $\sigma_2 < \sigma_3$, is explained by the use of the sensor of rope tension deviation with a PI-controller in the system. Unlike the integrator, PI-controller allows increasing the overall transfer coefficient of the system and improving the location stabilization accuracy of the descent underwater vehicle. From the results of mathematical modeling, we can conclude that precise stabilization of the descent underwater vehicle is possible using the sensor of rope tension deviation.

Acknowledgements   The reported study is supported by the Ministry of Education and Science of Russian Federation (project #2.3649.2017/4.6).

References

1. B.W. Nam, S.Y. Hong, Y.S. Kim, Effects of passive and active heave compensators on deepwater lifting operation. Eng. Lett. **23**(1), 33–37 (2013)
2. J. Kim, Thruster modeling and controller design for unmanned underwater vehicles (UUVs). Underwater Vehicles, pp. 235–250, 2008
3. G.E. Kuvshinov, K.V. Chupina, D.V. Radchenko, P.I. Chepurin, Analysis of towed underwater vehicle system conduct under rough sea conditions, in *Proceedings of the Sixth International Symposium on Underwater Technology (UT2009)*, 2009, pp. 193–200
4. S. Ueki, H. Doi, S, Miyajima, K. Hasegava, H. Satoh, Hoisting device with vertical motion compensation function, U.S. Patent 2009/0166309 July, 2, 2009

5. U.A. Korde, Active heave compensation on drill-ships in deep water. Ocean Eng. **25**(7), 541–561 (1998)
6. S.I. Sagatun, Active control of underwater installation. IEEE Trans. Control Syst. Technol. **10** (5), 743–748 (2002)
7. S. Rowe, B. Mackenzie, R. Snell, Deepwater installation of subsea hardware, in *Proceedings of the 10th Offshore Symposium*, Houston, TX, 2001, pp. 1–9
8. J. Neupert, T. Mahl, B. Haessig, O. Sawodny, K. Schneider, *A Heave, Compensation Approach for Offshore Cranes, American Control Conference*, Seattle, Washington, USA, 2008, pp. 538–543
9. Giuseppe Conte, Andrea Serranu, Robust control of a remotely operated underwater vehicle. Automatica **34**(2), 193–198 (1998)
10. V. Vesely, Robust control methods a systematic survey. J. Electr. Eng. **64**(1), 59–64 (2013)
11. J. Yu, Z. Yan, J. Wang, Q. Li, Robust stabilization of ship course via convex optimization. Asian J. Control **16**, 871–877 (2013)
12. Y.-W. Liang, S.-D. Xu, T.-C. Chu, Robust control of the robot manipulator via an improved sliding mode scheme, in *International Conference on Mechatronics and Automation—Harbin, China: Proceeding of conference*, 1, pp. 1593–1598, 2007
13. S.A. Gayvoronskiy, T. Ezangina, I. Khozhaev, L. Gunbo, Maximization of the robust stability degree of interval systems by means of a linear controller in the presence of limits. J. Phys. Conf. Ser. **803**(1), pp. 1–5 (012045) (2017)
14. S.A. Gayvoronskiy, T. Ezangina, I Khozhaev, The interval-parametric synthesis of a linear controller of speed control system of a descent submersible vehicle. IOP Conf. Ser. Mater. Sci. Eng. **93**, 1–7(012055) (2015)
15. S.V. Efimov, M.I. Pushkarev, Determining direct measures of performance based on the location of zeros and pole of the transfer function. Optoelectron. Instrum. Data Process. **47**, 297–302 (2011)
16. D.P. Kim, Synthesis regulator maximum degree of stability. Driv Technol. 52–57 (2003)
17. G.V. Rimsky, A.A. Nesenchuk, Root locus methods for robust control systems quality and stability investigations, in *Proceedings of Conference, 1996 IFAC 13th Triennial World Congress—San Francisco, USA*, pp. 469–474
18. A.A. Nesenchuk, Parametric synthesis of qualitative robust control systems using root locus fields, in *Proceedings of the 15th Triennial World Congress of The International Federation of Automatic Control (IFAC)*, Barcelona, Spain, 21–26 July 2002, pp. 387–387
19. A.V. Tatarinov, A.M. Tsirlin, Mathematical programming problems containing complex variables, and the maximum degree of stability of linear dynamical systems. Bull. RAS. Ser. Comput. Syst. **1**, 28–33 (1995)
20. O.S. Vadutov, S.A. Gayvoronskiy, Application of edge routing to stability analysis of interval polynomials. J. Comput. Syst. Sci. Int. **42**(6), 833–838 (2003)
21. S.A. Gayvoronskiy, T. Ezangina, I. Khozhaev, The analysis of different methods to stabilize the location of Descent Underwater Vehicles, in *Proceedings of the International MultiConference of Engineers and Computer Scientists 2017*. Lecture Notes in Engineering and Computer Science, Hong Kong, 15–17 Mar 2017, pp. 240–244

A Comparison of Different Quasi-Newton Acceleration Methods for Partitioned Multi-Physics Codes

Rob Haelterman, Alfred Bogaers and Joris Degroote

Abstract In many cases, different physical systems interact, which translates to coupled mathematical models. We only focus on methods to solve (strongly) coupled problems with a partitioned approach, i.e. where each of the physical problems is solved with a specialized code that we consider to be a black box solver. Running the black boxes one after another, until convergence is reached, is a standard but slow solution technique, known as non-linear Gauss-Seidel iteration. A recent interpretation of this approach as a root-finding problem has opened the door to acceleration techniques based on Quasi-Newton methods that can be "strapped onto" the original iteration loop without modification to the underlying code. In this paper, we analyze the performance of different acceleration techniques on different multi-physics problems.

Keywords Acceleration · Fluid-structure interaction · Iterative method
Partitioned method · Root-finding · Quasi-Newton method

1 Introduction

Multi-physics problems occur in various disciplines. One example is fluid-structure interaction (FSI), which refers to the coupling between a fluid flow and a moving or deforming structure. This kind of problems can be found in heart valves,

R. Haelterman (✉)
Department of Mathematics, Royal Military Academy, Renaissancelaan 30, 1000 Brussels,
Belgium
e-mail: robby.haelterman@rma.ac.be

A. Bogaers
Council for Scientific and Industrial Research, Advanced Mathematical Modelling
Modelling and Digital Sciences, Meiring Naudé Road, Brummeria, Pretoria, South Africa
e-mail: abogaers@csir.co.za

J. Degroote
Department of Flow Heat and Combustion Mechanics, Ghent University,
Sint-Pietersnieuwstraat 41, 9000 Ghent, Belgium
e-mail: joris.degroote@ugent.be

© Springer Nature Singapore Pte Ltd. 2018
S.-I. Ao et al. (eds.), *Transactions on Engineering Technologies*,
https://doi.org/10.1007/978-981-10-7488-2_11

flapping flags, flutter of electricity cables, etc. Problems with more than two fields also exist, such as fluid-structure-thermal interaction in sintering processes and gas turbine blades. Mathematically these can be written as the non-linear system of equations:

$$\begin{cases} f_1(x_1, x_2, \dots, x_n) = 0 \\ f_2(x_1, x_2, \dots, x_n) = 0 \\ \qquad\vdots \\ f_n(x_1, x_2, \dots, x_n) = 0 \end{cases} \qquad (1)$$

where $f_i : D_F \subset \mathbb{R}^m \to \mathbb{R}^{k_i}$, $x_j \in \mathbb{R}^{m_j}$ $(i, j \in \{1, 2, \dots, n\})$, $\sum_{i=1}^n k_i = m$, $\sum_{j=1}^n m_j = m$.

Each equation describes (the discretized equations of) a physical problem that is spatially or mathematically decomposed. E.g. $f_1(x_1, x_2) = 0$ could give the pressure x_1 on the wall of a flexible tube for a given geometry x_2, while $f_2(x_1, x_2) = 0$, could give the deformed geometry of the wall due to the pressure exerted on it by the fluid. We will assume the problem has the following characteristics [11, 21]:

1. Good solvers exist for each equation of the system. For this reason, we will use a partitioned solution method.
2. The analytic form of f_i $(i = 1, 2, \dots, n)$ is unknown.
3. The problem has a large dimensionality.
4. Evaluating f_i $(i = 1, 2, \dots, n)$ is computationally costly.

2   Fixed-Point Formulation and Quasi-Newton Acceleration

A typical solution method for (1) is the fixed-point iteration [22].

Fixed-point iteration

1. Startup: Take initial values $x_2^1, x_3^1, \dots x_n^1$. Set $s = 1$.
2. Loop until convergence:
   2.1. Solve $f_1(x_1, x_2^s, \dots, x_n^s) = 0$ for x_1, resulting in x_1^{s+1}.
   2.2. Solve $f_2(x_1^{s+1}, x_2, \dots, x_n^s) = 0$ for x_2, resulting in x_2^{s+1}.
   $\dots$
   2.n. Solve $f_n(x_1^{s+1}, x_2^{s+1}, \dots, x_n) = 0$ for x_n, resulting in x_n^{s+1}.
   2.n+1. Set $s = s + 1$.

If F' (the Jacobian of $F = (f_1, \dots, f_n)$) satisfies the condition

$$\forall i < j < n : [F']_{ij} = 0 \qquad (2)$$

then we can write the whole process 2.1. to 2.n, as $x_n^{s+1} = H(x_n^s)$ [20]. The problem can now be considered as finding the fixed point of H, or alternatively as finding the zero of $K(x) = H(x) - x$, where we assume that K has continuous first partial derivatives and a nonsingular Jacobian in a neighborhood of its single zero. It is on this root-finding problem that we apply Quasi-Newton (QN) acceleration.

Quasi-Newton acceleration

1. Startup: Take initial values $x_2^1, x_3^1, \ldots x_n^1$. Set $s = 1$.
2. Loop until convergence:
　　2.1. Solve $f_1(x_1, x_2^s, \ldots, x_n^s) = 0$ for x_1, resulting in x_1^{s+1}.
　　2.2. Solve $f_2(x_1^{s+1}, x_2, \ldots, x_n^s) = 0$ for x_2, resulting in x_2^{s+1}.
　　. . .
　　2.n. Solve $f_n(x_1^{s+1}, x_2^{s+1}, \ldots, x_n) = 0$ for x_n, resulting in $H(x_n^s)$.
　　2.n+1. Compute an approximate Jacobian $\hat{K}_s'$ of K (see below)
　　2.n+2. $x_n^{s+1} = x_n^s - (\hat{K}_s')^{-1} K(x_n^s)$
　　2.n+3. Set $s = s + 1$.

Alternatively a slightly different Quasi-Newton step $x_n^{s+1} = x_n^s - \hat{M}_s' K(x_n^s)$ can be used. Here M_s' serves as an approximation to the inverse of the Jacobian at step s, whereas $\hat{K}_s'$ is an approximation of the Jacobian itself. We will designate methods that approximate the Jacobian as Type I methods, and methods that approximate the inverse Jacobian as Type II methods [11].

3　Different Choices of Quasi-Newton Methods

We define $\delta x_s = x_n^{s+1} - x_n^s$, $\delta K_s = K(x_n^{s+1}) - K(x_n^s)$ and $\{\iota_j; j = 1, \ldots, m_n\}$ as the canonical basis for $\mathbb{R}^{m_n}$ and $\langle \cdot, \cdot \rangle$ as the standard Euclidean scalar product.

3.1　Non-Linear Gauss-Seidel (GS)

This method (also called, among others, "Iterative Substructuring Method" or "Picard iteration") is nothing else than the fixed-point iteration described at the beginning of Sect. 2. It is seldom considered to be a Quasi-Newton method, but can take this form if we set $(\hat{K}_{s+1}')^{-1} = -I$ [22].

3.2   Aitken's δ^2 Method ($A\delta^2$)

Aitken's δ^2 method [1] is a relaxation method and as such is again seldom seen as a Quasi-Newton method, but it can take its form if we define $(\hat{K}'_{s+1})^{-1} = -\frac{1}{\omega_{s+1}}I$ with

$$\omega_{s+1} = -\omega_s \frac{\langle K(x^s_n), K(x^{s+1}_n) - K(x^s_n)\rangle}{\langle K(x^{s+1}_n) - K(x^s_n), K(x^{s+1}_n) - K(x^s_n)\rangle}. \tag{3}$$

3.3   Broyden's Good Method (BG), Bad Method (BB) and the Switched Broyden Method (SB)

1. Broyden's first (or "good") method is a Quasi-Newton method that is part of the family of Least Change Secant Update (LCSU) methods [4, 5, 8, 9, 12], where the approximate Jacobian $\hat{K}'_{s+1}$ is chosen as the solution of $\min\{\|\hat{K}' - \hat{K}'_s\|_{Fr}\}$, s.t. $\hat{K}'\delta x_s = \delta K_s$, which leads to the following rank-one update:

$$\hat{K}'_{s+1} = \hat{K}'_s + \frac{(\delta K_s - \hat{K}'_s \delta x_s)\delta x_s^T}{\langle \delta x_s, \delta x_s\rangle} \quad \text{or} \tag{4}$$

$$(\hat{K}'_{s+1})^{-1} = (\hat{K}'_s)^{-1} + \frac{(\delta x_s - (\hat{K}'_s)^{-1}\delta K_s)\delta x_s^T(\hat{K}'_s)^{-1}}{\langle \delta x_s, (\hat{K}'_s)^{-1}\delta K_s\rangle}. \tag{5}$$

$(\hat{K}'_1)^{-1}$ is typically set to be $-I$, i.e. the first iteration is a GS iteration.

2. Broyden's second (or "bad") method is a Quasi-Newton method that uses an approximation $\hat{M}'$ of the inverse Jacobian. It is also part of the family of LCSU methods [4, 9, 12], where $\hat{M}'_{s+1}$ is chosen as the solution of $\min\{\|\hat{M}' - \hat{M}'_s\|_{Fr}\}$, s.t. $\hat{M}'\delta K_s = \delta x_s$, which leads to the following rank-one update:

$$\hat{M}'_{s+1} = \hat{M}'_s + \frac{(\delta x_s - \hat{M}'_s \delta K_s)\delta K_s^T}{\langle \delta K_s, \delta K_s\rangle} \quad \text{or} \tag{6}$$

$$\left(\hat{M}'_{s+1}\right)^{-1} = \left(\hat{M}'_s\right)^{-1} + \frac{\left(\delta K_s - \left(\hat{M}'_s\right)^{-1}\delta x_s\right)\delta K_s^T\left(\hat{M}'_s\right)^{-1}}{\langle \delta K_s, \left(\hat{M}'_s\right)^{-1}\delta x_s\rangle}. \tag{7}$$

Again, typically $\hat{M}'_1 = -I$ is chosen.

3. Broyden himself [4] admitted that the "bad" formulation of his algorithm didn't function properly. The reasons for the "good" or "bad" behavior are not well understood, and in some instances the bad method actually outperforms the good method. For this reason we follow an idea suggested in [24] that avoids the need to

choose between the two methods and create a switched version of BG/BB (called "SB") based on the following reasoning: as both BG and BB are secant methods we have $\hat{K}'_s \delta x_{s-1} = \delta K_{s-1}$ and $\left(\hat{M}'_s \right)^{-1} \delta x_{s-1} = \delta K_{s-1}$. Using (4) and (7), we get

$$\hat{K}'_{s+1} \delta x_{s-1} - \delta K_{s-1} = \frac{(\delta K_s - \hat{K}'_s \delta x_s) \delta x_s^T}{\langle \delta x_s, \delta x_s \rangle} \delta x_{s-1} \tag{8}$$

$$\left(\hat{M}'_{s+1} \right)^{-1} \delta x_{s-1} - \delta K_{s-1} = \frac{\left(\delta K_s - \left(\hat{M}'_s \right)^{-1} \delta x_s \right) \delta K_s^T \left(\hat{M}'_s \right)^{-1}}{\langle \delta K_s, \left(\hat{M}'_s \right)^{-1} \delta x_s \rangle} \delta x_{s-1}. \tag{9}$$

Equations (8) and (9) can be considered to be a secant error at the next approximation with respect to the previous iterates. Thus, BG has a smaller error than BB when

$$\frac{|\delta x_s^T \delta x_{s-1}|}{\langle \delta x_s, \delta x_s \rangle} < \frac{|\delta K_s^T \delta K_{s-1}|}{\left| \langle \delta K_s, \left(\hat{M}'_s \right)^{-1} \delta x_s \rangle \right|}. \tag{10}$$

The same reasoning can be built-up starting from (5) and (6). We then obtain for the switching condition:

$$\frac{|\delta x_s^T \delta x_{s-1}|}{|\langle \delta x_s, (\hat{K}'_s)^{-1} \delta K_s \rangle|} < \frac{|\delta K_s^T \delta K_{s-1}|}{\langle \delta K_s, \delta K_s \rangle}. \tag{11}$$

To avoid situations where (10) and (11) contradict, we will only use the latter. If (11) is met then the BG-update will be applied, otherwise the BB-update.

3.4 Column-Updating Method (CU), Inverse Column-Updating Method (ICU) and Switched Column-Updating Method (SCU)

1. The Column-Updating method is a Quasi-Newton method that was introduced by Martinez [25, 27, 28]. The rank-one update of this method is such that the column of the approximate Jacobian corresponding to the largest coordinate of the latest increment δx_s is replaced in order to satisfy the secant equation $\hat{K}' \delta x_s = \delta K_s$ at each iteration. This results in:

$$(\hat{K}'_{s+1})^{-1} = (\hat{K}'_s)^{-1} + \frac{(\delta x_s - (\hat{K}'_s)^{-1} \delta K_s) \iota_{j_{K,s}}^T (\hat{K}'_s)^{-1}}{\langle \iota_{j_{K,s}}, (\hat{K}'_s)^{-1} \delta K_s \rangle} \tag{12}$$

where $\iota_{j_{K,s}}$ is chosen such that $j_{K,s} = \mathrm{Argmax}\{|\langle \iota_j, \delta x_s\rangle|; j = 1, \dots, m_n\}$.
$(\hat{K}'_1)^{-1}$ is typically set to be $-I$,

2. The Inverse Column-Updating method (ICU) is a Quasi-Newton method that was introduced by Martinez and Zambaldi [23, 26]. It uses a rank-one update such that the column of the approximation of the inverse of the Jacobian corresponding to the largest coordinate of δK_s is replaced in order to satisfy the secant equation $\hat{M}' \delta K_s = \delta x_s$ at each iteration. This results in:

$$\hat{M}'_{s+1} = \hat{M}'_s + \frac{(\delta x_s - \hat{M}'_s \delta K_s)\iota^T_{j_{M,s}}}{\langle \iota_{j_{M,s}}, \delta K_s \rangle}, \tag{13}$$

where $\iota_{j_{M,s}}$ is chosen such that $j_{M,s} = \mathrm{Argmax}\{|\langle \iota_j, \delta K_s\rangle|; j = 1, \dots, m_n\}$. Again, typically $\hat{M}'_1 = -I$ is chosen.

3. As far as the authors are aware, the idea behind SB has not yet been applied to CU and ICU, despite being straightforward. A similar reasoning as for SB gives the following switching condition:

$$\frac{|\iota^T_{j_{K,s}} \delta x_{s-1}|}{|\langle \iota_{j_{K,s}}, (\hat{K}'_s)^{-1}\delta K_s\rangle|} < \frac{|\iota^T_{j_{M,s}} \delta K_{s-1}|}{\left|\langle \iota_{j_{M,s}}, \delta K_s\rangle\right|}. \tag{14}$$

If this condition is satisfied, then the CU-update is used, otherwise the ICU-update.

3.5 Quasi-Newton Least Squares (QN-LS) and Quasi-Newton Inverse Least Squares (QN-ILS)

1. In the Quasi-Newton Least Squares Method (QN-LS) [14, 16–18, 30, 31] the approximate Jacobian $\hat{K}'_{s+1}$ is chosen as the solution of $\min\{\|\hat{K}' - \hat{K}'_s\|_{Fr}\}$ s.t. $\hat{K}'\delta x_i = \delta K_i$, $\forall i \in \{1, \dots, s\}$, which leads to the following rank-one update:

$$(\hat{K}'_{s+1})^{-1} = (\hat{K}'_s)^{-1} + \frac{(\delta x_s - (\hat{K}'_s)^{-1}\delta K_s)\left((I - \mathscr{L}_s\mathscr{L}^T_s)\delta x_s\right)^T (\hat{K}'_s)^{-1}}{\langle (I - \mathscr{L}_s\mathscr{L}^T_s)\delta x_s, (\hat{K}'_s)^{-1}\delta K_s \rangle}, \tag{15}$$

with $\mathscr{L}_s = [\bar{L}_1| \dots |\bar{L}_s]$, where $\bar{L}_k$ is the kth left singular vector of $V_s = [v^s_1 \dots v^s_{s-1}]$, with $v^s_i = x^s_n - x^i_n$ ($i = 1, \dots, s - 1$). $(\hat{K}'_1)^{-1}$ is typically set to be $-I$. As $\mathscr{L}_1$ doesn't exist we replace it by the zero matrix.

2. The Quasi-Newton Inverse Least Squares Method (QN-ILS) [7, 15] is similar to QN-LS but constructs an approximation to the inverse Jacobian.

The approximate Jacobian $\hat{M}'_{s+1}$ is chosen as the solution of $\min\{\|\hat{M}' - \hat{M}'_s\|_{Fr}\}$, s.t. $\hat{M}'\delta K_i = \delta x_i$, $\forall i \in \{1, \ldots, s\}$, which leads to the following rank-one update:

$$\hat{M}'_{s+1} = \hat{M}'_s + \frac{(\delta x_s - \hat{M}'_s \delta K_s)\left((I - \tilde{\mathscr{L}}_s(\tilde{\mathscr{L}}_s)^T)\delta K_s\right)^T}{\langle(I - \tilde{\mathscr{L}}_s(\tilde{\mathscr{L}}_s)^T)\delta K_s, \delta K_s\rangle}, \tag{16}$$

with $\tilde{\mathscr{L}}_s = [\tilde{L}_1|\tilde{L}_2| \ldots |\tilde{L}_s]$, where $\tilde{L}_k$ is the kth left singular vector of $\tilde{V}_s = [\tilde{v}_1^s \ldots \tilde{v}_{s-1}^s]$, with $\tilde{v}_i^s = K(x_n^s) - K(x_n^i)$ $(i = 1, \ldots, s-1)$. $\hat{M}'_1$ is typically set to be $-I$. As $\tilde{\mathscr{L}}_1$ does not exist we replace it by the zero matrix.

3. A similar reasoning as for SB gives the following switching condition:

$$\frac{|\langle((I - \mathscr{L}_s\mathscr{L}_s^T)\delta x_s)^T \delta x_{s-1}|}{|\langle(I - \mathscr{L}_s\mathscr{L}_s^T)\delta x_s, (\hat{K}'_s)^{-1}\delta K_s\rangle|} < \frac{|\left((I - \tilde{\mathscr{L}}_s(\tilde{\mathscr{L}}_s)^T)\delta K_s\right)^T \delta K_{s-1}|}{|\langle(I - \tilde{\mathscr{L}}_s(\tilde{\mathscr{L}}_s)^T)\delta K_s, \delta K_s\rangle|} \tag{17}$$

In this expression we observe that $\left((I - \mathscr{L}_s\mathscr{L}_s^T)\delta x_s\right) \perp V_s$ and $\delta x_{s-1} \in \mathscr{R}(V_s)$. Consequently $\left((I - \mathscr{L}_s\mathscr{L}_s^T)\delta x_s\right)^T \delta x_{s-1} = 0$.

Similarly $\left((I - \tilde{\mathscr{L}}_s(\tilde{\mathscr{L}}_s)^T)\delta K_s\right)^T \delta K_{s-1} = 0$. This switch will thus never be triggered and for that reason this Switched Quasi-Newton Least Squares method is rejected.

3.6 Non-linear Eirola-Nevanlinna Type I Method (EN1), Type II Method (EN2) and Switched Eirola Nevanlinna Method (SEN)

1. It is clear from the different update formulas for the approximate (inverse) Jacobian of all the previous methods that they can only be applied starting with $\hat{K}'_2$. In other words, $\hat{K}'_1$ needs to be chosen. Conventionally, this is set to be equal to $-I$. Likewise, for $A\delta^2$, ω_1 needs to be chosen and is set to 1. As a result all of the methods given above will have an identical first iteration, i.e. $x_n^2 = H(x_n^1)$.
The nonlinear Eirola-Nevanlinna (EN) was proposed by [33] as the nonlinear counterpart to the linear EN algorithm [10] and is different as it computes $\hat{K}'_1$ based on a virtual choice of $\hat{K}'_0$, set to be equal to $-I$ (which can also be interpreted as setting the initial approximation of the Jacobian of H as zero), which is used to create a first approximation $\hat{K}'_1$. The method is given by

$$(\hat{K}'_{s+1})^{-1} = (\hat{K}'_s)^{-1} + \frac{(p_s - (\hat{K}'_s)^{-1}q_s)p_s^T(\hat{K}'_s)^{-1}}{\langle p_s, (\hat{K}'_{s-1})^{-1}q_s\rangle}, \tag{18}$$

where $p_s = -(\hat{K}'_s)^{-1}K(x_n^{s+1})$ and $q_s = K(x_n^{s+1} + p_s) - K(x_n^{s+1})$. Note that the EN algorithm requires two calls of K (or H) per iteration.

2. Eirola and Nevanlinna did not propose a Type II method, but by generalisation this can be written as [11]:

$$\hat{M}'_{s+1} = \hat{M}'_s + \frac{(p_s - (\hat{K}'_s)^{-1} q_s) q_s^T}{\langle q_s, q_s \rangle}, \tag{19}$$

where p_s and q_s are defined as in the EN1 method.

3. As far as the authors are aware, the idea behind SB has not been applied to EN1 and EN2. The switching condition now becomes

$$\frac{|p_s^T \delta x_{s-1}|}{\left| \langle p_s, (\hat{K}'_{s-1})^{-1} q_s \rangle \right|} < \frac{|q_s^T \delta K_{s-1}|}{\langle q_s, q_s \rangle}. \tag{20}$$

When this condition is satisfied, then EN1 (Eq. 18) is used, otherwise EN2 (Eq. 19).

4 Re-Use of Previous Information

When the problem is time-dependent, and assuming that the changes over one time-step are relatively small, then the approximate (inverse) Jacobian of the previous time-step might be a relatively good initial guess for the approximate (inverse) Jacobian at the next time-step.

One of the primary aims of this investigation is to compare the performance of the various QN methods when the Jacobian, at the start of each new time step, $k + 1$, is either reset to $-I$, such that $(\hat{K}'_1)^{-1}_{k+1} = -I$, (which we will call "Jacobian reset") or set equal to the final approximation from the previous time step, i.e. $(\hat{K}'_1)^{-1}_{k+1} = (\hat{K}'_s)^{-1}_k$, where k indicates the time step counter (which we will call "Jacobian re-use"). A similar procedure is used for Quasi-Newton methods using $\hat{M}'_1$ and for Aitken's method.

5 Fluid Structure Interaction Test-Cases

Quasi-Newton methods have received significant attention in recent years within the field of partitioned fluid-structure interactions (FSI). In this section we aim to investigate the various QN methods when applied to a number of incompressible, transient, FSI benchmark problems by coupling OpenFOAM for the fluid flow solution and Calculix for the structural deformation. A relaxation factor of $\omega = 0.001$ is used for the first iteration of the first time step, or for the first iteration in every time step whenever the Jacobian is reset such that

$$x_n^2 = x_n^1 + \omega \delta x_1.$$ (21)

This is done to avoid the possibility of an excessively large first displacement guess, which can lead to divergence for strongly coupled problems. x_n^1 at the start of each new time step is set equal to the final converged solution from the previous time step.

5.1 Dynamic Piston-Channel Problem

The piston-channel test problem layout is shown in Fig. 1a, and consists of a 10 m long fluid domain which is forced out of the channel by an accelerating unit by unit moving block. The problem is a surprisingly difficult problem to solve when using partitioned solution schemes, and has been investigated in a number of publications (see for example [2, 6]). The coupling strength is sufficiently strong that simple fixed-point iterations, such as Gauss-Seidel iterations, are insufficient to guarantee convergence.

While the test case is intrinsically a one dimensional problem, it is modelled here in three dimensions, with a fluid domain discretised using 10 linear elements and a single linear structural element. The fluid density and viscosity are 1 kg/m³ and 1.0 kg/(m s), respectively, where the solid is described by a linear elastic material with a Young's modulus of $E = 10.0$ Pa with a zero density and Poisson's ratio. A slip boundary condition is applied to the fluid wall boundaries and the velocity is prescribed as $u(t) = 0.2t$ on the left side of the block (see Fig. 1a). The simulation is solved here using time step sizes of $\Delta t = 0.02$ s for a convergence tolerance of $\epsilon = \frac{||K(x_n^s)||}{\sqrt{m_n}} = 10^{-8}$. The interface displacement and velocity, along with a 1D-solution is shown in Fig. 1b. The 1D-model is found by simplifying the problem to a 1D mass-spring system, where the solid piston represents a linear spring and the fluid domain a variable mass (see [2] for more information on the simplified 1D-model). The accuracy of the FSI simulation can naturally be improved by increasing the mesh resolution or decreasing the time step size.

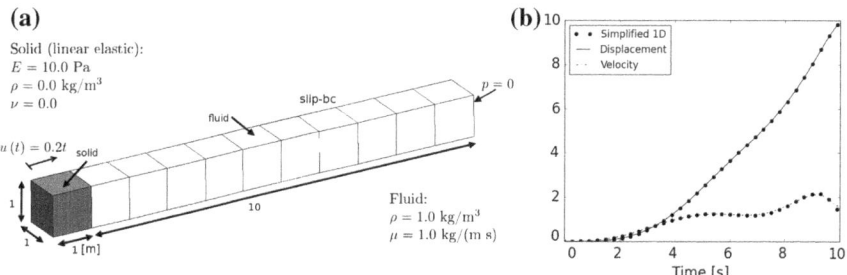

Fig. 1 **a** Piston-channel test problem layout. **b** The interface displacement and velocity for a time step size of $\Delta t = 0.02$s, compared to the simplified 1D mass-spring system

Table 1 The mean number of iterations required to reach convergence for the piston-channel test problem. Failure to converge is indicated by div() where the time step at which failure occurred is indicated in brackets; top performing method highlighted in bold

	Jacobian re-use	Jacobian reset
GS	N/A	div(1)
$A\delta^2$	div(1)	div(1)
BG	3.98	4.73
BB	3.98	**4.72**
SB	3.97	4.73
CU	3.98	div(75)
ICU	3.98	4.73
SCU	3.98	4.73
QN-LS	4.01	div(52)
QN-ILS	4.01	div(438)
EN1	3.91	div(108)
EN2	3.90	div(108)
SEN	**3.89**	div(108)

The average number of iterations required to reach convergence for each of the QN methods is summarised in Table 1. The results highlight the surprising complexity of the test problem, with both GS and Aitken failing to provide convergent results. The performance of all the other QN methods is virtually identical, with a significant improvement offered by retaining the Jacobian at the start of each new time step.

5.2 Dam Break with an Elastic Obstacle

The dam break problem consists of a collapsing column of water striking an elastic baffle, which has previously been analysed in [3, 32]. The problem layout is shown in Fig. 2a with a plot of the beam tip displacement shown in Fig. 2b. The FSI simulation is performed using a time step size of $\Delta t = 0.001$ s, 3670 linear fluid elements, and 14 quadratic, full integration finite-elements. The mean number of iterations to reach convergence for each of the QN methods is summarised in Table 2. The QN-LS family of methods is the top performing family of methods, with significant benefit seen in retaining the Jacobian from the previous time steps.

5.3 Wave Propagation in a Three Dimensional Elastic Tube

The 3D flexible tube problem was originally proposed in [13], inspired by the type of flow encountered in haemodynamics. The density ratios of the fluid and solid are near unity, which in conjunction with internal incompressible flow results in a very strongly coupled FSI problem that has received much attention in literature [2, 7, 13].

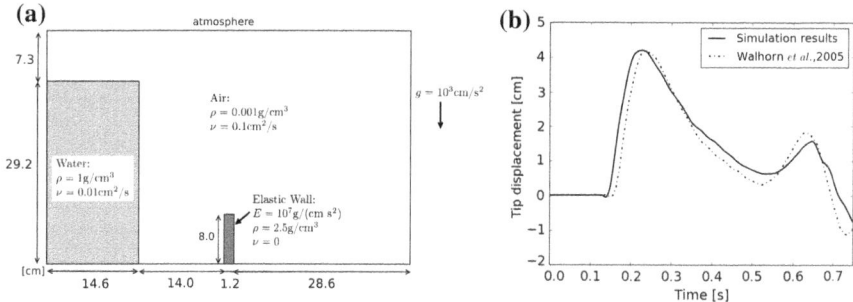

Fig. 2 **a** Dam break with elastic obstacle problem description. **b** Beam tip displacement results for the dam break test problem, compared to the results reproduced from [32]

Table 2 Mean number of iterations required by the QN methods for the dam break test problem. Failure to converge is indicated by div() where the time step at which failure occurred is indicated in brackets; top performing method highlighted in bold

	Jacobian re-use	Jacobian reset
GS	N/A	23.42
$A\delta^2$	div(279)	11.45
BG	4.17	7.09
BB	4.32	7.10
SB	4.14	7.07
CU	5.62	7.10
ICU	5.84	div(279)
SCU	5.80	7.29
QN-LS	**3.90**	6.52
QN-ILS	div(279)	**6.49**
EN1	4.37	8.05
EN2	4.58	8.05
SEN	4.58	8.04

The problem consists of a flexible tube of length $l = 5$ cm, with an inner and outer radius of $r_i = 0.5$ cm and $r_0 = 0.6$ cm, respectively. The flexible tube is modelled using a St. Venant-Kirchoff material model, with a Young's modulus of $E = 3 \times 10^6$ dynes/cm^2, density $\rho = 1.2$ g/cm^3 and Poisson's ratio of 0.3, where the fluid flow has a density of $\rho = 1.0$ g/cm^3 and a viscosity of $\mu = 0.03$ poise. The problem is modelled using 600 twenty-noded quadratic solid elements coupled with 6000 linear fluid flow elements resulting in an interface Jacobian size of $m_n = 1880$. The tube walls are fixed on both ends, and a smoothly varying pressure in the form of $p(t) = 1.3332 \times 10^4 (\sin(2\pi t/0.003 + 1.5\pi) + 1)/2$ is applied at the inlet over the first 0.003 s. The time step size for the simulation is $\Delta t = 0.0001$ s with a convergence tolerance $\epsilon = \frac{||K(x_n^s)||}{\sqrt{m_n}} = 10^{-8}$. The resulting pressure pulse propagation is illustratively shown for different time steps in Fig. 3.

The mean number of iterations to reach convergence is summarised in Table 3 for the various QN methods, and the number of coupling iterations required per time step, for each of the family of methods, is shown in Fig. 4. The QN-LS method was once again the top performing method.

5.4 2D Flexible Beam

The selected test case is a fluid-structure interaction problem consisting of flow around a fixed cylinder with an attached flexible beam. The beam undergoes large deformations induced by oscillating vortices formed by flow around the circular bluff

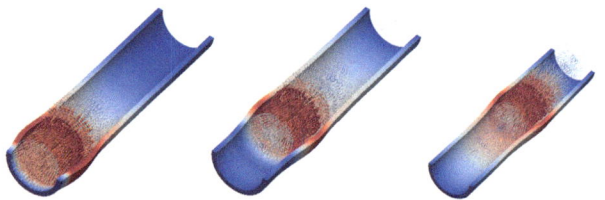

Fig. 3 Pressure pulse propagation at 0.003, 0.005 and 0.008 s (where the wall displacement is magnified by a factor 10)

Table 3 Comparison of the mean number of iterations required to reach convergence for the 3D flexible tube problem. Failure to converge is indicated by div() with the time step at which failure occurred in brackets; top performing method highlighted in bold

	Jacobian re-use	Jacobian reset
GS	N/A	div(1)
Aδ^2	div(1)	div(1)
BG	6.51	18.02
BB	7.68	div(47)
SB	6.75	18.02
CU	15.63	div(7)
ICU	div(14)	div(12)
SCU	13.89	div(10)
QN-LS	**5.64**	15.94
QN-ILS	5.95	**14.47**
EN1	6.91	div(52)
EN2	10.56	div(11)
SEN	7.37	div(2)

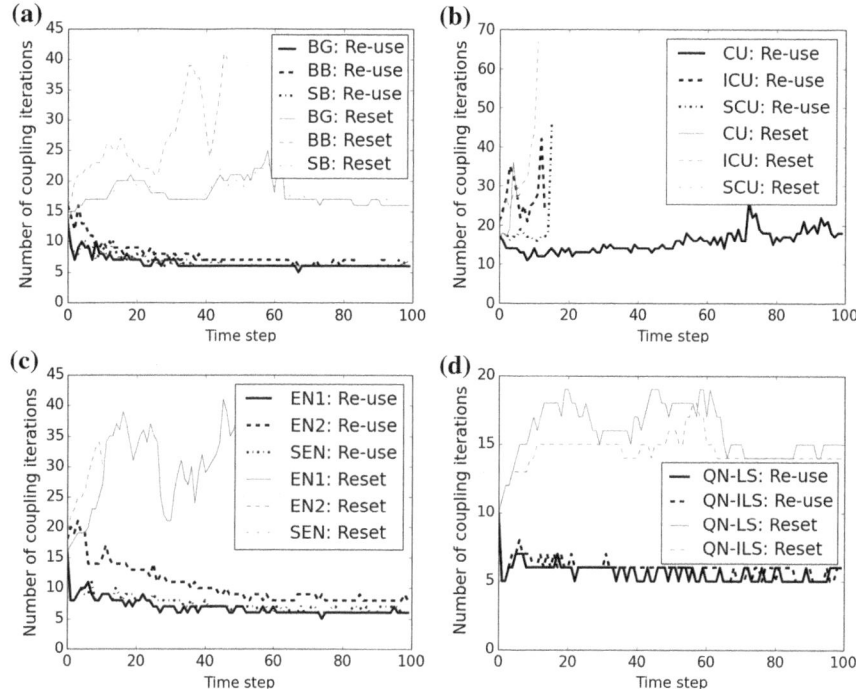

Fig. 4 A comparison of the number of coupling iterations required to reach convergence for the 3D flexible tube test problem

body. The problem was first proposed by Turek et al. [29], and has received substantial numerical verification. The problem layout and material properties are provided in Fig. 5a.

A parabolic inlet boundary condition, with mean flow velocity of $\bar{U} = 1$ m/s is slowly ramped up for $t < 0.5$ s. A snapshot of the beam tip displacement is illustratively shown in Fig. 5b. The convergence behavior for the various QN methods is summarized in Table 4.

Once again the QN-LS method is the top performing method. Besides for Broyden's method, the switching strategies provided no benefit for any of the other families of QN methods.

6 Other Application: Simplified Model of Plasma Heating by RF Waves in a Plasma

Quasi-Newton acceleration can also be applied to problems outside the field of fluid-structure interaction. The model that we present here is a simplified version of the set

(a)   **(b)**

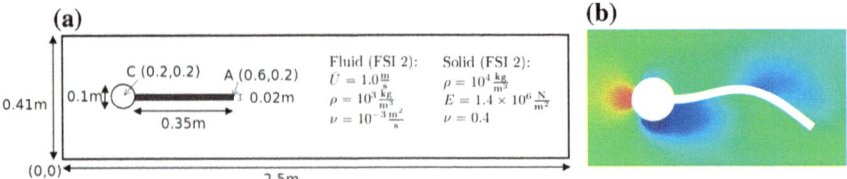

Fig. 5 **a** Flexible beam problem description and **b** snapshot of beam displacement and pressure contours at 8.7 s

Table 4 Comparison of the mean number of iterations required to reach convergence for the 2D flexible beam problem. Failure to converge is indicated by div() with the time step at which failure occurred in brackets; top performing method highlighted in bold

	Jacobian re-use	Jacobian reset
GS	N/A	div(1)
$A\delta^2$	10.31	14.00
BG	4.02	7.48
BB	4.25	7.42
SB	3.98	7.48
CU	5.26	7.99
ICU	5.39	7.94
SCU	6.34	8.05
QN-LS	**3.86**	6.69
QN-ILS	3.93	**6.63**
EN1	4.61	8.47
EN2	4.54	8.46
SEN	4.54	8.47

of codes commonly used to describe the steady state of plasma heating by radio frequency waves in a tokamak plasma, but still retains all the characteristics needed to validate our ideas as explained in Sect. 1. In an abstract form the governing equations can be written as:

$$\begin{cases} f_1(y_1, y_2, y_3, y_4, y_5, y_6, y_7) & = 0 \\ f_2(y_1, y_2, y_3, y_4, y_6, y_8, y_9, y_{10}) & = 0 \\ f_3(y_1, y_2, y_3, y_5, y_6, y_7, y_8, y_9, y_{10}) & = 0, \end{cases} \tag{22}$$

where f_1 represents a simplified 1-component wave equation, f_2 a simplified Fokker-Planck equation and f_3 a simplified 1D diffusion equation. y_1, y_2 and y_3 are the temperature profile of the species (majority ions, minority ions and electrons); y_4 the effective temperature of the minority ions; y_5, y_6 and y_7 are the power density profiles of the wave damping onto the species; y_8, y_9 and y_{10} are the collisionally redistributed minority power density fraction profiles onto the species. More details can be found in [19]. As the system in (22) does not satisfy condition (2), (y_1, y_2, y_3, y_4) need to be grouped as x_3, as explained in [20]. A convergence criterium of $\frac{\|x_n^{s+1} - x_n^s\|}{\|x_n^s\|} \leq 10^{-7}$ is used. The results are shown in Table 5 for different values of launched power P

Table 5 Simplified tokamak model. Number of function calls (of H) needed to reach convergence for various values of κ. $P = 2$ MW (left) and $P = 5$ MW (right). "div" = divergence or no convergence after 100 iterations. The top performing method is highlighted in bold

$P = 2$ MW. $\kappa(\cdot 10^{-2})$	3.5	4.0	4.5	5.0	7.5	10	25	50	75	100
G-S	div	div	95	88	67	55	30	19	14	14
$A\delta^2$	63	52	52	49	48	33	24	17	14	15
BG	div	33	30	29	28	23	18	14	**13**	12
BB	30	30	29	30	28	23	18	14	**13**	12
SB	31	div	28	25	26	21	18	14	**13**	12
CU	37	div	div	37	32	22	17	**12**	**13**	12
ICU	34	36	33	32	30	24	18	14	14	12
SCU	33	38	28	29	30	23	18	14	**13**	12
QN-LS	div	**24**	24	23	**21**	20	**16**	14	**13**	12
QN-ILS	**25**	25	**23**	**22**	**21**	19	**16**	14	**13**	12
EN1	36	34	40	30	30	24	22	16	16	16
EN2	34	36	34	34	32	26	22	16	16	16
SEN	35	37	34	34	30	24	22	16	16	16

$P = 5$ MW. $\kappa(\cdot 10^{-2})$	7.5	10	25	50	75	100
G-S	67	52	28	19	16	15
$A\delta^2$	39	36	23	19	16	16
BG	26	24	19	17	**14**	13
BB	25	24	20	17	**14**	13
SB	24	23	19	17	15	13
CU	36	div	21	17	**14**	**12**
ICU	27	30	21	18	15	14
SCU	26	27	21	17	**14**	14
QN-LS	22	23	**18**	**16**	**14**	13
QN-ILS	**21**	**22**	**18**	**16**	**14**	13
EN1	28	28	24	20	16	16
EN2	28	28	24	20	16	16
SEN	28	28	25	19	16	16

and diffusion coefficient κ. The QN methods clearly outperform G-S and Aitken's method. Of all the QN methods, the Least Squares methods most often give the best results (with a slight edge for QN-ILS over QN-LS), followed by the Broyden methods. While not equivocally so, the switching strategy often (slightly) improves the convergence speed of the underlying methods.

7 Conclusion

We have tested a wide variety of acceleration techniques on different multi-physics problems that are written as a fixed-point problem. While the choice of the best method remains problem dependent, it is clear that the best choice is the class of Quasi-Newton methods, of which, more often than not, the Least Squares methods come out on top. Re-using the Jacobian of all the QN methods at the beginning of the iterations of the next time step results in important reductions in the required number of iterations. With a few exceptions, a switching strategy, that hasn't drawn much attention in the past, is shown to offer only a slight boost of performance in exchange for a negligeable penalty in complexity. The class of Eirola-Nevanlinna methods, which are among the lesser known QN methods, have not shown their worth, and in the authors' opinion do not seem to warrant the complexity that they entail.

References

1. A.C. Aitken, On Bernouilli's numerical solution of algebraic equations. Proc. Roy. Soc. Edinb. **46**, 289–305 (1926)
2. A.E.J. Bogaers, S. Kok, B.D. Reddy, T. Franz, Quasi-Newton methods for implicit black-box FSI coupling. Comput. Methods Appl. Mech. Eng. **279**, 113–132 (2014)
3. A.E.J. Bogaers, S. Kok, B.D. Reddy, T. Franz, An evaluation of quasi-Newton methods for application to FSI problems involving free surface flow and solid body contact. Comput. Struct. **173**, 71–83 (2016)
4. C.G. Broyden, A class of methods for solving nonlinear simultaneous equations. Math. Comput. **19**, 577–593 (1965)
5. C.G. Broyden, Quasi-Newton methods and their applications to function minimization. Math. Comput. **21**, 368–381 (1967)
6. J. Degroote, K.-J. Bathe, J. Vierendeels, Performance of a new partitioned procedure versus a monolithic procedure in fluid-structure interaction. Comput. Struct. **87**(11), 793–801 (2009)
7. J. Degroote, R. Haelterman, S. Annerel, P. Bruggeman, J. Vierendeels, Performance of partitioned procedures in fluid-structure interaction. Comput. Struct. **88**(7), 446–457 (2010)
8. J.E. Dennis, J.J. Moré, Quasi-Newton methods: motivation and theory. SIAM Rev. **19**, 46–89 (1977)
9. J.E. Dennis, R.B. Schnabel, Least change secant updates for Quasi-Newton methods. SIAM Rev. **21**, 443–459 (1979)
10. T. Eirola, O. Nevanlinna, Accelerating with rank-one updates. Linear Algebra Appl. **121**, 511–520 (1989)

11. H.-R. Fang, Y. Saad, Two classes of multisecant methods for nonlinear acceleration. Numer. Linear Algebra Appl. **16**(3), 197–221 (2009)
12. A. Friedlander, M.A. Gomes-Ruggiero, D.N. Kozakevich, J.M. Martinez, S.A. dos Santos, Solving nonlinear systems of equations by means of Quasi-Newton methods with a nonmonotone strategy. Optim. Methods Softw. **8**, 25–51 (1997)
13. J.-F. Gerbeau, M. Vidrascu et al., A Quasi-Newton algorithm based on a reduced model for Fluid-structure interaction problems in blood flows. ESAIM: Math. Model. Numer. Anal. **37**(4), 631–647 (2003)
14. R. Haelterman, J. Degroote, D. Van Heule, J. Vierendeels, The Quasi-Newton least squares method: a new and fast secant method analyzed for linear systems. SIAM J. Numer. Anal. **47**(3), 2347–2368 (2009)
15. R. Haelterman, J. Degroote, D. Van Heule, J. Vierendeels, On the similarities between the Quasi-Newton inverse least squares method and GMRes. SIAM J. Numer. Anal. **47**(6), 4660–4679 (2010)
16. R. Haelterman, J. Petit, H. Bruyninckx, J. Vierendeels, On the non-singularity of the Quasi-Newton-least squares method. J. Comput. Appl. Math. **257**, 129–131 (2014)
17. R. Haelterman, B. Lauwens, F. Van Utterbeeck, H. Bruyninckx, J. Vierendeels, On the similarities between the Quasi-Newton least squares method and GMRes. J. Comput. Appl. Math. **273**, 25–28 (2015)
18. R. Haelterman, B. Lauwens, H. Bruyninckx, J. Petit, Equivalence of QNLS and BQNLS for affine problems. J. Comput. Appl. Math. **278**, 48–51 (2015)
19. R. Haelterman, D. Van Eester, D. Verleyen, Accelerating the solution of a physics model inside a Tokamak using the (Inverse) Column Updating Method. J. Comput. Appl. Math. **279**, 133–144 (2015)
20. R. Haelterman, D. Van Eester, S. Cracana, Does anderson always accelerate picard? in *14th Copper Mountain Conference on Iterative Methods* (Copper Mountain, USA, 2016)
21. R. Haelterman, A. Bogaers, J. Degroote, S. Cracana, Coupling of partitioned physics codes with Quasi-Newton methods, in *Lecture Notes in Engineering and Computer Science: Proceedings of The International MultiConference of Engineers and Computer Scientists 2017*, 15–17 Mar 2017, Hong Kong, pp. 750–755, 2017
22. C.T. Kelley, *Iterative methods for linear and nonlinear equations* (Frontiers in Applied Mathematics, SIAM, Philadelphia, 1995)
23. V.L.R. Lopes, J.M. Martinez, Convergence properties of the Inverse Column-Updating Method. Optim. Methods Softw. **6**, 127–144 (1995)
24. J.M. Martinez, L.S. Ochi, Sobre dois metodos de broyden. Mathemática Aplicada e Comput. **1**(2), 135–143 (1982)
25. J.M. Martinez, A Quasi-Newton method with modification of one column per iteration. Computing **33**, 353–362 (1984)
26. J.M. Martinez, M.C. Zambaldi, An Inverse Column-Updating Method for solving large-scale nonlinear systems of equations. Optim. Methods Softw. **1**, 129–140 (1992)
27. J.M. Martinez, On the convergence of the column-updating method. Comput. Appl. Math. **12**(2), 83–94 (1993)
28. J.M. Martinez, Practical Quasi-Newton method for solving nonlinear systems. J. Comput. Appl. Math. **124**, 97–122 (2000)
29. S. Turek, J. Hron, Proposal for numerical Benchmarking of Fluid-structure interaction between an elastic object and laminar incompressible flow, in *Fluid-Structure Interaction*, ed. by H.-J. Bungartz, M. Schäfer, Michael, Series "Modelling, Simulation, Optimisation" Vol. 53 (Springer, Berlin, 2006), pp. 371–385. ISSN:1439-7358
30. J. Vierendeels, Implicit coupling of partitioned fluid-structure interaction solvers using reduced-order models, in *Fluid-Structure Interaction, Modelling, Simulation, Optimization*, ed. by H.-J. Bungartz, M. Sch äfer, Lecture Notes in Computer Science Engineering, vol. 53 (Springer, Berlin, 2006), pp. 1–18
31. J. Vierendeels, L. Lanoye, J. Degroote, P. Verdonck, Implicit coupling of partitioned fluid-structure interaction problems with reduced order models. Comput. Struct. **85**, 970–976 (2007)

32. E. Walhorn, A. Kölke, B. Hübner, D. Dinkler, Fluid-structure coupling within a monolithic model involving free surface flows. Comput. Struct. **83**(25), 2100–2111 (2005)
33. U.M. Yang, A family of preconditioned iterative solvers for sparse linear systems. Ph.D. thesis, Department of Computer Science, University of Illinois, Urbana-Champaign, 1995

Implementation of International Safety Standard EN ISO 13849 into Machinery of Tyre Industry

Napat Wongpiriyayothar and Sakreya Chitwong

Abstract This paper presents application of international safety standard in risk assessment and risk reduction following by machinery directive. Many industries using machinery for manufacturing products have tendency to take risk from poor-quality of machinery design which may not be produced according to international safety standard. This can lead dangerous situation to machine user. The new standard EN ISO 13849-1 (Safety of machinery—Safety-related parts of control systems—Part 1: General principles for design, 2006) [2] which replaced the old standard EN 954-1 (Safety-related parts of control systems, 1996) [3] definitely in December 2011 made machine designer not be familiar with the new concept and feel confused due to most of concerned parameters shown in term of statistic value that there are difficulty in interpretation and understanding. In the present, there is still lack of examples of implementation international safety standard into machinery of specific industry, especially in tyre industry. Therefore the objective of this paper is made for implementation this standard into machinery of tyre industry in order to build a safe situation for machine user.

Keywords Average probability of dangerous failure per hour (PFHDavg) Common cause failure (CCF) · Diagnostic coverage (DC) · Mean time to dangerous failure (MTTFd) · Performance level (PL) · Safety-related parts of control system (SRP/CS)

N. Wongpiriyayothar (✉) · S. Chitwong
Department of Instrumentation and Control Engineering, King Mongkut's Institute of Technology Ladkrabang, Ladkrabang 10520, Bangkok, Thailand
e-mail: engineer_napat1@hotmail.com

S. Chitwong
e-mail: sakreya.ch@kmitl.ac.th

© Springer Nature Singapore Pte Ltd. 2018
S.-I. Ao et al. (eds.), *Transactions on Engineering Technologies*,
https://doi.org/10.1007/978-981-10-7488-2_12

153

1   Introduction

Tyre industry machinery processes starting from mixing process of raw material, preparing process of material, tyre building process, tyre curing process and final inspection process respectively which all processes caused unsafe situation to machine user, especially in tyre building process that most of unsafe situations came from *pinch point* and *rotation point* of automatic building unit in front of machine.

Most of machineries in tyre industry are automation machine using safety control circuits in order to prevent entering to moving part of machine and/or to prevent unexpected starting up of machine by generating stop function to hold machine in safe stage. The well-known safety rule explained about procedure of risk assessment and risk reduction is EN ISO 12100 [3] following five-step method, i.e. (1) determination of the limits of the machinery, (2) hazard identification, (3) risk estimation, (4) risk evaluation, and (5) risk reduction. In the risk reduction process consisting of three-step method which all suitable protective measures must be followed, i.e. first step, inherently safe design measures, second step, safeguarding and/or complementary protective measures, the last step, information for uses.

Therefore, the machine designer must eliminate hazard and/or reduce risks as much as possible by following three-step method of risk reduction process respectively in order to let machine perform safety function effectively and to provide safe situation for machine user.

This paper mainly focuses on the second step of risk reduction process that this part was regarded as belonging to standard EN ISO 13849 [1]. Thus, the purpose of this paper is to review and verify design of safety function of old machinery in tyre industry by comparing PLr (Performance level required) to PL (Performance level designed). If PL is greater than or equal to PLr (PL $\geq$ PLr), it means that machine can guarantee to perform the safe stage and meet requirement with design principles of international safety standard, on the other hand, if the verification result does not meet with requirement. What does machine designer need to do to eliminate hazard. These will be more explained in this paper.

2   Background

2.1   EN 954-1

In the past, the well-known machinery safety standard EN 954-1 [2] was introduced in 1996 which this version is familiar to most of machine designer and is used in many automation industrials broadly, especially in European region. This standard defined Safety Categories to manage fault under foreseeable condition to prevent loss of safety function. These categories are divided into five levels, termed Categories B, 1, 2, 3 and 4 which Cat-4 can provide the highest safety level with redundant configuration. The procedure of this standard is quite simple and easy to understand for machine designer due to most of concerned criteria that are presented in term of deterministic approach following these steps, i.e. firstly identify safety function required to eliminate hazard, secondly consider whether fault condition can lead to loss of safety function or not and finally select safety category to manage fault condition.

2.2 EN ISO 13849-1

Standard EN ISO 13849-1 [4] was introduced first time in 1999 (original version), then was revised in 2006 (second version, [1]). The purpose of this standard is to replace the old standard EN 954-1 [2] which is going to be retired in December 2011. The concept of this standard is not only focusing on the deterministic approach (Category), but also statistic approach (PL, MTTFd, CCF and DC). Moreover, there is determining the designated architectures of category to perform a safety function which may be implemented by one or more SRP/CS. Combination of SRP/CS to perform a safety function (see Fig. 1) consisting of input (SRP/CS$_a$), logic/processing (SRP/CS$_b$), output/power control elements (SRP/CS$_c$) and inter-connecting means (i_{ab}, i_{bc}). However, this standard is quite difficult to understand for machine designers due to most of concerned criteria and calculated parameters presented in term of both deterministic and statistic approach leading most of them to use some kind of commercial computerization program to provide the quick result without understanding the basic principle in calculation and source of those formulas having an effect on poor quality in risk reduction process. Thus, these will be introduced in the next part of this paper to be the guideline and overview for general principles of this standard.

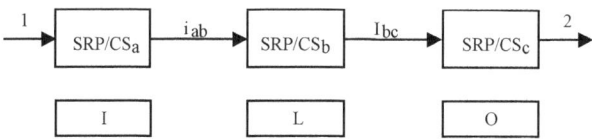

Fig. 1 Diagrammatic presentation of combination of safety-related parts of control systems for processing typical safety function

3 General Principles of EN ISO 13849-1

3.1 Overview

The purpose of this international safety standard is to provide machine designers, machine developers and machine manufacturers with an overall scope and guideline for design safety-related parts of control system (SRP/CS). The ability of safety-related parts of control systems is to perform a safety function under foreseeable conditions classified into five levels, called performance levels (PL) in term of PL a, b, c, d and e. These performance levels are defined in terms of probability of dangerous failure per hour (PFHD), (see Table 1). The probability of dangerous failure of the safety function depends on several factors, consisting of designated architecture of SRP/CS (Category), reliability of components (MTTFd, CCF), fault detection of mechanisms (DC), design process, operating stress, environmental conditions and operation procedures. In order to achieve PL, the concept of this standard based on the categorization of structures following specific design criteria and specific behaviors under fault conditions. These categories are classified into five levels, termed Categories B, 1, 2, 3 and 4. For example of SRS/CS (Input Elements: interlocking devices, electro-sensitive protective devices, pressure sensitive devices ... etc.), (Logic Elements: program logic controller devices (PLC), monitoring system, data processing unit ... etc.), and (output elements: contactors, relays, valves ... etc.).

This standard was developed to provide a clear cut concept in application of SRP/CS on machinery which can be assessed and audited by third party to certify whether safety function was designed correctly according to machinery directive or not.

3.2 Concept of EN ISO 13849-1

After we completed in risk assessment and risk reduction following EN ISO 12100 [3] if the result of risk reduction is required to implement protective measures on

Table 1 Classification of performance levels (PL)

PL	Average probability of dangerous failure per hour (PFHD) (1/h)
a	$\geq 10^{-5}$ to $<10^{-4}$
b	$\geq 3 \times 10^{-6}$ to $<10^{-5}$
c	$\geq 10^{-6}$ to $<3 \times 10^{-6}$
d	$\geq 10^{-7}$ to $<10^{-6}$
e	$\geq 10^{-8}$ to $<10^{-7}$

Note: Besides the average probability of dangerous failure per hour other measures are also necessary to achieve the PL

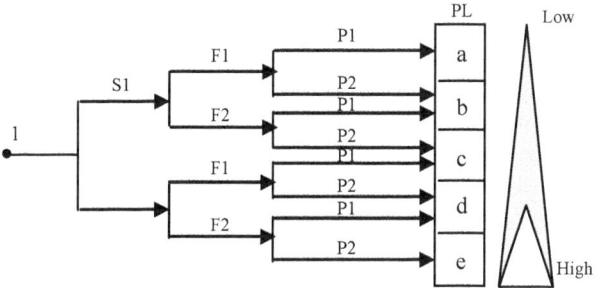

Fig. 2 Risk graph for determining required PLr for safety function

machinery in order to eliminate hazard and/or reduce risk. This will lead to part of EN ISO 13849-1 [1] which concerns with general principles for design of SRP/CS. Then, the iterative process for design of SRP/CS shall be followed according to EN ISO 13849-1 [1] (Page 13, Fig. 3) following these steps, i.e. (1) identify the safety functions to be performed by SRP/CSs, (2) determined the required performance level (PLr), (3) design safety function and identify SRP/CS to carry out safety function, (4) evaluate PL by considering Category, MTTFd, DC and CCF, (5) verify PL of safety function (PL ≥ PLr or not), (6) validate (meet with all requirements or not) sequentially. In step (5) and (6), if verification and validation step did not meet with requirement, this iterative process should be reconsidered.

Determination of PLr by Risk Graph Method

Risk Graph Method is part of standard EN ISO 13849-1 [1], using in determining PLr for each safety function to be carried out by SRP/CS (see Fig. 2). There are concerned parameters using in estimation of risk following these, i.e. Severity of injury represented by S "(S1, slight injury), (S2, serious injury)", Frequency/Exposure of hazard represented by F "(F1, seldom happened/exposure time is short), (F2, continuously happened/exposure time is long)", Possibility of avoiding hazard/limiting harm represented by P "(P1, possible under specific condition), (P2, impossible)", and point number 1 is starting point of this method. Thus, the result of this method will let us know the level of risk (low, medium or high) and required PLr in selection each SRP/CS to perform safety function. This method given here is to provide as the guideline concept to machine designer in estimation of risk.

3.3 Evaluation of PL by Category, MTTFd, DC and CCF

The ability of SRP/CS to perform safety function shall be expressed through PL and determined by estimation following these aspects: (1) Category, (2) MTTFd, (3) DC and (4) CCF.

(1) *Category*

System requirement and system behavior to withstand fault condition are explained in term of Categories. SRP/CS shall be met with requirement of one of the five categories, termed Categories B, 1, 2, 3 and 4 (see Figs. 3, 4, 5 and 6).

Category B is the basic category in which occurrence of fault can lead to the loss of safety function. This category provides the lowest safety level.

Category 1 is developed from Cat-B in which the occurrence of a fault can lead to the loss of safety function, but the ability to withstand fault is higher than Cat-B by using the concept of selection and implementation of well-tried components and well-tried safety principles.

Category 2 is required to apply Cat-B and Cat-1. In addition, the safety function shall be checked by machine control system periodically in which the occurrence of a fault can lead to the loss of safety function during checking period and the loss of safety function can be detected by the check.

Category 3 is required to apply Cat-B and Cat-1. In addition, safety-related parts shall be designed to ensure that single fault cannot lead the loss of safety function and single fault will be detected properly in case of reasonable practice in which the occurrence of the accumulated fault can lead to the loss of safety function.

Category 4 is required to apply Cat-B and Cat-1. In addition, safety-related parts shall be designed to ensure that single fault and accumulated fault cannot lead to the loss of safety function and the fault will be detected in time to prevent the loss of safety function. This category provides the highest safety level. For more explain in

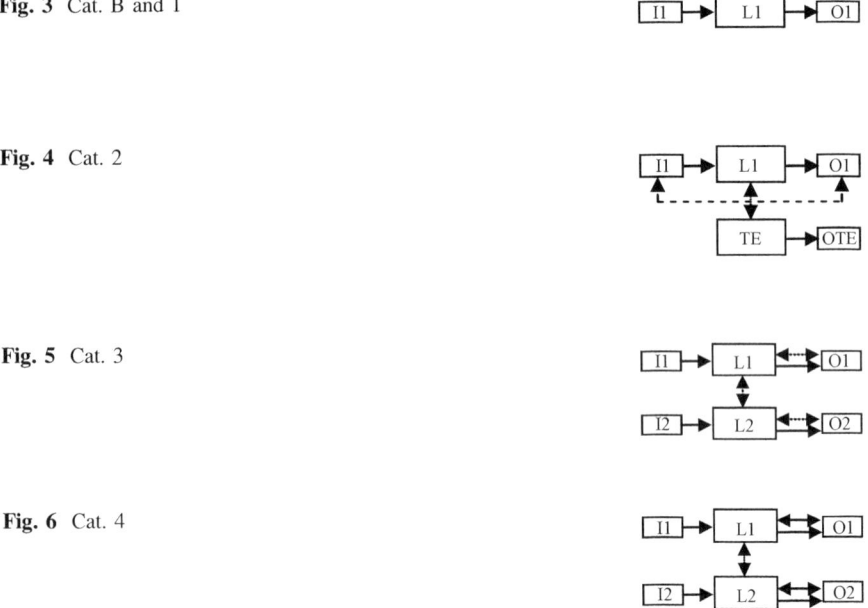

Fig. 3 Cat. B and 1

Fig. 4 Cat. 2

Fig. 5 Cat. 3

Fig. 6 Cat. 4

Table 2 Mean time to dangerous failure of each channel (MTTFd)

Denotation of each channel	Range of each channel
Low	3 years $\leq$ MTTFd $<$ 10 years
Medium	10 years $\leq$ MTTFd $<$ 30 years
High	30 years $\leq$ MTTFd $\leq$ 100 years

Table 3 Terminology

Symbol	Definition of abbreviate word
n_{op}	The mean number of annual operations
d_{op}	The mean operation, in days per year
h_{op}	The mean operation, in hours per day
B_{10d}	The mean number of cycles until 10% of components failure dangerously
t_{cycle}	The mean time between the beginning of two successive cycles of the component. (e.g. switching of a valve) in seconds per cycle

detail of categories are provided in EN ISO 13849-1 [1] (Page 38, Table 10— summary of requirements for categories).

(2) *MTTFd (Mean time to dangerous failure)*

MTTFd is classified into three levels (low, medium and high). This value describes the failure rate of component (reliability of component) in unit of years. The lowest MTTFd is 3 years and the highest MTTFd is 100 years to be taken into account (see Table 2).

Calculating or Evaluating MTTFd for single components

To evaluate the statistic value of MTTFd for each component, these value can be referred from standard value of components which are manufactured according to basic and well-tried safety principles as shown in EN ISO 13849-1 [1] (Page 50–56, Tables C.1–C.7) or can be calculated from B_{10d}, this is another statistic parameter provided by suppliers that they need to evaluate and declare into manufacturer data sheet. For terminology (see Table 3).

Calculation of MTTFd from B_{10d} can be referred from these formulas; "(1)" and "(2)".

$$MTTFd = \frac{B10d}{0.1 \times (nop)} \tag{1}$$

$$nop = \frac{(dop) \times (hop) \times 3,600(s/h)}{tcycle} \tag{2}$$

Table 4 Diagnostic coverage (DC)

Denotation	Range
None	DC < 60%
Low	60% ≤ DC < 90%
Medium	90% ≤ DC < 99%
High	99% ≤ DC

Calculating or Evaluating MTTFd for each channel

The MTTFd values of all single components which are part of the channel can be calculated by formula "(3)".

$$\frac{1}{MTTFd} = \frac{1}{MTTFd1} + \frac{1}{MTTFd2} + \cdots \frac{1}{MTTFdn} \tag{3}$$

(3) *DC (Diagnostic coverage)*

The diagnostic coverage is ratio between failure rate of dangerous failure that can be detected and failure rate of total dangerous failure (total dangerous failure consists of dangerous failure which can be detected and cannot be detected). The DC is presented in term of statistic value to measure effectiveness of diagnostics, classified into four levels (see Table 4). DC can be estimated from EN ISO 13849-1 [1] (Page 59–61, Table E.1).

For SRP/CS consisting of several parts, DC can be estimated by an average value of DC, so-called DC_{avg} and can be calculated by formula "(4)".

$$DCavg = \frac{\frac{DC1}{MTTFd1} + \frac{DC2}{MTTFd2} + \cdots \frac{DCn}{MTTFdn}}{\frac{1}{MTTFd1} + \frac{1}{MTTFd2} + \cdots \frac{1}{MTTFdn}} \tag{4}$$

(4) *CCF (Common cause failure)*

CCF concept is to provide a checklist to let machine designer take into account to evaluate whether common problem had already been solved or not following check list in EN13849-1 [1] (Page 63, Table F.1). Maximum of evaluation score is 100 points. If evaluation score is less than 65 points, means that does not meet with requirement. Thus, machine designer should select appropriate measures to improve this factor to get score higher than 65 points.

Evaluation of PL

After completed in considering of Category, MTTFd, DC and CCF, then machine designer can evaluate PL of SRP/CS by following EN ISO 13849-1 [1] (Page 81–82, Table K.1). To meet with requirement, machine designer has to verify that PL is greater than or equal to PLr. In case of PL is less than PLr, the iterative process should be reconsidered.

4   Example of Implementation

Life time to review and verify risk assessment of old machinery in tyre industry will be conducted every 5 years periodically in order to ensure whether safety function work properly according to concept of international safety standard or not by using Risk Graph Method to determine PLr and compare with PL. If PL is greater than or equal to PLr (PL ≥ PLr), this can guarantee that safety function meet with requirement but if PL is less than PLr (PL < PLr), then safety function and design feature of machinery must be reconsidered.

The old tyre building machine was reviewed following period of 5 years. For the required participants to verify risk assessment consist of machine designer, machine user (e.g. maintenance member, operation member, tooling change member and quality assurance member) and site safety officer to brainstorm any ideas in risk assessment and risk reduction to eliminate hazardous situation and/or reduce risk as much as possible.

5   Result of Risk Assessment and Risk Reduction

Risk of tyre building machine concerned with *pinch point* and *rotation point* of automatic building unit in front of machine (see Fig. 7). The result of risk assessment following risk graph method which was evaluated by concerned participants is S2, F2, and P1. Thus, PLd is required for eliminating hazardous situation (Fig. 8).

After completed in risk assessment process of old tyre building machine, we found 2 points of SRP/CS must be improved following second step of risk reduction process by implementing of safeguarding and complementary protective measures that consist of First point, upgrading system of emergency stop is needed due to the original design of this system was designed by category-1 (see Fig. 9) which provided only PLc (not meet requirement with PLd), and Second point, implementing system of safety light curtain is needed due to the original design of this system was designed without protective measures in front of automatic building

Fig. 7 Hazardous situation
from pinch and rotation point

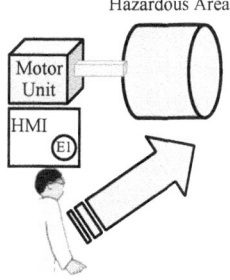

Hazardous Area

Fig. 8 Implemented of SRP/
CS to eliminate hazardous
situation

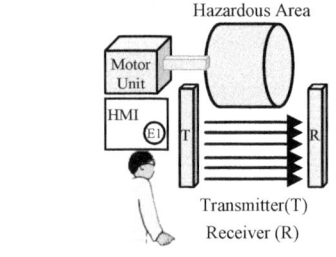

Fig. 9 Original design of
emergency stop system

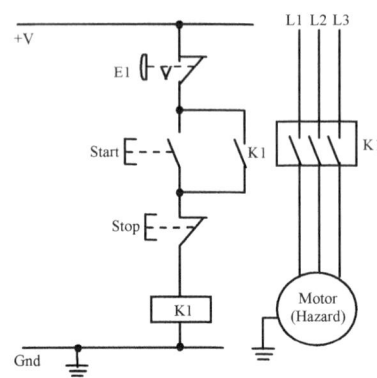

unit that can lead to unsafe situation when maintenance member access to dangerous zone to repair machine or operator access to verify specification of product or quality assurance member access to verify quality of product or tooling change member access to change equipment for producing the new size of tyre following daily production planning … etc. All of these behaviors have a chance to take risk from unexpected start-up of machine function and cause of serious injury eventually.

Therefore, these 2 points must be improved in order to eliminate hazardous situation and/or reduce risk.

6 Verification PL of Safety Function

6.1 Original Design of Emergency Stop System (see Fig. 9)

Safety function explanation:

- When emergency stop device E1 was activated, control voltage of contactor K1 will be interrupted and de-energized power out of movement part (Motor). Then, hazardous situation will be eliminated.

- This was designed by category-1 that cannot maintain all component failures. Safety function depends on reliability of components only. There is no implementing of fault detection that can lead to the loss of the safety function.

- The stopping function of emergency stop device is implementing of complementary protective measure to hazardous area.

Design feature:

- Meet requirement with category-B, implement of well-tried components and well-tried safety principles.

- Design of the closed-circuit current and earth connection regard to well-tried safety principles concept.

- Selection of emergency stop device E1 regards to well-tried components concept in according with IEC 60947-5-1 [5].

- Selection of contactor K1 regards to well-tried components concept in according with Table D.4 of EN ISO 13849-2 [6].

- Wiring control signal to contactor in according with stop category type 0 of EN 60204-1 [7].

Result of PFHD and PL:

- MTTFd was calculated by emergency stop E1 is standard emergency stop device according to Table C.1 of EN ISO 13849-1 [1], the life time of switching operation (B_{10d}) is 100,000 cycles and to be activated 3 times per day before starting each shift following standard operation procedure for testing safety device (3 shifts/day, 365 working day/year), Therefore n_{op} is 1,095 cycles/year and MTTFd is 913 years.

- MTTFd was calculated by contactor K1 according to Table C.1 of EN ISO 13849-1 [1], B_{10d} is 2,000,000 cycles and start/stop to be activated 6 times/day before starting/stopping of each shift (3 shifts/day, 365 working days/year), Therefore n_{op} is 2,190 cycles/year and MTTFd is 9,132 years.

- PL was defined by using $MTTFd_{avg}$ between E1 and K1 which is 830 years (consider at maximum value 100 years, high) and designated architecture which is category-1 according to Table K.1 of EN ISO 13849-1 [1], therefore the PFHD of this system is 1.14×10^{-6} per hour. *This corresponds to PLc.*

6.2 Upgrading Design of Emergency Stop System (see Fig. 10)

Safety function explanation:

- When emergency stop device E1 was activated, control voltage of contactor K1 and K2 will be interrupted and de-energized power out of movement part (Motor). Then, hazardous situation will be eliminated.

- This was designed by category-3 that both feedback signal of emergency stop E1 and feedback signal of redundant contactors K1, K2 were monitored by the monitoring safety relay (MSR1). But this cannot maintain an accumulation of undetected faults that can lead to the loss of the safety function.

- The stopping function of emergency stop device is implementing of complementary protective measure to hazardous area.

Design feature:

- Meet requirement with category-B, implement of well-tried components and well-tried safety principles.

- Design of the closed-circuit current and earth connection regard to well-tried safety principles concept.

- Selection of emergency stop device E1 regards to well-tried components concept in according with IEC 60947-5-1 [5].

- Selection of contactor K1, K2 regards to well-tried components concept in according with Table D.4 of EN ISO 13849-2 [6].

Fig. 10 Upgrading design of emergency stop system

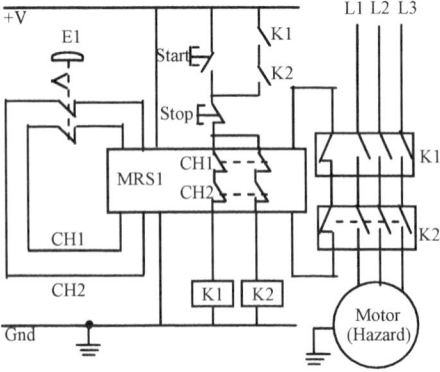

- The monitoring safety relay (MSR1) meet requirement with category-4, PLe, MTTFd is 4.35×10^{-9} per hour according to manufacturer datasheet.

Result of PFHD and PL:

- MTTFd calculated by emergency stop E1, is 913 years (Same concept as previous mentioned).

- MTTFd calculated by contactor K1, is 9,132 years. (Same concept as previous mentioned).

- MTTFd calculated by contactor K2, is 9,132 years. (Same concept as previous mentioned).

- DC_{avg} and CCF are relevant in category-3, Therefore DC_{avg} of E1 and K1, K2 are 90% according to Table E.1 of EN ISO13849-1 [1] and CCF of this system are 85 according to Table F.1 of EN ISO 13849-1 [1].

- PL was defined by using $MTTFd_{avg}$ between E1 and K1, K2 which is 761 years (consider at maximum value 100 years, high) and designated architecture which is category-3 and DC_{avg} is 90% (medium) according to Table K.1 of EN ISO 13849-1 [1], PFHD is 4.29×10^{-8} per hour. Following additional of subsystem MSR1 that PFHD is 4.35×10^{-9} per hour. Therefore the average PFHD of this system is 4.73×10^{-8} per hour. *This corresponds to PLe.*

6.3 Implementing of Protective Measure by Safety Light Curtain System (see Fig. 10)

Safety function explanation:

- When safety light curtain device (LC1) was activated, control voltage of contactor K1 and K2 will be interrupted and de-energized power out of movement part (Motor). Then, hazardous situation will be eliminated.

- This was designed by category-3 that both feedback signal of safety light curtain device (LC1) and feedback signal of redundant contactors K1, K2 were monitored by MSR1. But this cannot maintain an accumulation of undetected faults that can lead to the loss of the safety function.

- The stopping function of safety light curtain device (LC1) is implementing of complement protective measure to hazardous area.

Design feature:

- Meet requirement with category-B, implement of well-tried components and well-tried safety principles.

- Design of the closed-circuit current and earth connection regard to well-tried safety principles concept.

- Selection of contactor K1, K2 regards to well-tried components concept in according with Table D.4 of EN ISO 13849-2 [6].

- The monitoring safety relay (MSR1) meet requirement with category-4, PLe, MTTFd is 4.35×10^{-9} per hour according to manufacturer datasheet.

- The safety light curtain device (LC1) meet requirement with category-4, PLe, MTTFd is 7.93×10^{-9} per hour according to manufacturer datasheet.

Result of PFHD and PL:

- MTTFd calculated by contactor K1, is 9,132 years. (Same concept as previous mentioned).

- MTTFd calculated by contactor K2, is 9,132 years. (Same concept as previous mentioned).

- DC_{avg} and CCF are relevant in category-3, Therefore K1 and K2 are 90% according to Table E.1 of EN ISO 13849-1 [1] and CCF of this system are 85 according to Table F.1 of EN ISO 13849-1 [1].

- PL was defined by using $MTTFd_{avg}$ between K1 and K2 which is 4,566 years (consider at maximum value 100 years, high) and designated architecture which is category-3 and DC_{avg} is 90% (medium) according to Table K.1 of EN ISO 13849-1 [1], PFHD is 4.29×10^{-8} per hour. Following additional of subsystem

Fig. 11 Implementing of protective measure by safety light curtain

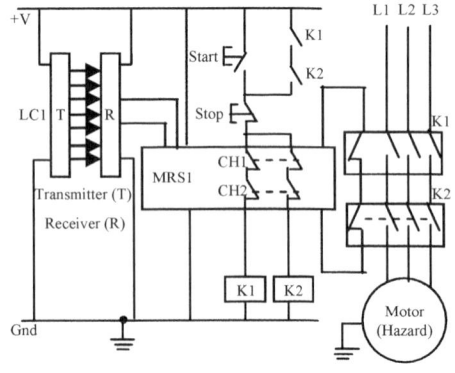

Table 5 Evaluation result of PL and PFHDavg of each system by Category, MTTFd, DC and CCF

Fig.	SRP/CS	CAT	B10d (c)	d_{op} (d/y)	Act SRP/CS (c/d)	n_{op} (c/y)	MTTFd (Year)	Avg	DC (%)	Avg	CCF (p)	PL	PFHD (1/h)	Avg	a
(9)	E1	1	1×10^5	365	3	1095	913	830	N/A	N/A	N/A	PLc	1.14×10^{-6}	1.14×10^{-6}	1
	K1	2	2×10^6	365	6	2190	9132		N/A		N/A				1
(10)	E1	3	1×10^5	365	3	1095	913	761	90	90	85	Ple	4.29×10^{-8}	4.73×10^{-8}	1
	K1	3	2×10^6	365	6	2190	9132		90						1
	K2	3	2×10^6	365	6	2190	9132		90						1
(11)	MRS1	4	N/A	N/A	N/A	N/A	355	355	N/A	N/A		Ple	4.35×10^{-9}		2
	LC1	4	N/A	N/A	N/A	N/A	20	20	N/A	N/A	85	Ple	7.93×10^{-9}	5.52×10^{-8}	2
	MRS1	4	N/A	N/A	N/A	N/A	355	355	N/A	N/A		Ple	4.35×10^{-9}		2
	K1	3	2×10^6	365	6	2190	9132	4566	90	90		Ple	4.29×10^{-8}		1
	K2	3	2×10^6	365	6	2190	9132		90						1

[a]Note: (1) Means that data of B10d refer from EN ISO 13849-1 [1] (Page 50, Table C.1) and calculation data of nop, MTTFd, DC, CCF and PFHD refer from method of standard EN ISO 13849-1 [1], (2) Means that data of MTTFd, PL and PFHD refer from manufacturer datasheet

Abbreviate word definition: (c) means cycles, (d/y) means day per year, (c/d) means cycles per day, (c/y) means cycles per year, (p) means point, (h) means hour

MSR1 that PFHD is 4.35×10^{-9} per hour and LC1 that PFHD is 7.93×10^{-9} per hour. Therefore the average PFHD of this system is 5.52×10^{-8} per hour. *This corresponds to PLe.*

7 Summary and Conclusion

From the result of risk assessment of the old tyre building machine, the result showed that PLd is required to eliminate hazardous situation and/or reduce risk. However, not only the original design of emergency stop system (see Fig. 9) that provide PLc is not enough to reduce risk, but also there are lacking of protective measure in front of hazardous area that can lead unsafe situation to machine user. Therefore the purpose of this paper is want to implement SRP/CS by upgrading design of emergency stop system (see Fig. 10) and implementing of protective measure by safety light curtain system (see Fig. 11) following international safety standard requirement. Both of these systems provide PLe (see Table 5) which is more than enough to reduce risk and can ensure that machine will be able to perform safe stage and build safe situation for machine user (see Fig. 8).

By the writer's opinions and experiences, all processes of risk assessment and risk reduction are not easy to achieve and get more effective result. The important parameters which need to be taken into account are experience and knowledge of participants who involved in this activity. If they are lack all of these, they cannot identify "where are the risk points which need to be eliminated" and cannot offer any improvement idea "how to develop SRP/CS to eliminate hazardous situation". Therefore, in order to get more effective result the chairman and/or project leader should require concerned participants who have an experience related with machinery up to 5 years in different domains to do this activity. Machine designer is not only the person in charge of this activity, but other domains also are essential to exchange any different point of view in eliminating risk and optimization of investment cost should be considered also.

Acknowledgements I would like to sincerely thank my thesis advisor Assoc. Prof. Sakreya C. and also professors of the KMITL's university for their participation in supporting my work and helping me get results of better quality.

I would like to thank my friends and my family for providing me with unfailing support and continuous encouragement throughout my years of study. This accomplishment would not have been possible without them.

References

1. ISO 13849-1:2006, Safety of machinery—safety-related parts of control systems—part 1: general principles for design
2. CEN EN 954-1:1996, Safety-related parts of control systems
3. ISO 12100:2010, Safety of machinery—general principles for design—risk assessment and risk reduction
4. ISO 13849-1:1999, Safety of machinery—safety-related parts of control systems—part 1: general principles for design
5. IEC 60947-5-1:2003, Low-voltage switchgear and control gear–part 5: control circuit devices and switching elements–section 1: electromechanical control circuit devices
6. EN ISO 13849-2:2010, Safety of machinery—safety-related parts of control systems—part 2: validation
7. EN 60204-1:2009, Safety of machinery—electrical equipment of machines—part 1: general requirements

Investigation on Bidirectional DC/DC Converters for Energy Management and Control

Chao-Tsung Ma

Abstract This paper investigates a DSP based dual active full bridge phase shift (DAFBFS) DC/DC power converter for performing advanced energy management and control functions in renewable energy based power generation systems, e.g., wind and solar power generation systems. The hardware system of the proposed DAFBFS DC/DC converter includes two full-bridge circuit units, a coupling inductor and a high-frequency transformer especially designed for fast charging and discharging control of a battery energy storage system (BESS). The proposed DAFBFS converter has a number of merits, i.e., electrical isolation, high voltage gain, fast response in real-time current regulation and simplicity in designing its controllers with a single control variable. In this paper, issues regarding the operating principles, mathematical modeling and controllers of the DAFBFS converter are discussed. To achieve a better efficiency and enhance functional flexibility in hardware implementation, a fully digital control scheme with a TI DSP as the core controller is developed and experimentally verified. Typical simulation and experimental results are presented to demonstrate the performance of the proposed control scheme.

Keywords Battery energy storage system · Dc/Dc converter
Digital controller · Digital signal processor · Energy management
Renewable power generation

1 Introduction

In recent years, the renewable energy based distributed power generation (DG) has become the most economic option for adding new power generation capacity, especially for countries that mainly depend on nuclear energy, natural gas, coal and

C.-T. Ma (✉)
Department of EE, CEECS, National United University, #2 Lien-Da Rd,
Nan-Shih Li, Miaoli City 36063, Taiwan, ROC
e-mail: ctma@nuu.edu.tw

© Springer Nature Singapore Pte Ltd. 2018
S.-I. Ao et al. (eds.), *Transactions on Engineering Technologies*,
https://doi.org/10.1007/978-981-10-7488-2_13

fuel oil for power generation, such as Taiwan and other countries in Asia [1–3]. It has been well accepted that a self-sufficient, domestic renewable energy based power generation system has a number of advantages, e.g., it increases energy security, eliminates the need for expensive fuel imports, and reduces the burden on national budgets. It can also create substantial economic and societal benefits, such as increasing national competitiveness and job opportunities apart from its global environmental benefits. In Taiwan, the goal of DG using renewable energy based power generation devices, e.g., wind turbine generators (WTG) and photovoltaic (PV) modules, is to meet the continuously growing electric power demand without increasing greenhouse gas emissions and to gradually reduce the share of nuclear energy based power generation. It should be noted that due to the intrinsic feature of power fluctuation in most renewable generation systems, e.g., WTG and PV the dispatching of their output power is quite difficult and even impossible [4, 5]. Therefore, the addition of certain battery energy storage systems (BESS) with real-time charging and discharging capabilities is indeed necessary to increase the power management and control flexibility of future power systems embedded with a high penetration level of DG.

Over the last decade, the development of high-performance BESS [6, 7] and the related advanced power management and control technologies have enabled the fast development of renewable energy applications and so-called intelligent, hybrid DC/AC micro-grids [8, 9]. In the design of grid-level BESS, a number of conventional batteries can be chosen; however, the all vanadium redox flow battery (VRFB) has a number of intrinsic merits; e.g., its output power and capacity can be independently designed; the charging and discharging functions can be carried out simultaneously; it can be fully discharged without any harm; it has a high energy-conversion efficiency; it takes into account the safety and environmental protection concerns and above all, its system maintenance cost is relatively low [10, 11]. With the above advantages, the VRFB system is very suitable for applications in a wide variety of DG to optimize the integration of energy management and power quality control techniques. In addition, any advanced BESS must be equipped with advanced power converters and the related system control algorithms to achieve satisfactory performance [12–15].

Considering the required energy management and control functions in wind and solar power generation systems, a digital controller based dual active full bridge phase shift (DAFBFS) DC/DC power converter with fast current regulation capability is investigated in this paper. To confirm the correctness of the proposed control scheme, several simulations cases are carried out on PSIM software and an 1.2 kW hardware system with the TMS320F28335 as core controller and the related hardware interface circuits are also constructed for practical evaluations. The arrangement of sections in this paper is as follows. In Sect. 2, the main topology of isolated DC/DC power converters are briefly reviewed. The operating principles and control schemes of the proposed DAFBFS power converter are then addressed in Sect. 3. In Sect. 4, a fast charging and discharging control case is stated and typical results obtained from simulation studies and experimental tests are presented with discussions. Section 5 gives a summary on the technical issues concerned and the performance of the DAFBFS power converter with the proposed control scheme.

2 The Topologies of Dc/Dc Converters

In the open literature, there are a lot of research and discussions on DC/DC converter systems and their control methods [16, 17]. Depending on whether an isolation transformer is utilized, bi-directional DC/DC converters can be simply classified into non-isolated and isolated converter topologies. Figure 1 shows the classification of bi-directional DC/DC converters for industrial applications. In this section, some of the isolated bi-directional DC/DC converters presented in the literatures are briefly reviewed.

Figure 2 shows the typical structure of an isolated bidirectional DC/DC converter [18]. As can be seen in Fig. 2, a high-frequency transformer connects the two sets of inverters with their own filtering capacitors and inductors. Because the transformer is a non-ideal component the effects of leakage inductance on voltage spikes appear, making it not suitable for high power applications.

In paper [19], the authors proposed a revised version of the circuit shown in Fig. 2 to improve the above mentioned drawback by adding a RCD circuit to the output side of the secondary bridge. However, the use of RCD circuit will lead to additional losses, resulting in the decrease of overall efficiency of the converter. Therefore, the authors of paper [20] proposed a low-loss damping circuit as shown in Fig. 3, which can be operated in high-power application conditions and at the same time with relatively high system efficiency.

To further improve the overall performance of the above mentioned converters, the damping circuit in Fig. 3 was replaced by an active clamping circuit by the

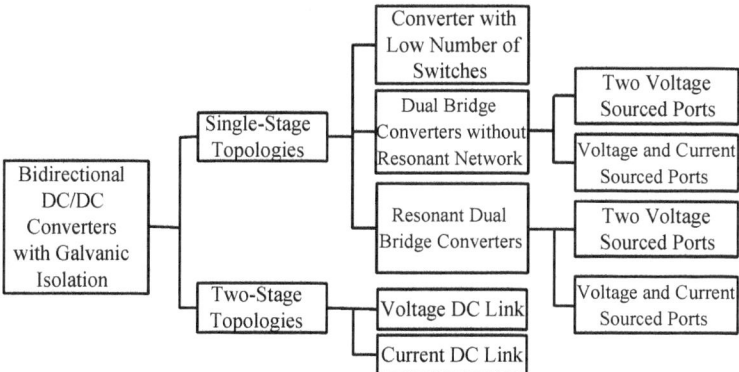

Fig. 1 The classification of bi-directional DC/DC converters

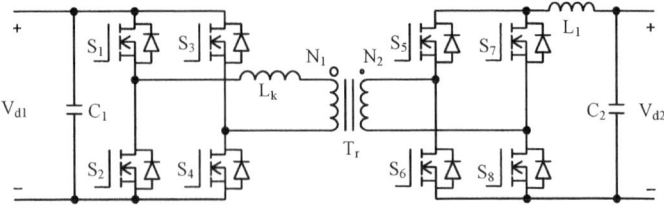

Fig. 2 The isolated bidirectional DC/DC converter

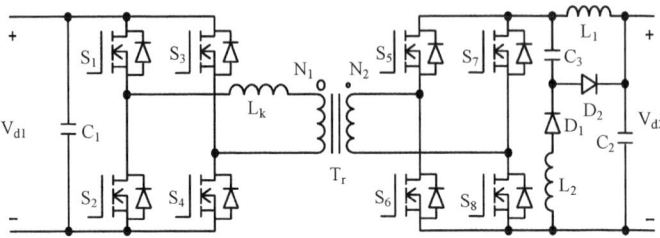

Fig. 3 The low-loss isolated bidirectional DC/DC converter circuitry

authors of [21]. This arrangement has the following advantages: (1) zero-voltage switching for the first two switches on the secondary side; (2) zero voltage and zero current switching for all switches on the primary side of the transformer; (3) no circulating currents. Theoretically, bidirectional DC/DC converters with active clamped buffer circuits have higher conversion efficiency compared to conventional full-bridge phase-shift converters; however, power switches of this circuit are required to withstand a relatively higher voltage stress than its secondary voltage, Vd2. To eliminate this shortcoming, the authors of paper [22] proposed an method to improve the circuit performance by changing the primary-side full-bridge into a half-bridge topology. This arrangement has successively achieved the following advantages: (1) zero voltage switching for all switches; (2) constant clamping voltage on all switches; (3) simplicity in designing driving circuits.

3   The Operating Principles of the Proposed DC/DC Converter

3.1   The Operating Principle of DAFBFS Converters

The transformer isolated DAFBFS DC/DC converter investigated in this paper uses an energy transferring inductor and the output capacitors of its power switches to achieve zero voltage switching. This converter topology is especially suitable for application scenarios in DG systems since it provides power isolation, high voltage

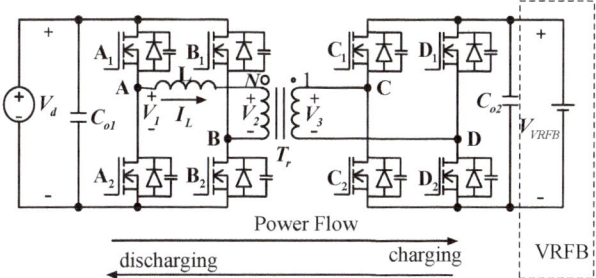

Fig. 4 The isolated DAFBFS DC/DC converter with a simplified VRFB model

conversion ratio, and high efficiency. The DAFBFS DC/DC converter with a simplified VRFB model is shown in Fig. 4. The primary side of the transformer connecting to a full-bridge converter of four active switches, whose DC side is normally connected to a DC bus as considered the high-voltage side. The secondary side full-bridge converter is connected to a VRFB as the low-voltage side. In operation, the DAFBFS DC/DC converter is controlled by a set of bipolar PWM switching signals with a 50% conduction period. By controlling the active switches of the primary and secondary sides, a phase shift, θ, can be produced between the two voltages across the energy transferring inductor, V1 and V2. The resonance inductor, L, here, is acting as a transferring media for a bidirectional energy transmission, thereby controlling the direction and magnitude of power flows with the changing of θ. Figure 5 shows the theoretical waveforms of the operations of charging and discharging of the VRFB.

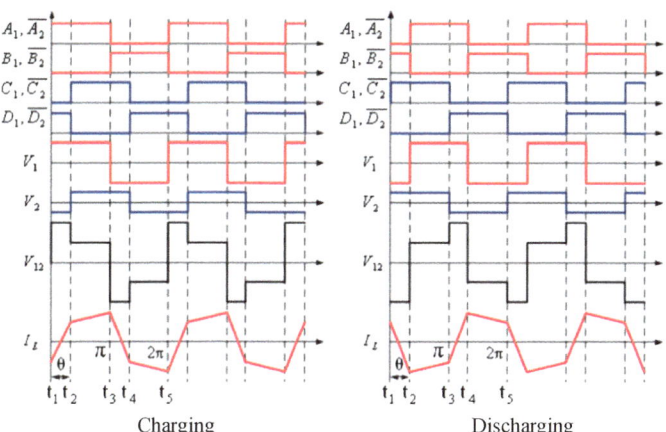

Fig. 5 The theoretical waveforms of the DAFBFS converter charging and discharging of the VRFB

3.2 The Analysis of Power Flow in DAFBFS Converters

Theoretically, the DAFBFS DC/DC converter mainly relies on shifting the phase θ to control the quantity and direction of power flow. According to this principle, the relationship between the phase shift, θ, and the power flow, P_T, can be easily deduced. Assuming that all components are ideal, the transferred power, P_T, is equal to the output power, P_o, expressed as follows.

$$P_T = P_o = \frac{1}{\pi} \int_0^\pi I_L(\delta) \cdot V_2(\delta) d\delta \tag{1}$$

As shown in Fig. 6, the inductor current $I_T(\delta)$ can be divided into 0 to θ and θ to π two intervals, so that the current at the point of 0 is labeled as I_{L1i}, at the point of θ is labeled I_{L2i}, and at the point of π is $-I_{L1i}$.

The inductor current, $I_{L1}(\delta)$, from 0 to θ can be defined as follows:

$$I_{L1}(\delta) = \frac{V_d + NV_o}{\omega L} \delta + I_{L1i} \tag{2}$$

The inductor current, $I_{L2}(\delta)$, from 0 to θ can then be defined as follows:

$$I_{L2}(\delta) = \frac{V_d - NV_o}{\omega L} (\delta - \theta) + I_{L2i} \tag{3}$$

According to the relationship between (2) and (3) one can have the following Eqs. (4) and (5).

$$-I_{L1i}(\delta) = \frac{V_d - NV_o}{\omega L} (\pi - \theta) + I_{L2i} \tag{4}$$

$$\frac{V_d + NV_o}{\omega L} \theta + I_{L1i} = I_{L2i} \tag{5}$$

Fig. 6 The key waveforms in a DAFBFS converter

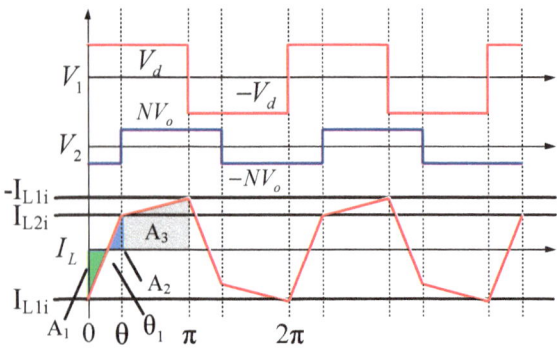

Solving the Eqs. (4) and (5), the two inductor currents can be obtained.

$$I_{L1i} = \frac{-(V_d + NV_o)\theta - (V_d - NV_o)(\pi - \theta)}{2\omega L} \tag{6}$$

$$I_{L2i} = \frac{(V_d + NV_o)\theta - (V_d - NV_o)(\pi - \theta)}{2\omega L} \tag{7}$$

Using (1) and the related parameters in the 0 to π interval, one can have the following power flow equation.

$$P_T = \frac{NV_o}{\pi} \left[\int_0^{\theta_1} I_L(\delta)d\delta - \int_{\theta_1}^{\theta} I_L(\delta)d\delta + \int_{\theta}^{\pi} I_L(\delta)d\delta \right] \tag{8}$$

After some mathematical derivations, (8) can be rewritten as:

$$P_T = \frac{NV_o}{\pi} \left(-I_{L2i} \cdot \theta + \frac{-I_{L1i} + I_{L2i}}{2}\pi \right) \tag{9}$$

Substituting equivalent parameters into (9), the final expression of transmission power, P_T , is reached as follows.

$$P_T = P_o = \frac{NV_d V_o}{\omega L} \left(\theta - \frac{\theta^2}{\pi} \right) \tag{10}$$

From the above (10), one can see that when the phase shift angle is greater than 0 degrees, the power will be transmitted from the primary side to the secondary side and when the phase shift angle is less than 0 degrees, the power flow is reversed. The maximum range of operating phase shift is from $-\pi/2$ to $\pi/2$.

3.3 Control of the DAFBFS DC/DC Converter

Because the battery unit used in this paper is connected to the secondary side of the DC/DC converter and a controlled DC bus is connected to the primary side of DAFBFS DC/DC converter it can be viewed as two ideal DC voltage sources connected to both sides of the converter. The DAFBFS DC/DC converter can then control the power flow in both directions simply by changing θ. In this control scheme, both voltage sources remains almost constant, only the current magnitude and direction are changed, thereby changing the power magnitude and direction as desired. As can be seen from Fig. 4 that the voltages V1 and V2 on both sides of the inductor can be varied by properly control the power switches in the primary and secondary sides to achieve the desired phase difference θ. According to the above description, the following equations can be directly derived.

$$V_L = L\frac{di_L}{dt} = V_1 - V_2 \tag{11}$$

Since the V1 and V2 voltages are square waves and with a phase difference of θ, (11) can be rewritten as follows.

$$V_L = \frac{4}{\pi}V_d \sin \omega t - \frac{4}{\pi}NV_o \sin (\omega t + \theta) \tag{12}$$

Adding the perturbation terms to both I_L and θ in (11) and (12), one has

$$L\frac{d(I_L + \hat{i}_L)}{dt} = \frac{4}{\pi}V_d \sin \omega t - \frac{4}{\pi}NV_o \sin(\omega t + \theta + \theta) \tag{13}$$

By removing the steady-state terms in (13) and performing some mathematical manipulations, the transfer function of the converter system can be derived in s-domain as shown in (14).

$$\frac{\hat{I}_L(s)}{\theta(s)} = \frac{\hat{I}_d(s)}{\theta(s)} = -\frac{\frac{4}{\pi}NV_o \cos (\theta_0)}{sL} \tag{14}$$

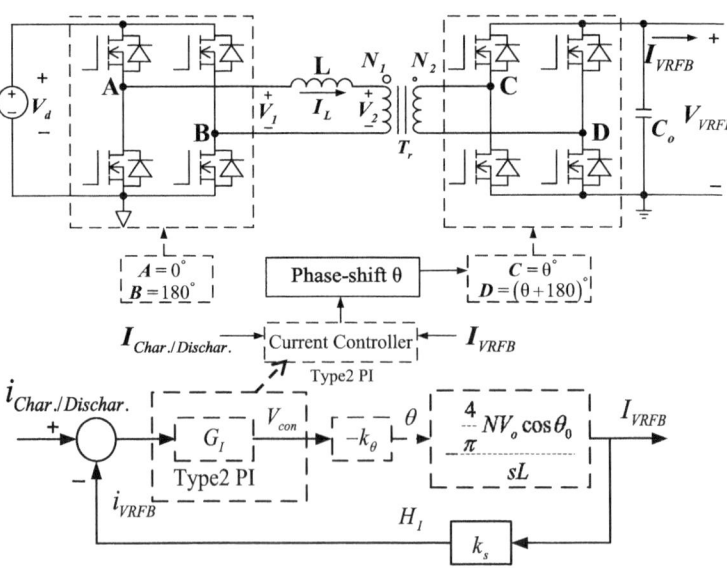

Fig. 7 The overall charging and discharging control system of the DAFBFS DC/DC converter

With the consideration of sensing parameters, the final mathematical model for designing current controllers is obtained as shown in (15).

$$H_I(s) = \frac{\frac{4}{\pi} k_\theta k_s N V_o \cos{(\theta_0)}}{sL} \qquad (15)$$

As the controller plays an important role in any converter operations, in this paper the type-II compensator and k-factor method are chosen to design the required current controller GI (s). In general, the phase margin of the current controller is usually set at 45°–70° and the crossover frequency is about 1/4–1/10 times the switching frequency. In this design case, the phase margin is set to 65° and the crossover frequency is set to 1/8 times the switching frequency that is 50 kHz. After some mathematical derivations, the parameters of the controller can be obtained and the control block diagram of a type-II PI controller, GI(s), used for simulation studies and the DSP based experimental tests carried out in this paper is shown in Fig. 7.

4   The Simulation and Experimental Results

In this section, a fast charging and discharging control function of the DAFBFS DC/DC converters is firstly verified with simulations on the PowerSIM (PSIM) software, followed by a practical test using a small-scale experimental hardware setup with the same operation conditions. The test system parameters are: DC Bus

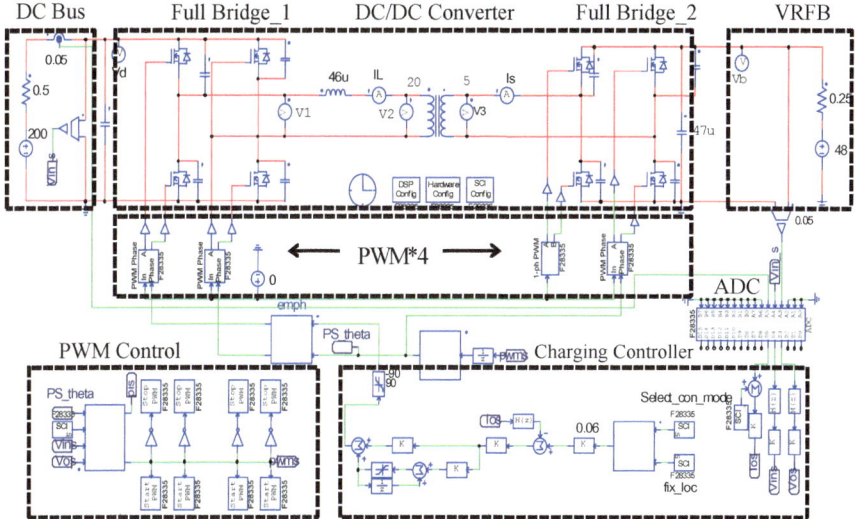

Fig. 8 The PSIM model of the DAFBFS DC/DC converter

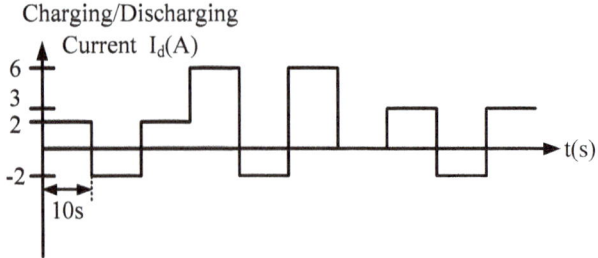

Fig. 9 A list of continuous charging and discharging commands

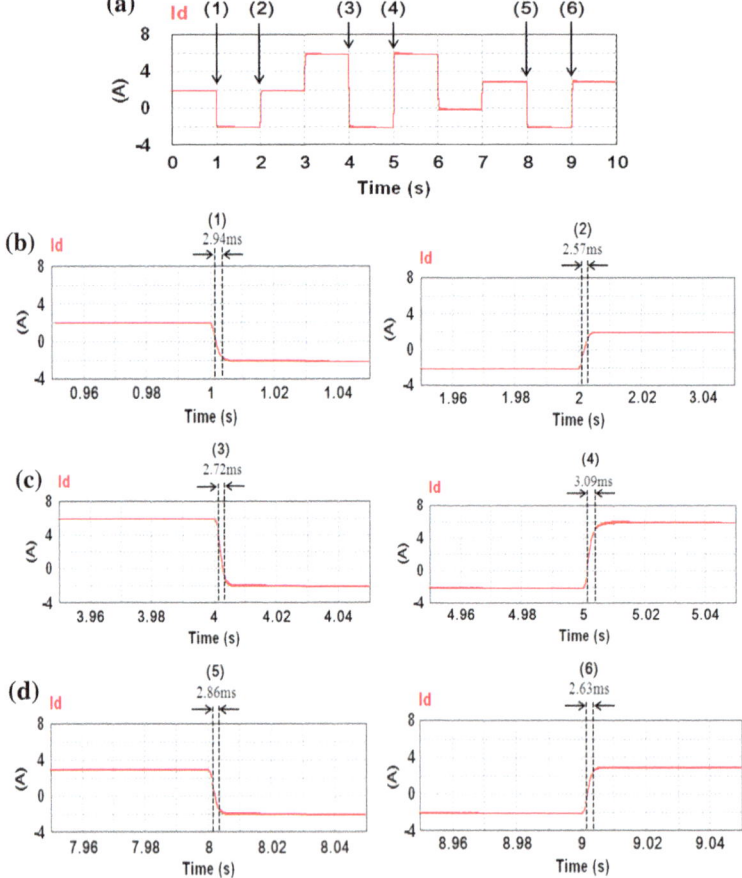

Fig. 10 **a** A list of continuous charging and discharging commands (simulated). **b** Results of step changes in current command, between 2 and −2 A with the falling time of 2.94 ms and rising time of 2.57 ms. **c** Results of step changes in current command, between 6 and −2 A with the falling time of 2.72 ms and rising time of 3.09 ms. **d** Results of step changes in current command, between 3 and −3 A with the falling time of 2.86 ms and rising time of 2.63 ms

Fig. 11 The hardware setup of an 1.2 kW DAFBFS converter system

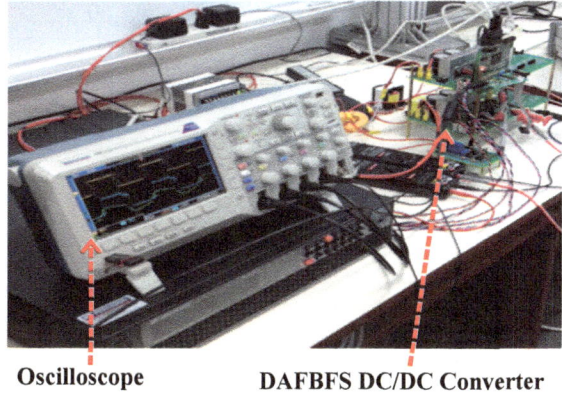

Oscilloscope   **DAFBFS DC/DC Converter**

has a voltage of 200 V, the rated voltage of VRFB is 48 V. The power rating of the DAFBFS converter is 1.2 kW. The final PSIM model for simulation studies is shown in Fig. 8.

4.1 Simulation Results of DAFBFS Converters

In this simulation case, the DAFBFS DC/DC converter acts as a charging and discharging controller for the VRFB energy storage system and a list of continuous charging and discharging commands is shown in Fig. 9. Figure 10a–d show a set of simulation results.

4.2 Experimental Results of DAFBFS Converters

To confirm the correctness of the proposed control scheme, an 1.2 kW hardware system with the TMS320F28335 as core controller and the related hardware interface circuits are constructed as shown in Fig. 11. Figure 12a–d show a set of measured results. The system parameters and operating conditions are the same as that used in the simulation case presented in Sect. 4.1.

5 Conclusion

The output power fluctuation of renewable energy based DG systems is unavoidable. Without adequate control schemes, the power fluctuation phenomena is prone to make the grid voltage and frequency unstable and degrade the quality of electric

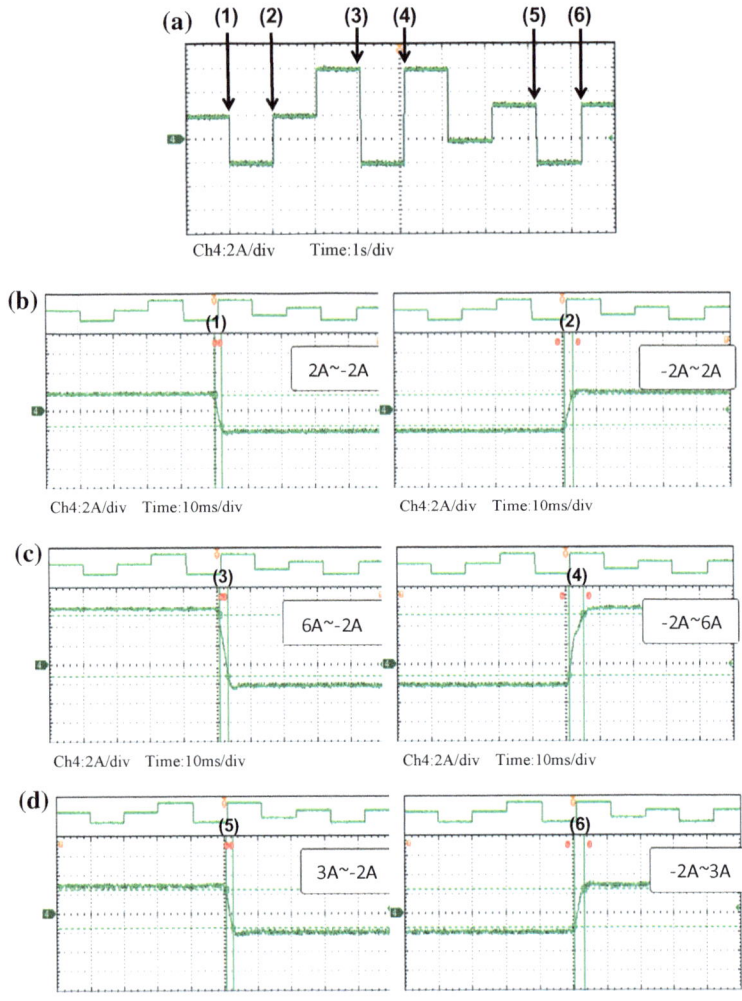

Fig. 12 **a** A list of continuous charging and discharging commands (measured). **b** Results of step changes in current command, between 2 and −2 A with the measured falling time of 2 ms and the rising time of 2.4 ms. **c** Results of step changes in current command, between 6 and −2 A with the measured falling time of 2.6 ms and the rising time of 4.2 ms. **d** Results of step changes in current command, between 3 and −2 A with the measured falling time of 2 ms and the rising time of 2.8 ms

power supply system. It has been proved that using a properly designed BESS with fast-response power converter systems and appropriate power control strategies can effectively solve the problem. In the aspect of BESS and fast-response power converter systems, this paper has briefly reviewed the potential topologies of bidirectional DC/DC power converters and proposed a simple phase-shifting power control scheme. The overall performance of a selected DAFBFS DC/DC power converter acting as the charging and discharging interface of a BESS is then fully studied. The feasibility and effectiveness of the proposed fast current control method based on a single phase-shifting parameter has been verified with simulation studies and tests with a small-scale, DSP based experimental hardware setup. Typical results obtained from both simulation and practical measured waveforms have been presented with brief discussions.

Acknowledgements This work was supported in part by the Ministry of Science and Technology, R.O.C. under Grant MOST 105-2221-E-239-022.

References

1. L.F. Ochoa, G.P. Harrison, Minimizing energy losses: optimal accommodation and smart operation of renewable distributed generation. IEEE Trans. Power Syst. **26**(1), 198–205 (2011)
2. D.Q. Hung, N. Mithulananthan, DG allocation in primary distribution systems considering loss reduction, in *Handbook of Renewable Energy Technology* (World Scientific Publishers, Singapore, 2011), pp. 587–628
3. I. El-Samahy, E. El-Saadany, The effect of DG on power quality in a deregulated environment. Proc. IEEE Power Eng. Soc. Gen. Meet 2969–2976 (2005)
4. L. Xiangjun, H. Dong, L. Xiaokang, Battery energy storage station (BESS)-based smoothing control of photovoltaic (PV) and wind power generation fluctuations. IEEE Trans. Sustainable Energy **4**(2), 464–473 (2013)
5. R. Sebastián, Application of a battery energy storage for frequency regulation and peak shaving in a wind diesel power system. IET Gener. Transm. Distrib. **10**(3), 764–770 (2016)
6. M. Farhadi, O. Mohammed, Energy storage technologies for high-power applications. IEEE Trans Ind Appl **52**(3), 1953–1961 (2016)
7. M. Zidar, P.S. Georgilakis, N.D. Hatziargyriou, T. Capuder, D. Skrlec, Review of energy storage allocation in power distribution networks: applications, methods and future research. IET Gener. Transm. Distrib. **10**(3), 645–652 (2016)
8. A. Chaouachi, R.M. Kamel, R. Andoulsi, K. Nagasaka, Multiobjective intelligent energy management for a microgrid. IEEE Trans. Ind. Electron. **60**(4), 1688–1699 (2013)
9. M. Kumar, S.C. Srivastava, S.N. Singh, M. Ramamoorty, Development of a control strategy for interconnection of islanded direct current microgrids. IET Renew. Power Gener. **9**(3), 284–296 (2015)
10. M. Guarnieri, P. Mattavelli, G. Petrone, G. Spagnuolo, Vanadium redox flow batteries: potentials and challenges of an emerging storage technology. IEEE Ind. Electron. Mag. **10**(4), 20–31 (2016)
11. D.A. Riccardo, L. Baumann, A. Damiano, E. Boggasch, A vanadium-redox-flow-battery model for evaluation of distributed storage implementation in residential energy systems. IEEE Trans Energy Convers **30**(2), 421–430 (2015)
12. C.-T. Ma, Design and implementation of a bidirectional DC/DC converter for BESS operations, in *Lecture Notes in Engineering and Computer Science: Proceedings of the International MultiConference of Engineers and Computer Scientists 2017*, Hong Kong, 15–17 Mar 2017, pp. 666–671
13. J.S. Wang, S.X. Li, PID decoupling controller design for electroslag remelting process using cuckoo search algorithm with self-tuning dynamic searching mechanism. Eng. Lett. **25**(2), 125–133 (2017)

14. J.K. Obichere, M. Jovanovic, S. Ademi, improved power factor controller for wind generator and applications. Eng. Lett. **24**(2), 125–131 (2016)
15. H. Karaca, E. Bektas, Selective harmonic elimination using genetic algorithm for multilevel inverter with reduced number of power switches. Eng. Lett. **24**(2), 138–143 (2016)
16. S. Bhattacharya, T. Zhao, G. Wang, S. Dutta, S. Baek, Y. Du, B. Parkhideh, X. Zhou, A.Q. Huang, Design and development of generation-isilicon based solid state transformer, in *The Proceedings of the 25th Annual IEEE Applied Power Electronics Conference and Exposition* (Palm Springs, CA, 2010), pp. 1666–1673
17. S. Inoue, H. Akagi, A bidirectional isolated dc-dc converter as a core circuit of the next-generation medium-voltage power conversion system. IEEE Trans. Power Electron. **22** (2), 535–542 (2007)
18. O. Garcia, L.A. Flores, J.A. Oliver, J.A. Cobos, J. de la Pena, Bidirectional DC-DC converter for hybrid vehicles, in *IEEE Power Electronics Specialists Conference*, June 2005, pp. 1881–1886
19. L. Zhu, A novel soft-commutating isolated boost full-bridge ZVS-PWM DC-DC converter for bi-directional high power applications, in *IEEE Power Electronics Specialists Conference*, June 2004, pp. 2141–2146
20. E.S. Kim, K.Y. Joe, H.Y. Choi, Y.H. Kim, Y.H. Cho, An improved soft switching bi-directional PSPWM FB DC/DC converter, in *IEEE Industrial Electronics Society Conference*, Sept 1998, pp. 740–743
21. R. Li, A. Pottharst, N.K. Witting, M. Dellnitz, O. Znamenshchykov, R. Feldmann, Design and implementation of a hybrid energy supply system for railway vehicles, in *Twentieth Annual IEEE Applied Power Electronics Conference and Exposition*, Mar 2005, pp. 474–480
22. S.J. Jang, T.W. Lee, W.C. Lee, C.Y. Won, Bidirectional DC to DC converters for fuel cell generation system, in *IEEE Power Electronics Specialists Conference*, June 2004, pp. 4722–4728

Distributed Lock Relation for Scalable-Delay-Insensitive Circuit Implementation Based on Signal Transition Graph Specification

Pitchayapatchaya Srikram and Arthit Thongtak

Abstract Signal transition graph specification has a potential to describe behavior of hardware system in term of concurrent, sequential and one instance of the same events. One typical idea is for asynchronous control circuits, which is a variety of delay assumption design by means of signal transition graph specification. This paper proposes a distributed lock relation to determine the completion path for multiple-cycle signals. We select the tardy internal-completion signal to be the volunteer signal based on Scalable-Delay-Insensitive (SDI) model. The effectiveness of the proposed methodology is evaluated by cost of area, which is number of internal input signals and literal logic gates.

Keywords Asynchronous control circuit · Asynchronous logic synthesis
Lock relation · Multiple-cycle signal · Scalable-insensitive delay model
Signal transition graphs

1 Introduction

The recent advancement in asynchronous control circuits design, the absence of clock circuits, there has been considerable in an wire fork that lead to design sophisticated asynchronous circuits under unbounded gate and wire delay assumption. As regards the Quasi-Delay-Insensitive (QDI) model was designed on the basis of wire forks, on which of the propagation delay of each output is poised on delay, called an isochronic fork. Since, the main challenge of design faced in

P. Srikram
Rajamangala University of Technology Thunyaburi, Pathumthani, Thailand
e-mail: Pitchayapatchaya.s@en.rmutt.ac.th

A. Thongtak (✉)
Chulalongkorn University, Bangkok, Thailand
e-mail: Arthit.t@chula.ac.th

© Springer Nature Singapore Pte Ltd. 2018
S.-I. Ao et al. (eds.), *Transactions on Engineering Technologies*,
https://doi.org/10.1007/978-981-10-7488-2_14

185

designed isochronic fork with utilizing time information. It has been shown that two paths may travel through n gates before acknowledge another, is called extended isochronic forks or denoted as Q^nDI [1], likewise difference propagation delays, called an asymmetric isochronic-forks assumption [2]. The delay assumption of isochronic forks is still not completely understood; However, scalable-delay-insensitive (SDI) model is one of major design considerations that alleviated the ascetic completion signal, this assumes on relative delay ratio between any two component is bounded by K value that guarantees the correct circuit operation [3]. Although considerable a research has been done on asynchronous combinational circuit on data path, rather than describe design in term of asynchronous control circuit in the class of asynchronous sequential circuits. In current practice the design of asynchronous SDI control circuits has been present SDI optimization [4]. As with the SDI optimization, the approach is delved and modified each a concurrent transition model on wholly of Signal Transition Graph (STG)—there is concurrent relation of primary-input signal transition and non-primary signal transition. Despite of this, the underlying STG satisfies its properties previously, After SDI optimization, the approach is satisfies STG properties such as persistence and complete state coding, but it is not preserved safeness and liveness in the event that is contained multiple token. One solution has been presented the determined concurrent transition model whether it can be optimized based on SDI model by using lock relation [5]. The result of above approach indicated that is reduced cost of circuits; However, all the previously mention approaches suffer from some limitations for handled some multiple-cycle signal at STG domain.

This article is introduced a distributed-lock relation in order to simplified multiple-cycle signal. This investigation is taken the form of a case-study of the design of asynchronous control circuit by using the novel SDI optimization. To design such circuit at STG domain with multiple-cycle signal, since each transition on the wire fork needs to be acknowledged explicitly, in the other word, it is a casual relation. This is exemplified in the research undertaken by determination of the volunteer signal, which is a tardy acknowledgement of completion path signal. This proposed methodology is demonstrated through experimental result as an example on optimization and implementation of asynchronous control circuits on SDI model. The procedure of this method is illustrated in Fig. 1, which is compared to the previous method. As well as, this approach is not implemented circuit on SDI model, if the given STG has not to satisfy the properties of STG implementability. The synthesis of circuit from STG is based on S. Park method [6].

The core contribution of this paper can be summarized as follow: In the next section is briefly introduced the related work, an overview of the STG notation used in this paper, the basic of lock relation properties. In addition, this section is presented the distributed lock relation specification in detail and the definition of Scalable-Delay-Insensitive (SDI) model. Then, Sect. 3 discusses distributed-lock relation based design style of empirical SDI model. Section 4 is proposed method of SDI implementation and optimization. Then, the experimental results and conclusions are given in Sects. 5 and 6, respectively.

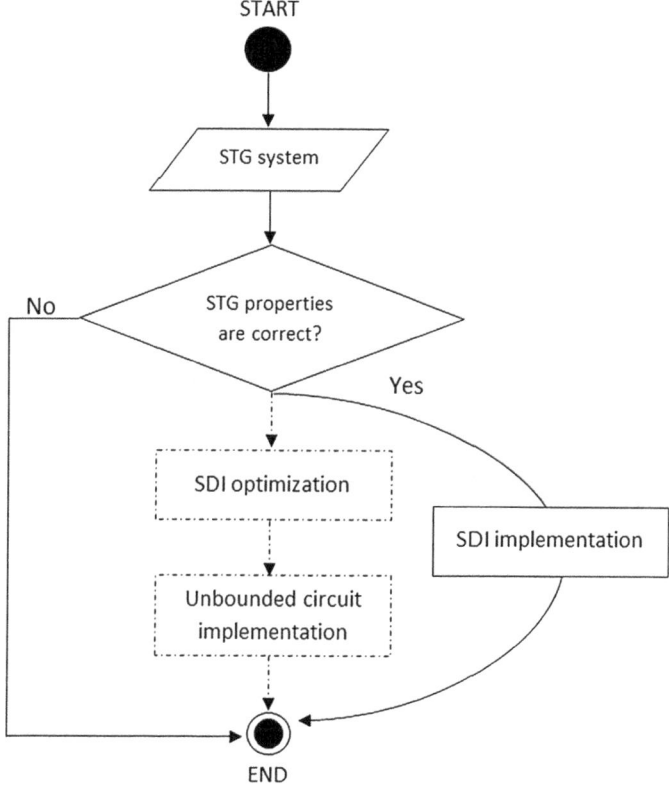

Fig. 1 Overall procedure for SDI implementation

2 Related Work

The following is the brief description of overview of signal transition graph, lock relation and Scalable-Delay-Insensitive model definition.

2.1 Signal Transition Graph (STG) Notation: An Overview

The signal Transition Graph (STG) is formalization of characteristic of asynchronous control circuits in described [7]. This is interpreted as live safe free-choice Petri nets. The signal transition is represented as rising and falling, It seem, event-driven on timing diagrams, if whose transitions are rising and falling immediately, is called single-cycle signal as a^+ and a^-. The transition $a*$ denotes either a^+ or a^-, while $\overline{a*}$ denotes the complement of $a*$. Otherwise, the same of signal is rising and falling various times. This is called multiple-cycle signal as a^+/t and a^-/t. The arcs are represented casual relation i.e. ai is trigger signal for bo, when bo is achieved transition, it means that bo is acknowledge signal transition for ai. Marking ($\bullet$, token) is described the enabled transition which $\mu 0$ is initial marking.

However, the STG is implementable, this is be satisfied following properties:

- **Safeness** if, it is a strongly-connected graph and every simple cycle in STG has exactly one marking (token, μ).
- **Liveness**, the initial marking can be reached enabled all the transition in STG.
- **Persistency**, the signal transition must not be disabled by another signal transition.
- **Complete state coding** (CSC), all the signal transition is produce the difference binary code.
- **Simple cycle**, the signal transition is sequenced as $t1 \ldots titj \ldots t1$ in STG.

2.2 Formal Definition of Lock Relation

Lock relation describe the interleaving signal transitions between couple signal, there is causal relation by live STG which is formalized two tuple $G = \langle V, E \rangle$. Previous research has shown that the lock relation is sufficient condition for circuit implementation and verification of STG properties which is implementable circuit [6, 8].

- **Full-lock** if, the two signals a and b on simple cycle, are interleaved its transition that $a* \rightarrow b* \rightarrow \overline{a*} \rightarrow \overline{b*}$.
- **Semi-lock** if, the two signals a and b on simple cycle, are interleaved its transition that $b* \rightarrow a* \rightarrow \overline{b*}$ or $a* \rightarrow b* \rightarrow \overline{a*}$.
- **Associate-lock** if, A is minimal set of full-lock relation of two signals as $a1$ and $a2$, and the any transition of set A is fully-locked with signal b, this is such that $\exists a1, a2 \in a1 \rightarrow b* \rightarrow a2 \rightarrow \overline{b*}$ on simple cycle and also called *transitive-lock relation*.

The STG is satisfied lock-relation, this is enable implementing hazard-free circuits. According to lock relation is interpreted not only single-cycle signal but also multiple-cycle signal in described. It has been is presented solving state coding problem on STG domain by transitive-lock relation on which contains multiple-cycle signal [9].

One is splitting multiple-cycle into virtual single-cycle signal. Meanwhile, this paper is developed multiple-cycle signal specification by a distributed-lock relation.

- **Distributed-lock** if, all candidate transition of multiple-cycle signal a/t are full-lock relation with signal b and c, then signal b and c are fully locked. This is denoted as $a/tL(bLc)$, $t\ni\{1, 2, \ldots, n+1\}$ [10].

Proof If the multiple-cycle signal a/t is full-lock relation with signal b on the simple cycle such as in (1):

$$\exists a/t: a/t^* \rightarrow b^* \rightarrow \overline{a/t^*} \rightarrow a/(t+1)^* \rightarrow \overline{b^*} \rightarrow \overline{a/(t+1)^*}$$
$$\text{Or,} \quad \exists a/t: a/t^* \rightarrow \overline{a/t^*} \rightarrow b^* \rightarrow a/(t+1)^* \rightarrow \overline{a/(t+1)^*} \rightarrow \overline{b^*} \tag{1}$$

While, the multiple cycle signal a/t is full-lock relation with signal c on the simple cycle such as in (2):

$$\exists a/t: a/t^* \rightarrow c^* \rightarrow \overline{a/t^*} \rightarrow a/(t+1)^* \rightarrow \overline{c^*} \rightarrow \overline{a/(t+1)^*}$$
$$\text{Or,} \quad \exists a/t: a/t^* \rightarrow \overline{a/t^*} \rightarrow c^* \rightarrow a/(t+1)^* \rightarrow \overline{a/(t+1)^*} \rightarrow \overline{c^*} \tag{2}$$

Thus (1) and (2), there are formalized as seen in the continuity Eq. (3):

$$\{[(a/t \rightarrow b^*) \cup (a/t \rightarrow \overline{b^*})] \cup [(a/t \rightarrow c^*) \cup (a/t \rightarrow \overline{c^*})]\} \tag{3}$$

Then (3), there is summarized as in (4) and the final result of summarization of (4) is formalized as in (5):

$$(a/tLb^*) \cup (a/tLc^*): b\ni\{b^*, \overline{b^*}\} \ and \ c\ni\{c^*, \overline{c^*}\} \tag{4}$$

$$a/tL(bLc), a/t\ni\{a/t^*, \overline{a/t^*}, \ldots,$$
$$a/t(t+1)^*, \overline{a/t(t+1)^*}\}, n\ni\{1, 2, \ldots, n+1\} \tag{5}$$

∎

2.3 Scalable-Delay-Insensitive (SDI) Model

The SDI model [11] refers to an interconnection of two components (C1, C2). There is $t0$ signal that causes to $t1$ and $t2$ signal as illustrated in Fig. 2. The propagation delay paths of two circuit components from $t0$ to $t1$ denote delay as D1 and another from $t0$ to $t2$ denote delay as D2, Let De1 and De2 refer to estimated delay for two paths accordingly, and Da1 and Da2 refer to actual delay that occur in two paths.

K is constant of margin for correctness operation circuit based on SDI model, is called the maximum variation ratio. For the present case under consider signal

Fig. 2 The SDI assumption
model

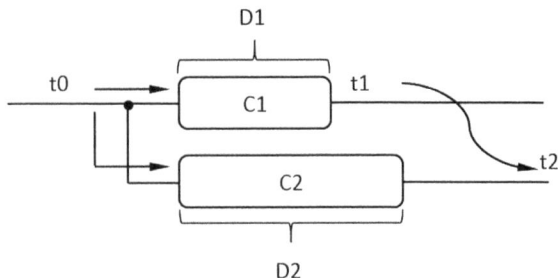

transition *t1* is specified to precede *t2*. Consequently, The K value is given for De1
such that the relation K • De1 < De2, then Da1 < Da2. On the Contrary, if K value
is given for De2 such that the relation K • De2 < De1, then Da2 < Da1, since
signal transition *t2* is preceded *t1*.

3 Distributed-Lock Relation-Based Design Style of SDI Model

In this section, this is approach to implement such circuit based on SDI design style
from STG, which hold the multiple-cycle signal. As explained earlier, the isochronic forks is caused simultaneous pair of signal transition, then the conversional
asynchronous control circuits confined to the each non-primary input for practi-
cality and ease of circuit implementation on unbounded delay model, which is not
only caused by the primary-input signal, but also the other acknowledgement with a
subsequent non-primary input signal as example illustrated in Fig. 3a, signal *t0* is
trigger signal for signal *t1* and *t2*, both signal *t1* and *t2*, is caused by each other's
acknowledgement signal transition (*t1'* and *t2'*) between two paths.

In term of STG, this characteristic refers to causal relation, which is interpreted
STG as illustrated in the Fig. 3b. There are causal relation among *t0*, *t1* and *t2*. On
the single-cycle signal *t0* is caused to accomplished signal *t1* and *t2*. As well as,
Even if the circuit operation is multiple-cycle signal of *t0*, it is refers that *t0* can
accomplish signal *t1* and *t2* with multiple-cycle signal. As mention above, this
schema is given by corresponding distributed-lock relation, as illustrated in Fig. 3c.

However, the timing assumption on SDI model refers that any two paths in
which wire fork, one non-final transition signal is accomplished, the other is also
accomplished by K value i.e. *t1* precede *t2*, this means that if *t2* is accomplished, *t1*
is also accomplished. On the other hands, *t2* is volunteer successor acknowledge the
t1. Therefore, *t2* is called volunteer signal. In term of delay, *t2* is tardy

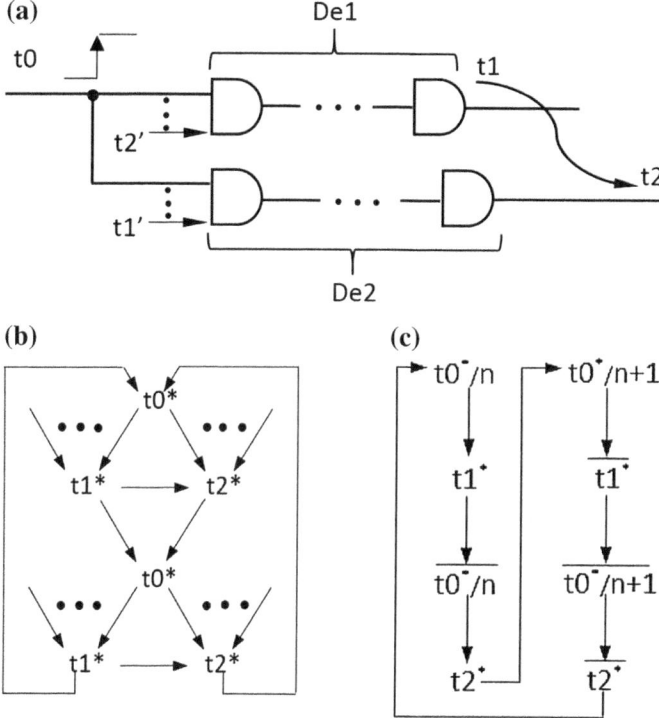

Fig. 3 The SDI circuit model **a** and **b** simplifying STG which are behaviorally equivalent and corresponding STG with multiple-cycle signal

internal-completion signal. Therefore, the distributed-lock relation which guarantees the completion path, there can be operation based on SDI model as discussed in the method Sect. 4.

4 SDI Optimization and Implementation

Since the distributed-lock relation can be handled STG, which is obtained the multiple-cycle signal based on SDI model, as described in the previous section. This section is presented the method to implement and optimize circuit based on SDI model. A case-study approach was adopted to determine the distributed-lock relation on which the primary input signal is multiple-cycle signal and the full-lock relation set of non-primary input signal. However the terminologies of signal, these is describe on the algorithm as illustrate on Table 1.

Table 1 Terminologies for signal and arcs on empirical SDI design

Notation	Description
$Vip_{(i)}(*/t)$	Set of primary input with multiple-cycle signal, $Vip_{(i)}(*/1), Vip_{(i)}(*/2), \ldots, Vip_{(n+1)}(*/n+1)$
$Vnp_{(i)}$	Set of primary input with single-cycle signal, $Vnp_{(1)}, Vnp_{(2)}, \ldots, Vnp_{(n+1)}$
$Vip_{(i)}$	Set of primary input with single-cycle signal, $Vip_{(1)}, Vip_{(2)}, \ldots, Vip_{(n+1)}$
$V_{(i)}$	Set of both primary input signal and non-primary input signal with single-cycle signal, $V_{(1)}, V_{(2)}, \ldots, V_{(n+1)}$
$Vop_{(i)}$	Set of non-primary input on which are candidate of distributed-lock relation, $Vop_{(1)}, Vop_{(2)}, \ldots, Vop_{(n+1)}$
$\bullet enp_{(i)}$	Set of incoming marking on which is arc of non-primary input, $[\bullet enp_{(i)} \vert V_{(i)} \longrightarrow enp_{(i)} Vnp_{(i)}]$
$enp'_{(i)}\bullet$	Set of outgoing token on which is arc of non-primary input, $[enp_{(i)} \bullet \vert Vnp_{(i)} \longrightarrow enp_{(i)} V_{(i)}]$

where i and t is denoted as $i, t \ni \{1, 2, 3, \ldots, n+1\}$

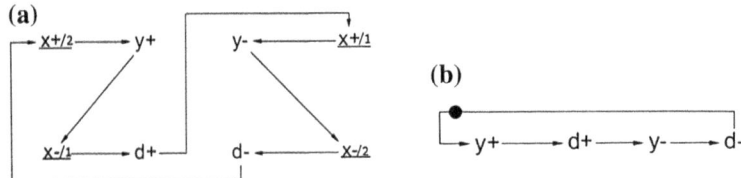

Fig. 4 The STG on which **a** is distributed-lock relation and **b** is to determine volunteer signal

As regarding the methodology, which is consists of two steps; First step is to determine a volunteer signal, then established to a partial transition graph as described on the following.

4.1 To Determine a Volunteer Signal

This algorithm is described to determine a volunteer signal on the distributed-lock relation as shown in Algorithm I and given as procedure example as illustrate in Fig. 4.

Algorithm I: To determine a Volunteer signal

Input	: The distributed-lock relation
Output	: Volunteer signal (v)

1: **begin**

2: **while** distributed-lock $\ni < V_{ip}(*/t), V_{np(1)}, V_{np(2)} >$ **do**

3:      **Let** $< V_{np(1)}, V_{np(2)} > \in$ set of{full-lock on simple cycle $\subseteq \mu 0$ }

4:      Check :volunteer signal

5:      **if** ($\mu 0 \mid \bullet enp_{(1)} \cup enp_{(2)}') \bullet \neq \varnothing$)

6:        **then** Let volunteer signal (v)$:= V_{np(2)}$

7:      **endif**

8: **return** (volunteer signal (v))

9: **end**

Since the distributed-lock relation set of $<X^*/t, y^*, d^*>$ as illustrated in Fig. 4a then extract full-lock relation set of $<y^*, d^*>$ including the initial marking as illustrated in Fig. 4b. The propagation timing assumption is given that if initial state is accomplished, then the final state is also accomplished. This is referred that if y^* is accomplish signal, then d^* will be also accomplished. Therefore, y^* is a volunteer signal.

This algorithm is satisfied for multiple-distributed lock relation on STGs, if the multiple-cycle input signal is not redundant to another set of distributed-lock relation. The case of redundant primary-input signal set of distributed-lock relation; However, the case-study of SDI model can guarantee only the pair of non-primary input signal on which caused by the same input signal. Therefore, the designer have to decide on the guaranteed path selection. As illustrated in Fig. 5, the signal X^*/t is distributed-lock relation not only set of y^* and d^* but also the set of o^* and d^*. The guaranteed path can be either set of distributed-lock relation $<X^*/t, y^*, d^*>$ or set of distributed-lock relation $<X^*/t, o^*, d^*>$.

4.2 *Established Partial Transition Graph for Each Non-primary Signal on Distributed-Lock Relation*

Due to each non-primary input signal needs to be acknowledged explicitly; however, the volunteer signal is implied as successor acknowledged path. This is lead it

Fig. 5 The multiple distributed-lock relation

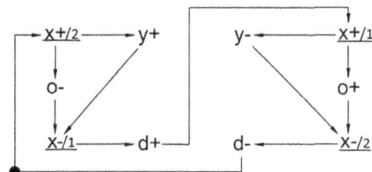

Fig. 6 The partial transition graph

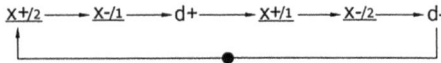

to be negligible in input signal, Therefore, As shown in Algorithm II, which is describe the established partial transition graph for each non-primary input on the distributed-lock relation which is eliminated the volunteer signal of whose others' input signal from the given STG. As given STG example as illustrated in Fig. 4a, the volunteer signal is y^*. Thus, y^* was eliminate of ds' input signal as illustrated in Fig. 6.

Algorithm II: Established Partial Transition Graph

Input   : The given STG (G)
Output : The partial STG for each non-primary input on distributed-lock relation $g(Vop(i))$
1:   **Begin**
2:   **do** The partial transition graph for each output signal: $\{g(V_{op(i)}) \ni \bigcup_{i=1}^{n} V_{ip(i)}\}$
3:      Let $g(v) \ni \{G\,(\bigcup_{i=1}^{n} V_{ip(i)}(*/t) \cup \bigcup_{i=1}^{n} V_{ip(i)} \cup \bigcup_{i=1}^{n} V_{np(i)})\}$
4:      Let $g(V_{np(i)}) \ni \{G\,(\bigcup_{i=1}^{n} V_{ip(i)}(*/t) \cup \bigcup_{i=1}^{n} V_{ip(i)} \cup (\bigcup_{i=1}^{n} V_{np(i)} - v))\}$
5:   **return** (the partial transition graph : $g(v), g(V_{np(1)})$)
6:   **End**

---

5   Experimental and Result

As was pointed out introduction to this research paper, our proposed method is significantly to improve area cost of asynchronous control circuit with multiple-cycle signal. In this section has demonstrated a through experimental result as providing STG which can be instantiated with optimization and practical asynchronous control circuits in SDI implementation as illustrate in Fig. 7a.

As illustrate in Fig. 7a the providing STG has a multiple-cycle input signal (Ai^*/t), non-primary input signal (Co^*, Xo^*) and multiple-cycle output signal (Bo^*/t). Thus, there is a set of distributed-lock relation $<Ai^*/t, Co^*, Xo^*>$, Then its membership is extracted it by representation on simple cycle as illustrate in Fig. 7b. As the distributed-lock relation is determined the volunteer signal as Co^* as illustrated in Fig. 7c.

For SDI implementation, the partial transition graph was established for Xo^* and Co^* circuit implementation as illustrated in Fig. 8a, b, respectively. The partial transition graph for Xo^* which signal Co^* was eliminated for the non-primary input signal of signal Xo^*.

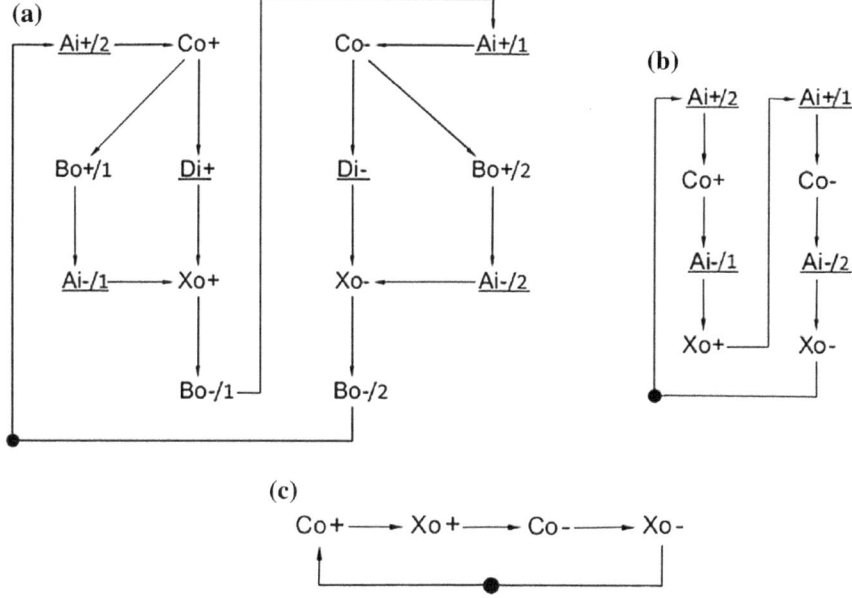

Fig. 7 As Example **a** the providing STG **b** distributed-lock relation **c** the volunteer signal on simple cycle

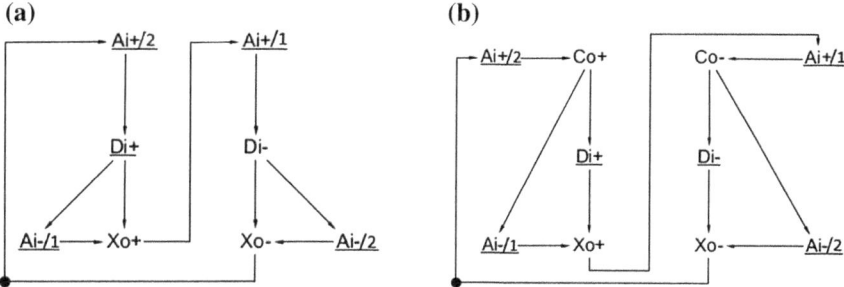

Fig. 8 The partial transition graph from STG on Fig. 7a, are **a** for Xo circuit implementation and **b** for Co circuit implementation

As the result of the providing STG, the circuit implementation based on QDI model and SDI model, which is synthesized by using S. Park's method as illustrated in Fig. 9a, and Fig. 9b, respectively. This is indicated that SDI circuit implementation is smaller than QDI circuit implementation.

Since, the effectiveness of the proposed methodology is evaluated by comparing as result cost area of circuit implementation, which is literal counts of non-primary input, number of conventional logic gate and number of c-element on which obtain 4 NAND gate. The table below illustrate the circuit implementation from the other

(a)   **(b)**

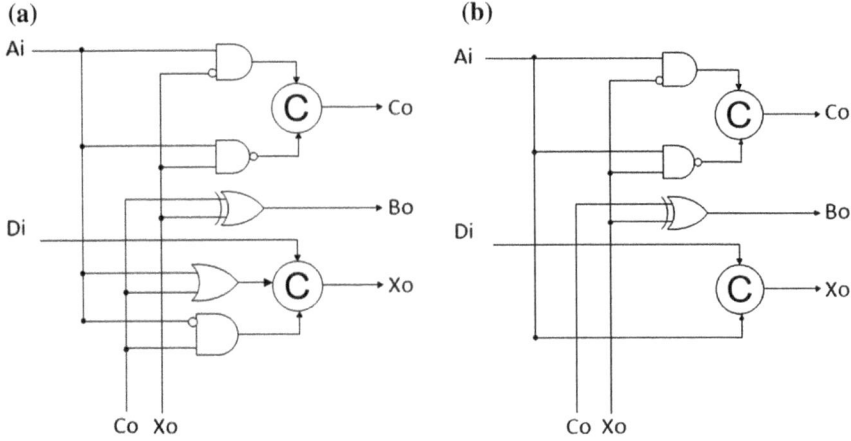

Fig. 9 As result of wholly signal of circuit implementation **a** is based on QDI and **b** SDI

Table 2 Comparing with QDI implementation

Example	The presented method		QDI implementation	
	#literal	#gate/c-element	#literal	#gate/c-element
ebergen Fig. 9	4	3/2	6	5/2
converta	10	6/3	12	7/3
wrdata	10	1/3	10	1/3
wrdatab	27	10/5	27	10/5

Table 3 Comparing with the other method

Example	The presented method		The other method [5]	
	#literal	#gate/c-element	#literal	#gate/c-element
ebergen Fig. 9	4	3/2	N/A	N/A
converta	10	6/3	N/A	N/A
wrdata	10	1/3	8	2/3
wrdatab	27	10/5	27	10/5

STG benchmarks which is compared with QDI model and the presented method as illustrated in Table 2.

As the Table 3 illustrate, there is a significant interesting result of circuit implementation between the other method and the presented method. There is indicated that the presented method is perform for handled STG with multiple-cycle signal.

This can summarize that the presented method improves the cost area of circuit implementation, where the STG obtain a distributed-lock relation.

6 Conclusion and Future Work

In this paper, we have presented the novel design of SDI optimization and implementation for multiple-cycle signals in STG. We also introduce distributed-lock relation which leverage on simplifying multiple-cycle signal on the STG. Our method determines the guaranteed path which is undertaken K value using distributed-lock relation as corresponding SDI model. In order to determine volunteer signal—the acknowledge successor relation in guaranteeing another completion signal. Then, this method is established to the partial transition graph for other non-primary input signal on distributed-lock relation, on which eliminated volunteer signal. As the result of this study shows that the improving area of circuit implementation is depended on distributed-lock relation of whose STG. In the future, we plan to establish to the element circuit for direct synthesized SDI circuit with single-cycle and multiple-cycle signal.

References

1. K. van Berkel, A. Peeters, F. Beest, Stretching Quasi delay insensitivity by means of extended isochronic forks, in *Proceedings of 2nd Working Conference Asynchronous Design Methodologies* (1995), pp. 99–106
2. K van Berkel, Beware the isochronic fork, Technical Report UR 003/91 (Philips Research Laboratories, Netherland, 1991)
3. T. Nanya, A. Takamura, M. Kuwako, M. Imai, M. Ozawa, M. Ozcan, R. Morizawa, H. Nakamura, Scalable-delay-insensitive design: a high-performance approach to dependable asynchronous systems (Invited paper), in *Proceedings of International Symposium on Future of Intellectual Integrated Electronics*, March 1999, pp. 531–540
4. P. Srikram, A. Thongtak, A design of asynchronous control circuits based on SDI model, in *Proceeding of The International MultiConference of Engineering and Computer Scientists 2014, IMECS 2014*. Lecture Notes in Engineering and Computer Science, Hong Kong. 12–14 March 2014, pp. 728–732
5. P. Srikram, A. Thongtak, Scalable-dalay-insensitive optimization based on lock relation, in *Proceeding of International Technical Conference on Circuits Systems, Computer and Communication, ITC-CSCC 2015*, Korea. 29 June–2 July 2015, pp. 786–789
6. S.B. Park, Synthesis of asynchronous VLSI circuits from signal transition graph specifications, Ph.D. Thesis (Tokyo Institute of Technology, Japan, 1996)
7. J. Sparso, S. Furber, *Principles of Asynchronous Circuit Design: A System Perspective* (Springer Publishing Company, 2010)
8. W. Lawsunnee, A. Thongtak, W. Vatanawood, Signal persistence checking of asynchronous system implementation using SPIN, in *Proceedings of the International MultiConference of Engineering and Computer Scientists, IMECS 2015*. Lecture Notes in Engineering and Computer, Hong Kong. 18–20 March 2015, pp. 604–609
9. K.J. Lin, J.W. Kuo, C.S. Lin, Direct synthesis of hazard-free asynchronous circuits from STGs based on lock relation and MG-decomposition approach, in *Proceedings of the European Conference on Design Automation, European Test Conference, The European Event in ASIC Design, EDAC-ETC-EUROASIC 1994*, Paris, France. 28 Feb–3 March 1994, pp. 178–183

10. P. Srikram, A. Thongtak, Distributed-lock relation for scalable-delay-optimization in multiple-cycle STG specifications, in *Proceeding of The International MultiConference of Engineering and Computer Scientists 2017, IMECS 2017*. Lecture Notes in Engineering and Computer Science, Hong Kong. 12–14 March 2014, pp. 672–677
11. M. Imai, M. Ozcan, T. Nanya, Evaluation of delay variation in asynchronous circuits based on the scalable-delay-insensitive model, in *Proceedings of the 10th International Symposium on Asynchronous Circuits and Systems, ASYNC'04*, Greece. 19–23 April 2004, pp. 62–71

A Neural Network Based Soft Sensors Scheme for Spark-Ignitions Engines

Yujia Zhai, Ka Lok Man, Sanghyuk Lee and Fei Xue

Abstract With the coming of massive application on autonomous vehicles, the safeness has been one of the features with highest development priority, which are considered in the design of automotive control systems. The development of intelligent sensors is an effective way to achieve this goal. For spark-ignition engines, the regualation of air fuel ratio and the control of engine speed are the keys to obtain reliable engine performance. This paper proposes a neural network (NN) based soft sensor scheme for air/fuel ratio sensor and crankshaft speed sensor, which are two important measurements for the control in spark-ignition engines. The modeling results show that satisfactory modeling performance can be obtained with moderate computational load.

Keywords Air fuel ratio · Automotive control systems · Engine performance
Intelligent sensors · Neural networks · Spark-ignition

1 Introduction

Advanced algorithms, 3D mapping, LiDAR (stands for light detection and ranging), and radar and camera sensors have been identified as the key technologies for autonomous vehicles [1–3]. This is because, to make appropriate maneuvers on

Y. Zhai (✉) · S. Lee · F. Xue
Department of Electrical and Electronic Engineering,
Xi'an Jiaotong-Liverpool University, 111 Ren Ai Road, Suzhou, China
e-mail: iemaeema@163.com

S. Lee
e-mail: sanghyuk.lee@xjtlu.edu.cn

F. Xue
e-mail: fei.xue@xjtlu.edu.cn

K. L. Man
Department of Computer Science, Xi'an Jiaotong-Liverpool University,
111 Ren Ai Road, Suzhou, China
e-mail: kalok2006@gmail.com

© Springer Nature Singapore Pte Ltd. 2018   199
S.-I. Ao et al. (eds.), *Transactions on Engineering Technologies*,
https://doi.org/10.1007/978-981-10-7488-2_15

road, autonomous vehicles should be equipped with control systems that are capable of analyzing sensory data and making correct decisions like a human driver. Consequently, the philosophy of control systems design in AV becomes different. Traditional vehicles work in a passive mode. It means that control systems consider drivers input as disturbance. Each electronic control unit (ECU) has its own objective to achieve. It is human drivers responsibility to make correct controls [4, 5]. For autonomous vehicles, ECUs gain full controls on all the actuators in systems. One of the advantages for this change is that the satisfactory engine performance in transient is easily obtained as the actuators for both air and fuel are controlled by ECUs. The speed control and efficient fuel injections could be easier to achieve as the air and fuel dynamics are is highly predictable. However, the challenging question is, without human interference, how an AV can coordinate all the control modules, such as path and speed planning, engine management, body dynamics, and etc., to ensure the maximum performance on effectiveness, efficiency, and stability.

In a typical fault tolerant control system, there are two parts—a fault detection and isolation tool (FDI) that detect the fault with the utilization of physical (additional hardware sensors) or analytical (soft sensors) redundancy. The data by fault analysis is delivered to a mechanism that can reconfigure the control data. The self-reconfigured controller can then maintain the specified control performance with faults. For the water management of fuel cell, Lebreton developed an active fault-tolerant control that is tested on a polymer electrolyte membrane fuel cell system. The artificial neural network based fault-tolerance system showed very promising results [6]. Providing redundancy in both sensor setup and data processing would be a common practice for safety reasons, especially when shifting functionality from a research level to series production level. Duel-sensor systems can be found in some of control application in automobiles, which provide sensor information redundancy on hardware-level. A comprehensive research study between active and passive approaches for fault-tolerant control/regulation systems can be found in Jiangs work [7]. Based on the understanding of engine dynamics, it is also possible to construct soft sensors for some critical application to ensure the robustness of control systems. For example, if any part in sensor systems goes wrong on road, it is highly desirable to have a backup plan on driving control, for the purpose of passengers safety.

In this paper, soft sensors for air fuel ratio and engine speed are realized by using neural network black-box model. The balance between performance and computational burden are considered for practical application. Section 2 introduces an benchmark engine model for spark-ignition (SI) engine; radial basis function neural network (RBFNN) models adopted in this research are explained in Sect. 3; simulation results of soft sensors are shown in Sect. 4; the conclusion is drawn in Sect. 5.

2 Mean Value Engine Model

In both industrial practice and scientific research, it has been more popular to use engine simulation models to make engine system analysis and design because it is much more economical than using a real engine test bed. The engine model adopted in this paper is referred to as the mean value engine model (MVEM) developed by Hendricks [8], which is a widely used benchmark for engine modeling and control. The three distinct subsystems of this model are the fuel injection, manifold filling and engine speed dynamics and those systems are modeled independently.

2.1 Manifold Filling Dynamics

The intake manifold filling dynamics are analysed from the viewpoint of the air mass conservation inside the intake manifold. It includes two nonlinear differential equations, one for the manifold pressure and the other for the manifold temperature. The manifold pressure is mainly a function of the air mass flow past throttle plate, the air mass flow into the intake port, the exhaust gas re-circulation (EGR) mass flow, the EGR temperature and the manifold temperature. It is described as

$$
\begin{aligned}
\dot{p}_i &= \frac{\kappa R}{V_i}(-\dot{m}_{ap}T_i + \dot{m}_{at}T_a + \dot{m}_{EGR}T_{EGR}) \\
&= f_p(\alpha, p_i, T_a T_i, n, m_{EGR}, T_{EGR})
\end{aligned}
\tag{1}
$$

The manifold temperature dynamics are described by the following differential equation

$$
\begin{aligned}
\dot{T}_i &= \frac{RT_i}{p_i V_i}[-\dot{m}_{ap}(\kappa - 1)T_i + \dot{m}_{at}(\kappa T_a - T_i) + \dot{m}_{EGR}(\kappa T_{EGR} - T_i)] \\
&= f_T(\alpha, p_i, T_a, T_i, n, m_{EGR}, T_{EGR})
\end{aligned}
\tag{2}
$$

The air mass flow past throttle plate $\dot{m}_{at}$ is related with the throttle position and the manifold pressure. The air mass flow into the intake port $\dot{m}_{ap}$ is represented by a well-known speed-density equation:

$$
\dot{m}_{at}(u, p_i) = m_{at1}\frac{p_a}{\sqrt{T_a}}\beta_1(u)\beta_2(p_r) + m_{at0}
\tag{3}
$$

$$
\dot{m}_{ap}(n, p_i) = \frac{V_d}{120RT_i}(\eta_i \cdot p_i)n
\tag{4}
$$

where

$$
\beta_1(u) = 1 - \cos(u) - \frac{u_0^2}{2!}
\tag{5}
$$

$$
\beta_2(p_r) = \begin{cases} \sqrt{1 - (\frac{p_r - p_c}{1 - p_c})^2} & if \quad p_r \geq p_c \\ 1 & if \quad p_r < p_c \end{cases}
\tag{6}
$$

$$p_r = \frac{p_i}{p_a} \tag{7}$$

and m_{at0}, m_{at1}, u_0, p_c are constants. Additionally, instead of directly model the volumetric efficiency η_i, it is easier to generate the quantity $\eta_i \cdot p_i$ which is called normalised air charge. The normalised air charge can be obtained by the steady state engine test and is approximated with the polynomial Eq. (8)

$$\eta_i \cdot p_i = s_i(n)p_i + y_i(n) \tag{8}$$

where $s_i(n)$ and $y_i(n)$ are positive, weak functions of the crankshaft speed and $y_i \ll s_i$.

2.2 Crankshaft Speed Dynamics

The crankshaft speed is derived based on the conservation of the rotational energy on the crankshaft. Its state equation can be written as

$$\dot{n} = -\frac{1}{In}(P_f(p_i, n) + P_p(p_i, n) + P_b(n)) + \frac{1}{In}H_u\eta_i(p_i, n, \lambda)\dot{m}_f(t - \Delta\tau_d) \tag{9}$$

$$= f_n(p_i, n, m_f, \theta, \lambda)$$

Both the friction power P_f and the pumping power P_p are related with the manifold pressure p_i and the crankshaft speed n. The load power P_b is a function of the crankshaft speed n only. The indicated efficiency η_i is a function of the manifold pressure p_i, the crankshaft speed n and the air fuel ratio λ.

2.3 Fuel Injection Dynamics

It has been found that the fuel jet from the injector can be characterised into two portions. One portion mixes with the air stream and enters the cylinder directly; the other portion deposits as fuel film on the surfaces of the intake system components, and mixes with the air stream through the reentrainment/evaporation process during subsequent engine cycles. This is known as wall-wetting.

According to Hendrick's identification experiments with SI engine, the fuel flow dynamics could be described as following equations [8]

$$\ddot{m}_{ff} = \frac{1}{\tau_f}(-\dot{m}_{ff} + X_f\dot{m}_{fi}) \tag{10}$$

$$\dot{m}_{fv} = (1 - X_f)\dot{m}_{fi} \tag{11}$$

$$\dot{m}_f = \dot{m}_{fv} + \dot{m}_{ff} \tag{12}$$

where the model is based on keeping track of the fuel mass flow. The parameters in the model are the time constant for fuel evaporation, τ_f, and the proportion of the fuel which is deposited on the intake manifold or close to the intake valves, X_f. These parameters are operating point dependent and thus the model is nonlinear in spite of its linear form. The MVEM provided by Elbert Hendrick has been validated using the real time data acquired from the engine test bed that equipped with Ford 1.6 L engine. The parameters for this model could be approximately expressed in the terms of the states of the model as

$$\tau_f(p_i, n) = 1.35(-0.672n + 1.68)(p_i - 0.825)^2 + (-0.06n + 0.15) + 0.56 \tag{13}$$

$$X_f(p_i, n) = -0.277p_i - 0.055n + 0.68 \tag{14}$$

2.4 MVEM Under AFR Measurement Delay

The AFR could be calculated using Eq. (15)

$$\lambda = \frac{\dot{m}_{ap}}{\dot{m}_f} \tag{15}$$

Nowadays, in the practical application of automotive industry, oxygen sensors are used in the fuel injection system. They determine if the air fuel ratio exiting a gas-combustion engine is rich (with unburnt fuel vapour) or lean (with excess oxygen), then, a closed-loop feedback controller,usually a proportional-integral(PI) controller, adjusts fuel injection rate m_{fi} according to real-time sensor data rather than operating with a open-loop fuel map. Therefore, the time delay of injection systems should also be considered. Manzie's research [9, 10] has shown there are three causes of time delay for injection systems: the two engine cycle delay between the injection fuel and the expulsion from the exhaust valves, the propagation delay for the exhaust gases to reach the oxygen sensor and the sensor output delay. It has been found that the engine speed has more influence on these delays than the manifold pressure. Therefore, the following equation can be used to represent the delays of injection systems.

$$t_d = 0.045 + \frac{10\pi}{n} \tag{16}$$

The time delay on air fuel ratio measurement has not been considered in original MVEM. A module used for air fuel ratio measurement is added into original MVEM for the research purpose of AFR control, which is based on Eq. (16). The expanded MVEM is shown in Fig. 1.

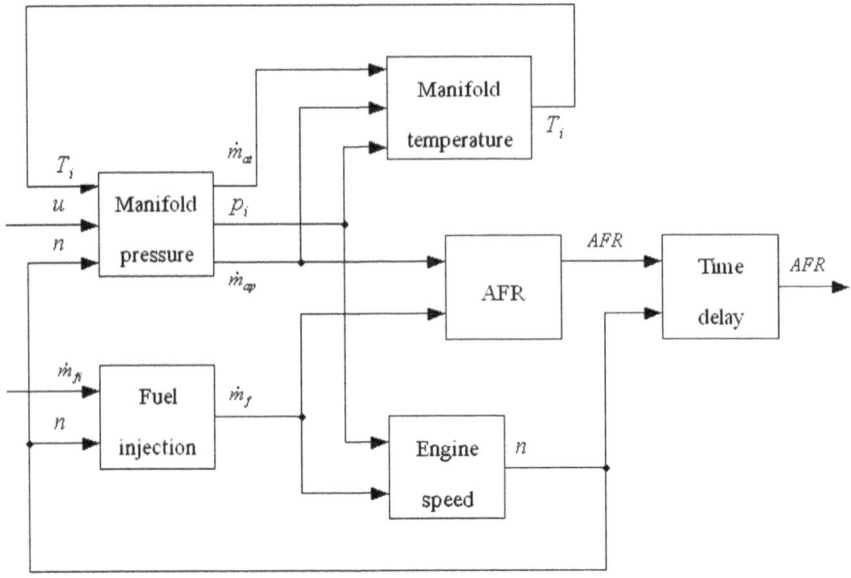

Fig. 1 Mean value engine model with AFR measurement delay

3 NN Based Engine Modeling

The radial basis function neural network (RBFNN) consists of three layers: input layer, hidden layer and output layer, as shown in Fig. 2, where $x = [x_1, x_2, \ldots, x_n]^T \in \mathscr{R}^n$ is the input vector, $h = [h_1, h_2, \ldots, h_q]^T \in \mathscr{R}^q$ is the hidden layer output vector, $W(k) \in \mathscr{R}^{p \times q}$ is the weight matrix with entry w_{ij}, which is the weight linking the jth node in the hidden layer to the ith node in the output layer, and $\hat{y} = [\hat{y}_1, \hat{y}_2, \ldots, \hat{y}_p] \in \mathscr{R}^p$ is the output vector of the RBFNN.

In mathematical terms, we have the following equations to describe the RBFNN.

$$\hat{y}(k) = W \cdot h(k) \tag{17}$$

$$h(k) = f[z(k)] \tag{18}$$

$$z_i(k) = \sqrt{[x(k) - c_i]^T [x(k) - c_i]} \tag{19}$$
$$= \| x(k) - c_i \|$$

where $i = 1, 2, \ldots, q$. $c_i \in R^n$ is the ith centre in the input space, and $f[.]$ is the non-linear activation function in hidden layer. The gaussian basis function given by

$$f[z(k), \sigma] = e^{\frac{-z^2(k)}{\sigma^2}} \tag{20}$$

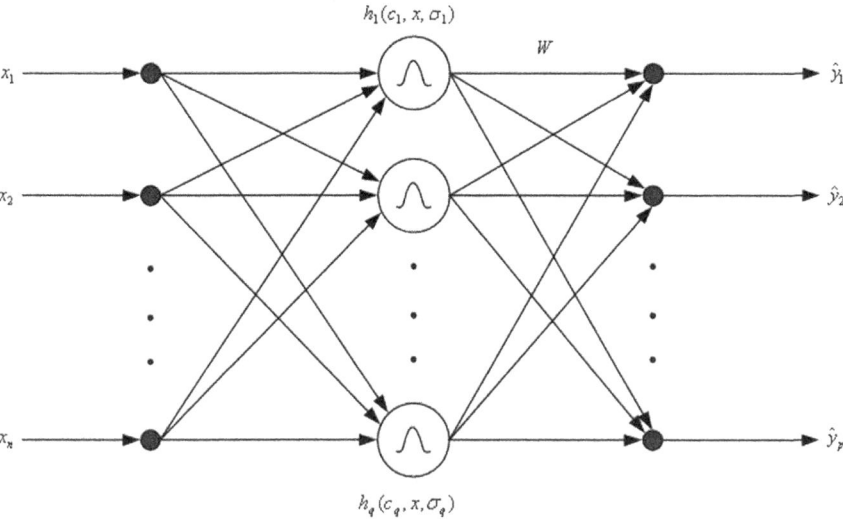

Fig. 2 RBFNN structure

is chosen in this research, where σ is a positive scalar called width, which is a distance scaling parameter to determine over what distance in the input space the unit will have a significant output.

The radial basis function neural network models are used in this research to predict system outputs. The procedure of RBFNN modelling and prediction is to determine network inputs according to system dynamics; data collection and scaling; network training and validation; using the network to do prediction. The network training includes determining the number of centres, q, appropriate centres and widths, c_i and σ_i, $i = 1, \ldots, q$ from the training data set; obtaining the weights W by training data, and validating the network by the test data [11].

3.1 K-Means Algorithm

The K-means clustering algorithm is used to choose the centres of RBFNN from a set of training data in this research. Its objective is to minimise the sum squared distances from each input data to its closest centre so that the data is adequately covered by the activation functions $f[.]$.

Procedure:

Step 1: Choose q initial cluster centres $c_1(1), c_2(1), \ldots, c_q(1)$.
Step 2: At the kth iteration step, distribute the sample $\{x\}$ into $S_j(k)$ among the q cluster domains. $S_j(k)$ denotes the set of samples whose cluster is $c_j(k)$

$$x \in S_j(k) \qquad if \; \| x - c_j(k) \| < \| x - c_i(k) \| \qquad (21)$$

where $j = 1, 2, \ldots, q$ and $i = 1, 2, \ldots, j - 1, j + 1, q$.

Step 3: Update the cluster centres.

$$c_j(k + 1) = \frac{1}{N_j} \sum_{j}^{N_j} S_j(k) \qquad (22)$$

where N_j is the number of elements in $S_j(k)$.

Step 4: Repeat *Step 2* to *Step 3* until $c_j(k + 1) = c_j(k)$.

3.2 *ρ-Neighbourhood Method*

The RBFNN width σ of each unit is computed by the ρ-Nearest neighbours method. The guideline is that the excitation of each node should overlap with some other nodes(usually closest) so that a smooth interpolation surface between nodes is obtained. To achieve this each hidden node must activate at least one other hidden node to a significant degree. Therefore, the width is selected so that σ is greater than the distance to the nearest unit centre.

$$\sigma_i = \left[\frac{1}{\rho} \sum_{j=1}^{\rho} \| c_i - c_j \|^2 \right]^{\frac{1}{2}} \qquad (23)$$

where $i = 1, 2, \ldots, q, c_j$ is the ρ-nearest neighbours of c_i. For nonlinear function approximation ρ depends on the problem and requires to be experimented.

3.3 *Recursive Least Squares Algorithm*

Recursive least squares (RLS) algorithm is a recursive form of the Least Squares (LS) algorithm. It evaluates for each new sample the parameter matrix W newly. The basic idea of RLS algorithm is to compute the new parameter estimate $W(k)$ at discrete time steps k by adding some correction information to the previous parameter estimate $W(k - 1)$ at time instant $k - 1$. It is used to find the RBFNN weights W, which can be summarised as follows [11]:

$$y_p(k) = y_c(k) - W(k - 1)h(k) \qquad (24)$$

$$g_z(k) = \frac{P_z(k-1)h(k)}{\mu + h^T(k)P_z(k-1)h(k)} \tag{25}$$

$$P_z(k) = \mu^{-1}[P_z(k-1) - g_z(k)h^T(k)P_z(k-1)] \tag{26}$$

$$W(k) = W(k-1) + g_z(k)y_p(k) \tag{27}$$

where $W(k)$ and $h(k)$ represent the RBFNN weights and activation function outputs at iteration k, $y_c(k)$ is the process output vector, P_z and g_z are middle terms. μ here is called *forgetting factor* ranging from 0 to 1 and is chosen to be 1 for off-line training. The parameters g_z, w and P_z are updated orderly for each sample with the change of the activation function output $h(k)$.

4 Soft Sensor for Air Fuel Ratio and Engine Speed

In this research, the SI engine model introduced in Sect. 2 is used for engine data collection. The sampling time is chosen as 0.01 s, which is the same value used in control systems in automotive application. There are totally 5000 samples that are collected. The first 4000 samples are used for training and the left 1000 samples for validation. In this section, the design of the structure of soft sensors for both AFR and speed are explained, and the performance are shown based on simulations.

The wideband O2 sensor, also called wide-range air fuel (WRAF) sensor, is widely equipped in modern automotive vehicles to replace the traditional Zirconia oxygen sensor that produces only binary sequence of air fuel ratio. Engine control management system can control the air/fuel mixture at stoichiometric ratio inside the combustion chamber, using the measured signal by WRAF. Since sensors are usually located in exhaust stream, a certain time-delay on the AFR measurement can not be avoided. Due to the harsh working condition and aging effect, the measured AFR can be biased by control circuits of WRAF sensor. Speed sensor is another important component for the fuel injection system. The over-reading and under-reading on engine speed value can significantly influence the performance of SI engines. It has been reported in many practical applications that the quality control measures can be obtained by using of soft sensor and the stringent requirements imposed on hardware-based sensors can be reduced significantly.

Following this idea, a soft sensor for both AFR and engine speed is constructed by using RBFNN model in this research. After studying on SI engine dynamics, the dynamics of AFR can be represented by the following equation:

$$[\hat{\lambda}, \hat{n}] = g(P, T, \theta, \dot{m}_{fi}) \tag{28}$$

Here, g is a nonlinear function by RBFNN, which is used to mapping the input and output data of SI engines. Therefore, the RBFNN based soft sensor for AFR and

Fig. 3 The structure of
RBFNN soft sensor for AFR
and engine speed

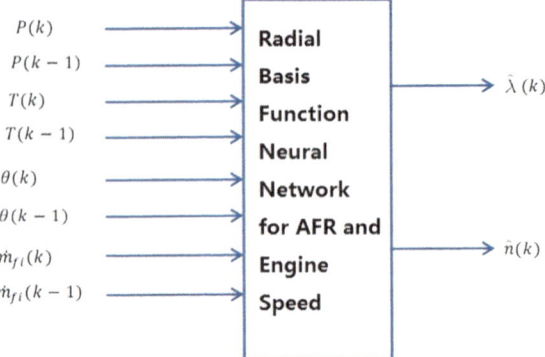

engine speed can be realized using the measured variables, such as throttle angle θ, fuel injection rate $\dot{m}_{fi}$, intake manifold pressure P, intake manifold temperature T. Then, the AFR in combustion chamber and engine speed can be inferred as accordingly. Considering the nonlinearity and time-delay in engine dynamics, a second order structure of RBFNN is chosen to construct such soft sensor as shown in Eq. 29:

$$[\hat{\lambda}, \hat{n}] = g[P(k), T(k), \theta(k), \dot{m}_{fi}(k), P(k-1), T(k-1), \theta(k-1), \dot{m}_{fi}(k-1)] \quad (29)$$

Its structure shown in Fig. 3. All data for the inputs and outputs has been scaled from 0 to 1 using linear scaling equation.

The performance of the soft sensor is shown in Figs. 4 and 5 The mean absolute error (MAE) of the shown 500 samples is 0.008 and 0.036, respectively. The satisfactory performance can be achieved using the developed soft sensor scheme for

Fig. 4 The performance of
AFR soft sensor

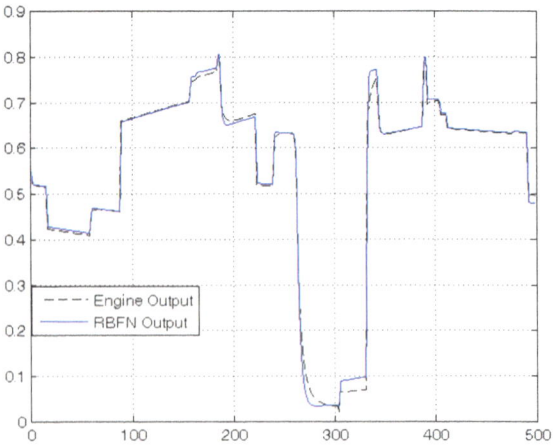

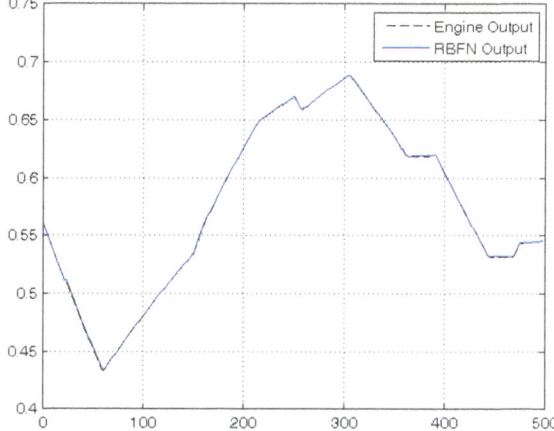

Fig. 5 The performance of engine speed soft sensor

both AFR and engine speed. The information redundancy provided by the soft sensor can be used as a key component for the fault-tolerant module in engine control system, which is critically important for autonomous vehicles in future.

5 Conclusion

- The soft sensor for air/fuel ratio and engine speed, which is based on neural network model, has been developed in this research. Redundancy using different principles for sensing and processing is important in modern automotive control systems.
- The modeling performance shows that it could provide reliable information, and such soft sensors can work together with hardware sensors for engine control to achieve the high robustness that is essential for future autonomous vehicles operations.
- Given the computational power of current ECUs, the developed soft sensor utilizes RBFNN with 30 nodes, which is a reasonable size for practical application. It might also be useful for on-board fault diagnosis systems.

Acknowledgements This research was financially supported by the Centre for Smart Grid and Information Convergence (CeSGIC) at Xian Jiaotong-Liverpool University. The authors would like to thank all the parties concerned.

References

1. Y. Zhai, K.L. Man, S. Lee, F. Xue, A neural network based soft sensor for air fuel ratio dynamics in SI engines lecture notes in engineering and computer science, in *Proceedings of the International MultiConference of Engineers and Computer Scientists*, (Hong Kong, 2017), 15–17 March 2017, pp. 719–722
2. Institution of mechanical engineers, ford reveals autonomous taxi plan, http://www.imeche.org/news/news-article/ford-reveals-autonomous-taxi-plan
3. AutoblogGreen, Tesla D is, as expected, an AWD Model S but new autopilot features surprise, http://www.autoblog.com/2014/10/09/tesla-d-awd-model-s-new-autopilot-surprise/
4. W. Zhu, J. Miao, J. Hu, L. Qing, Vehicle detection in driving simulation using extreme learning machine. Neurocomput. **128**, 160–165 (2014), https://doi.org/10.1016/j.neucom.2013.05.052
5. P. Reiner, B.M. Wilamowski, Efficient incremental construction of RBF networks using quasi-gradient method, in *Special Issue on Information Processing and Machine Learning for Applications of Engineering*, Vol. 150, Part B, 20 Feb 2015, pp. 349–356
6. C. Lebreton, M. Benne, C. Damour, N. Yousfi-Steiner, B. Grondin-Perez, D, Hissel, J.-P. Chabriat, Fault tolerant control strategy applied to pemfc water management. Int. J. Hydrogen Energy **40**, 10636–10646 (2015)
7. J. Jiang, X. Yu, Fault-tolerant control systems: a comparative study between active and passive approaches. Ann. Rev. Control **36**, 60–72 (2012)
8. E. Hendricks, D. Engler, M.A Fam, Generic mean value engine model for spark ignition engines, in *Proceedings of 41st Simulation Conference*, (DTU Lyngby, Denmark, SIMS, 2000)
9. C. Manzie, M. Palaniswami, H. Watson, Gaussian networks for fuel injection control. Proc. Inst. Mech. Eng. J. Automobile Eng. **215**(10), 1053–1068 (2001)
10. C. Manzie, M. Palaniswami, D. Ralph, H. Watson, X. Yi, Model predictive control of a fuel injection system with a radial basis function network observer. J. Dyn. Syst. Measur. Control Trans. ASME **124**(4), 648–658 (2002)
11. O. Nelles, *Nonlinear System Identification* (Springer, 2001)

A Tourist Spot Search System Based on Paragraph Vector Model of Location and Category Tags Using User Reviews

Daisuke Kitayama, Tomofumi Yoshida, Shinsuke Nakajima and Kazutoshi Sumiya

Abstract Tourist spots have certain functions, such as "spot suitable for seeing the night view" or "spot suitable for meeting point". In this paper, we propose a method for searching tourist spots that uses the similarity of their functional features. In our method, we extract distributed representations of a tourist spot as a paragraph vector of user reviews for the tourist spot. We extract distributed representations of location and category in the same way. Next, we extract the functional feature of a tourist spot by combining distributed representations of the tourist spot, locations, and categories. Finally, we search tourist spots using the extracted functional feature and distributed representations of other locations and categories. In this phase, the most important thing is the way in which locations and categories are represented. We can employ an extraction method using a paragraph vector for user reviews based on a location or a category (the conventional method). On the other hand, we can use the average vector of distributed representations of all spots in a location or a category. In this paper, we compared our method (average vector) with the conventional method (paragraph vector). By this experiment, we evaluated the effectiveness of our method for searching tourist spots.

Keywords Geographical information retrieval · Location based social network
Paragraph vector model · Text processing · Tourist information · User reviews

D. Kitayama (✉) · T. Yoshida
Kogakuin University, Tokyo, Japan
e-mail: kitayama@cc.kogakuin.ac.jp

T. Yoshida
e-mail: yoshidat@datalinks.co.jp

T. Yoshida
Datalinks Corporation, New York, USA

S. Nakajima
Kyoto Sangyo University, Kyoto, Japan
e-mail: nakajima@cc.kyoto-su.ac.jp

K. Sumiya
Kwansei Gakuin University, Hyōgo, Japan
e-mail: sumiya@kwansei.jp

© Springer Nature Singapore Pte Ltd. 2018   211
S.-I. Ao et al. (eds.), *Transactions on Engineering Technologies*,
https://doi.org/10.1007/978-981-10-7488-2_16

1   Introduction

In recent years, tourist information Web sites such as TripAdvisor [1] and Jalan [2] have come to be widely used. Tourists can post user reviews of a tourist spot on these Web sites. Therefore, there is much tourist information and many user reviews on the Web. In general, tourist spots have metadata of location tags and category tags on tourist information Web sites. Figure 1 shows metadata of a spot "Hachiko". This spot has "Shibya" as the location tag and "Monuments and Statues" as the category tag. And then, it has some user reviews. Users can search tourist spots using keywords and these tags to narrow down the list of spots. When wishing to see a historic spot in Kyoto, Japan, most users select the "Kyoto" tag and the "Shrine and Temple" tag.

However, not all tourist requirements are represented by the metadata given to the spot. In other words, users want to search using features other than location and category. When a user is planning a trip with the theme of night viewing, "Tokyo Tower" and "Mt. Hakodate" are suitable spots. "Tokyo Tower" is a communications and observation tower in Tokyo, Japan, and it is one of the most famous night viewing spots in Japan. "Mt. Hakodate" is a mountain in Hokkaido, Japan; it too is one of the most famous night viewing spots in Japan. These spots have the same function, "spot suitable for seeing a night view," although the location tags and category tags of these spots differ.

We call this function of a spot its functional feature. The functional feature does not depend on location and category. For example, shrines and temples are generally in quiet places. However, the shrines and temples category does not indicate the function "he/she can be quiet and calm". In fact, shrines and temples are also present in downtown areas. We show concrete example of functional feature using Fig. 2. The "Hachiko" is the famous meeting point in Japan. In this time, "meeting point" is a functional feature and it is not depend on location tag "Shibuya" and category tag "Monuments and Statues". Therefore, the conventional search method using location tags and category tags is insufficient for narrowing the search by functional feature.

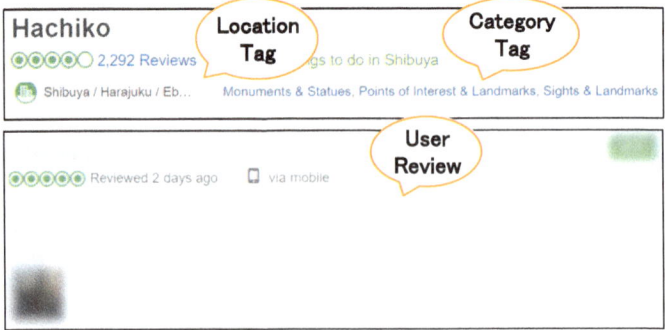

Fig. 1   Metadata of tourist information

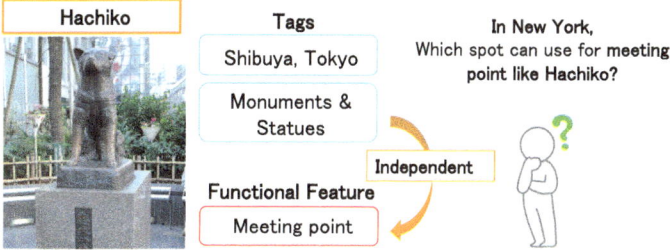

Fig. 2 Concept of functional feature

In the above way, we focus on the functional features of tourist spots, which are not extracted from the given metadata and tags [3]. We propose a method for searching tourist spots based on the similarity of their functional features. By this method, a user can search for a suitable spot simply by selecting "well-known" as the spot's functional feature even if the user wants to find spots in an area unfamiliar to the user.

In this paper, we use the paragraph vector model, an efficient method for vectorizing sentences and documents, on user reviews of each tourist spot to generate distributed representation (paragraph vector) of each tourist spot. We extract distributed representations of location tag and category tag in the same way. Then, we generate a query vector that shows the functional feature of a tourist spot by following steps. First, we subtract the location vector and category vector from the target spot vector; this generated vector is considered to represent the functional feature of the spot. Next, the user selects the search location and category. Finally, we add the location vector and the category vector selected by the user to the vector generated in the first step.

In this way, we can generate a query vector for searching for spots that have the same functional feature but in other locations and categories. For example, if a user would like to find a spot in New York having the same functional feature "point suitable for meeting" as "Hachiko," a famous statue of a dog in Tokyo, our method generates a query vector such as *Hachiko − Tokyo + New York*, and then calculates the cosine similarity between the query vector and the paragraph vector for each tourist spot. In an ideal retrieval result, our method will retrieve "Grand Central Terminal Clock," a famous meeting point in New York (See Fig. 3). Figure 4 shows an overview of our proposed search method.

The remainder of this paper is organized as follows. Use of the paragraph vector model for tourist spots and its problems are presented in Sect. 2. Section 3 describes the proposed method and an evaluation experiment. Related work is mentioned in Sect. 4. We conclude the paper in Sect. 5.

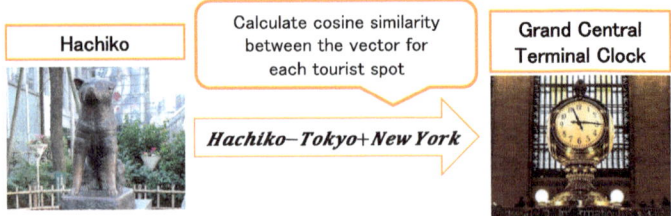

Fig. 3 A tourist spot search system based on paragraph vector model of location and category tags

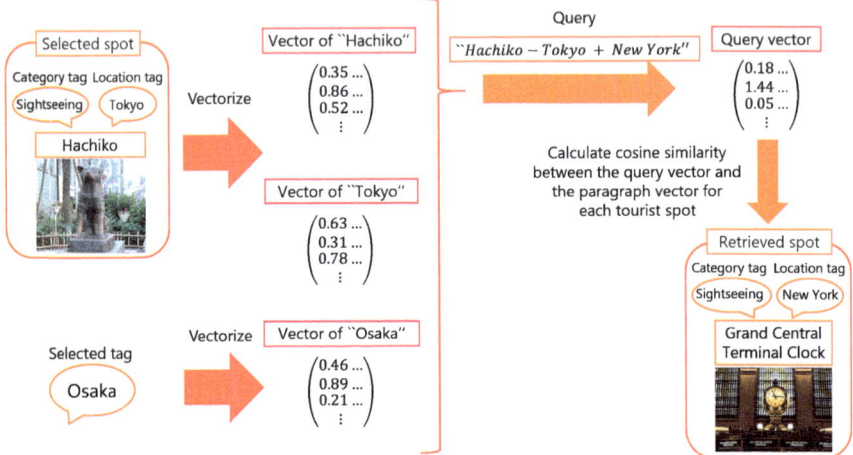

Fig. 4 Overview of our proposed search method

2 Use of Paragraph Vector Model for Tourist Spots and Its Problems

2.1 Paragraph Vector for Tourist Spot

The paragraph vector model was proposed by Le and Mikolov [4]. The idea for this model was based on the study by Mikolov et al. [5], known as word2vec [6]. Based on the distributional hypothesis given by Harris in 1954 [7], the paragraph vector model learns fixed-length feature representations, called paragraph vectors, from variable-length texts such as sentences, paragraphs, and documents by using a neural network. In this paper, we denote the minimum unit of learning data for the paragraph vector model as a "document". Thus, the paragraph vector model learns the distributed representations of each document.

Prior to evaluating our proposed method, we first analyzed how the paragraph vector for each tourist spot represents the feature of the corresponding tourist spot.

Table 1 Jalan dataset

Number of spots	43,759
Number of reviews	1,481,831
Types of location tag	Prefecture, Area, City, Town
Types of category tag	Parent category, Child category

Table 2 Package information and learning parameters

Version of Python	3.5.2
Version of gensim	0.12.4
Learning model	DBOW
Window size	8
Number of dimensions	300

We hypothesized that the paragraph vector for a given tourist spot represents the following three features of that tourist spot: location feature, category feature, and functional feature. Thus, we examined whether the similarity between the paragraph vectors for tourist spots that have similar location features or category features increased regardless of the similarity of their functional features by preliminary experiment 1, described in Sect. 2.2. Secondly, we analyzed the paragraph vector for the location tag and category tag by preliminary experiment 2, described in Sect. 2.3. On the basis of the results of the two preliminary experiments, we propose a method for calculating the paragraph vector for the location tag and category tag by using the average vector of the paragraph vectors for all spots that have the same location tag or the same category tag.

In addition, we describe the setup for our preliminary experiments. Table 1 shows the dataset for the preliminary experiments, which was extracted from Jalan, one of the leading tourist information Web sites in Japan. Each tourist spot has four location tags, two category tags, and user reviews. We used these user reviews as training data for the paragraph vector model. Furthermore, to generate the paragraph vectors, we used the implementation of the paragraph vector model by gensim [8], an open-source natural language processing toolkit implemented in the Python language. Table 2 shows the environment and parameter settings for gensim. Values of parameters that are not mentioned in Table 2 follow gensim's default settings [9].

2.2 Preliminary Experiment 1: Analyzing the Paragraph Vectors for Tourist Spots

As the first preliminary experiment, we analyzed how a paragraph vector represents the features of a tourist spot. In this experiment, we treated all user reviews for

Table 3 Tourist spots with a high similarity to "Statue of Hachiko"

Name of spot	Similarity	Location 1	Location 2	Location 3	Category
Statue of Hachiko	–	Tokyo	Shibuya area	Shibuya-ward	Sightseeing
Moyai Statue	0.7825	Tokyo	Shibuya area	Shibuya-ward	Sightseeing
Hachiko Family Relief	0.6197	Tokyo	Shibuya area	Shibuya-ward	Others
Saigo Takamori Statue	0.5719	Tokyo	Asakusa area	Taito-ward	Sightseeing
Umeda BIGMAN	0.4992	Osaka	Umeda area	Osaka city	Sightseeing
Statue of Hachiko (Odate station)	0.4675	Akita	Towada area	Odate city	Sightseeing

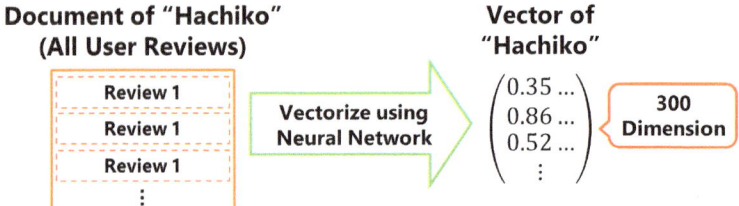

Fig. 5 The paragraph vector of the spot

one tourist spot as one document and generated the paragraph vector for each spot. Figure 5 shows the paragraph vector of "Hachiko". Table 3 shows the five tourist spots having the highest cosine similarity between their respective paragraph vectors and the paragraph vector for "Hachiko".

"Hachiko" is one of the most famous meeting points in Japan. This functional feature is hard to guess from its location tags, "Tokyo" and "Shibuya," and the category tags, "Sightseeing" and "Historic Sites". In this table, several famous meeting points in Japan such as "Moyai Statue," "Saigo Takamori Statue," and "Umeda BIGMAN" scores a high cosine similarity to "Hachiko". Many user reviews of these spots mention the reviewers' meeting experiences; therefore, the paragraph vector model generated distributed representations of these spots that contained the functional feature "point suitable for meeting" from the user review contexts.

However, many user reviews also mention the location feature and category feature of the tourist spot; thus, the cosine similarity between the paragraph vector for "Hachiko" and the paragraph vector for other spots also increases when their location features or category features are similar to each other. For example, "Moyai Statue," as well as "NANAKO Statue" and "QFRONT" (which, although not displayed in Table 3, came in sixth and ninth place, respectively, in cosine similarity with "Hachiko"), are located near the Shibuya Station, which is also located near "Hachiko". Moreover, many of the category tags in Table 3 are "Sightseeing" or "Historic Sites," the same category tags as for "Hachiko". On the basis of the above, we confirm our first hypothesis that the similarity between paragraph vectors for tourist spots that have similar location features or category features increases regardless of the similarity of their functional features.

2.3 Preliminary Experiment 2: Analyzing the Paragraph Vectors for Location Tags and Category Tags

As the second preliminary experiment, we analyzed the paragraph vectors for location tags and category tags. In this experiment, we treated all user reviews of all tourist spots having the same tag as one document for each tag and generated the paragraph vector for each tag. For this, we treated "location 3" as the location tag and "category 1" as the category tag for each spot. For example, when we generated the paragraph vector for "Kyoto city," we treated all reviews of spots whose location tag was "Kyoto city" as a "Kyoto city" document and generated a distributed representation of "Kyoto city" using the paragraph vector model. As an exception, we treated special wards of Tokyo such as "Shibuya-ku" and "Chiyoda-ku" as one location tag, "Tokyo special wards".

We analyzed whether these paragraph vectors for location tags and category tags are efficient for generating query vectors. Table 4 shows the five tourist spots having the highest cosine similarity between their respective paragraph vectors and the query vector *Hachiko − Tokyo special wards + Osaka city*. In this table, a famous amusement park in Osaka city, "Universal Studios Japan," scores a high cosine similarity to the query vector, although an amusement park is not a typical spot for a meeting point. The main reason for this result is that the cosine similarity between the paragraph vector for "Osaka city" and that for "Universal Studios Japan" is 0.7893, the highest similarity to "Osaka city" of any of the spots listed. Therefore, the cosine

Table 4 Tourist spots with a high similarity to the query vector "Hachiko − Tokyo special wards + Osaka city"

Name of spot	Similarity	Location	Category
Moyai Statue	0.5192	Tokyo special wards	Sightseeing
Universal Studios Japan	0.5047	Osaka city	Amusement
Hachiko Family Relief	0.3680	Tokyo special wards	Others
Umeda BIGMAN	0.3530	Osaka city	Sightseeing
Saigo Takamori Statue	0.3487	Tokyo special wards	Sightseeing

similarity between the query vector and "Osaka city" has increased. On the basis of the above, we confirm that treating all user reviews of all tourist spots having the same tag as one document does not generate an efficient query vector.

3 Experiment

We compared extracted tourist spots using query vectors generated by the proposed method and by some simple methods, and we examined the characteristics of each method. We generated paragraph vectors in the same way as for preliminary experiment 1 (Sect. 2.2). We used the following three methods for generating query vectors.

Filter method: We extracted tourist spots that have high cosine similarity to the input spot and added the same location/category tags.

Direct method: We generated a paragraph vector for location/category tags using a set of reviews for a spot having the exact same target location/category tags. This is the same method used in preliminary experiment 2 (Sect. 2.3).

Proposed method: We generated a paragraph vector for location/category tags by calculating the average of the paragraph vectors for spots that have the target location/category tags. This method is explained in Sect. 2.1.

Fig. 6 shows overview of each method. We used two queries, the first one to assess the effectiveness of the location tag and the second one to assess the effectiveness of the category tag:

- *Shinjuku Gyoen National Garden − Tokyo special wards + Osaka city*
- *Meiji Shrine − Shrine + Sightseeing*

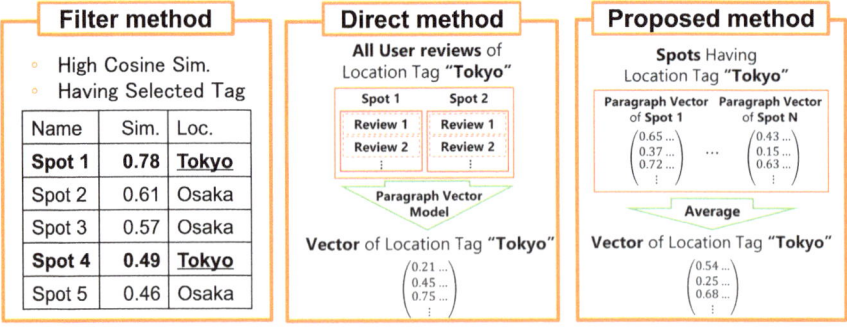

Fig. 6 Overview of each method

3.1 Query 1: Shinjuku Gyoen National Garden − Tokyo Special Wards + Osaka City

We used *Shinjuku Gyoen National Garden − Tokyo special wards + Osaka city* as the generated query vector based on location tags. We call this Q_1. "Shinjuku Gyoen National Garden" has "Tokyo special wards" as its location tag and "Sightseeing" as its category tag. With this query, we expect that a nature-rich park in Osaka city will be extracted. Table 5 shows the five spots having the highest cosine similarity to "Shinjuku Gyoen National Garden". Tables 6, 7, and 8 show the top five spots in the results based on the respective methods.

First, in examining Tables 7 and 8, we observe that nature-rich parks around Osaka such as "Osaka Castle Nishinomaru Garden" and "Kyoto Prefectural Botanical Garden" in the results for the proposed method are ranked more highly than the results for the direct method. Therefore, we consider that the proposed method generates a paragraph vector for location tags that is more suitable than that of the direct method. However, nature parks around Tokyo such as "Cherry Blossoms at Shinjuku Gyoen National Garden" and "Imperial Palace East Gardens" remain in the results for the proposed method. The norm of the paragraph vector for location tags generated by the proposed method tends to be small because it is averaged (we can say that the same is true for category tags). Therefore, we expected that the contribution of the generated paragraph vector for location/category tags to the query vector would be small. In almost all cases, the results for the proposed method are a fine adjustment of the results according to similarity (see Tables 5 and 8). We need to develop a suitable method to amplify the generated paragraph vector for location/category tags.

Second, the filter method was able to extract some nature-rich parks in Osaka such as "Nagai Botanical Garden" and "Nakanoshima rose garden," which are similar to "Shinjuku Gyoen National Garden" (see Table 6). These spots were not able to be extracted by the other methods. The similarity of the paragraph vector for the spot itself includes the similarity of its functional feature. As a result, the filter method with location tags went very well (see Table 6). When a user can select location tags appropriately, we should use the filter method. On the other hand, when a user wants to search for spots on or around selected location tags, our method can extract tourist spots at selected location tags and around the selected location.

3.2 Query 2: Meiji Shrine − Shrine + Sightseeing

We used *Meiji Shrine − Shrine + Sightseeing* as the generated query vector based on category tags. We call this Q_2. Meiji Shrine has Tokyo special wards as its location tag and Shrine as its category tag. With this query, we expect that historical spots with tourist facilities will be extracted rather than a place of worship.

Table 5 Top five spots based on cosine similarity to "Shinjuku Gyoen National Garden"

Name of spot	Similarity	Location	Category
Cherry Blossoms at Shinjuku Gyoen National Garden	0.7022	Tokyo special wards	Animal/plant
Imperial Palace East Gardens	0.4939	Tokyo special wards	Sightseeing
Kyoto Prefectural Botanical Garden	0.4774	Kyoto city	Sightseeing
Yoyogi Park	0.4339	Tokyo special wards	Sightseeing
Otani Natural Park	0.4306	Yamato Takada city (Nara)	Sightseeing

Table 6 Top five spots based on cosine similarity to Q_1 using filter method

Name of spot	Similarity	Location	Category
Osaka Castle Nishinomaru Garden	0.4003	Osaka city	Sightseeing
Nagai Botanical Garden	0.3500	Osaka city	Sightseeing
Nakanoshima Rose Garden	0.3284	Osaka city	Sightseeing
Fujita Manor Park	0.3224	Osaka city	Sightseeing
Tsurumi Ryokuchi	0.3214	Osaka city	Sightseeing

Table 7 Top five spots based on cosine similarity to Q_1 using direct method

Name of spot	Similarity	Location	Category
Universal Studios Japan	0.5409	Osaka city	Amusement
Cherry Blossoms at Shinjuku Gyoen National Garden	0.3448	Tokyo special wards	Animal/Plant
Universal City Walk Osaka	0.2958	Osaka city	Others
Dahlia of Kurohime Highlands	0.2642	Shinanomachi (Nagano)	Animal/Plant
Cherry Blossoms at Akashi Park	0.2621	Akashi city	Animal/Plant

Table 8 Top five spots based on cosine similarity to Q_1 using proposed method

Name of spot	Similarity	Location	Category
Cherry Blossoms at Shinjuku Gyoen National Garden	0.6754	Tokyo special wards	Animal/Plant
Kyoto Prefectural Botanical Garden	0.4791	Kyoto city	Sightseeing
Imperial Palace East Gardens	0.4564	Tokyo special wards	Sightseeing
Osaka Castle Nishinomaru Garden	0.4370	Osaka city	Sightseeing
Otani Natural Park	0.4123	Yamato Takada city (Nara)	Sightseeing

Table 9 shows the five spots having the highest cosine similarity to "Meiji Shrine". Tables 10, 11, and 12 show the top five spots in the results based on the respective methods. According to the results in Table 10, the spots extracted by the filter method (such as "Auxiliary Shrine of Ishigami Jingu" and "Mausoleum of Emperor Jinmu") do not have the category "Shrine". However, they are spots that are practically equivalent to shrines.

We considered that these inappropriate results were due to the adding of the feature tag "Sightseeing". In general, the category to which a spot belongs cannot be determined uniquely, and there are many spots that can belong to a plurality of category tags. Therefore, it is difficult to effectively extract spots using the filter method based on category tags.

On the other hand, as seen in Table 11, the paragraph vector for the category tag learned by the direct method extracted tourist attractions near "Meiji Shrine," such as "Yoyogi Park". In addition, "Nogi Shrine," located in Nogizaka near "Meiji Shrine," was extracted. These spots frequently include reviews such as "I walked from Harajuku to Akasaka by accident," meaning that they visited sightseeing spots near other sightseeing spots. Therefore, we believe that the generated paragraph vector represents the meaning of the category tags.

The results for the proposed method are similar to the cosine similarity results (see Tables 9 and 12). This has the same tendencies as in the case of the location tag. Although they are not displayed in Table 12, the similarity of spots near "Meiji Shrine" and highly related to "Sightseeing" tags such as "Yoyogi First Gymnasium" and "Nogi Shrine" have risen, in addition to "Yoyogi Park". That is, only the spots related to the added/subtracted tag have changed. We regard this as a more rigorous result compared with the direct method.

Experimental results show that the query vector using the paragraph vector for the location/category tag generated by the proposed method has the following three features.

Table 9 Top five spots based on cosine similarity to "Meiji Shrine"

Name of spot	Similarity	Location	Category
Meiji Shrine Inner Garden	0.5512	Tokyo special wards	Sightseeing
Atsuta Shrine	0.5034	Nagoya city	Shrine
Miyazaki Shrine	0.4855	Miyazaki city	Shrine
Hokkaido Shrine	0.4761	Sapporo city	Shrine
Yajima Shrine	0.4377	Miyazaki city	Shrine

Table 10 Top five spots based on cosine similarity to Q_2 using filter method

Name of spot	Similarity	Location	Category
Meiji Shrine Inner Garden	0.5512	Tokyo special wards	Sightseeing
Auxiliary Shrine of Ishigami Jingu	0.3704	Tenri city	Sightseeing
Yoyogi Park	0.3589	Tokyo special wards	Sightseeing
Mausoleum of Emperor Jinmu	0.3535	Kashihara city (Nara)	Sightseeing
Kashihara Forest Garden	0.3417	Kashihara city (Nara)	Sightseeing

Table 11 Top five spots based on cosine similarity to Q_2 using direct method

Name of spot	Similarity	Location	Category
Meiji Shrine Inner Garden	0.4704	Tokyo special wards	Sightseeing
Nogi Shrine	0.3256	Tokyo special wards	Shrine
Cornerstone of Peace Memorial	0.3227	Itoman city	Others
Yoyogi Park	0.3210	Tokyo special wards	Sightseeing
Dr. Hebron House Memorial	0.3057	Yokohama city	Sightseeing

- It is particularly effective in cases that include tags that are difficult to classify uniquely, such as category tags.
- The paragraph vectors for the location/category tag based on average vectors make a small contribution to the query vector. Therefore, we need to develop an amplification method for the paragraph vectors for the location/category tag.
- It is possible to extract only spots that are highly relevant to the target tags via addition and subtraction using the generated paragraph vectors.

Table 12 Top five spots based on cosine similarity to Q_2 using proposed method

Name of spot	Similarity	Location	Category
Meiji Shrine inner garden	0.5389	Tokyo special wards	Sightseeing
Atsuta Shrine	0.4291	Nagoya city	Shrine
Hokkaido Shrine	0.4193	Sapporo city	Shrine
Yoyogi Park	0.4168	Tokyo special wards	Sightseeing
Miyazaki Shrine	0.4103	Miyazaki city	Shrine

4 Related Work

There are many studies on tourism information extraction. First, we mention a method for automatically extracting sightseeing spots using geotagged content posted on a location-based social network service. Crandall et al. [10] proposed a method for extracting a popular spot and landmark. They extracted clusters as a spot by using the large number of photos, the spot's location information, and social tags. Moreover, they showed that the trajectory of a photography route could be obtained from data of the same photographer. Hirota et al. [11] proposed visualization of multiple points for which there are considerable numbers of geotagged photos, and extraction of a landmark's shape by using the shooting directions. Further, Oku and Hattori [12] proposed a method to extract the features of a sightseeing spot based on the tweets in that area. Next, we mention the expansion method for extracting tourist information. Fujii et al. [13] are working on a method to extract information based on real experience from blogs and community Q&As, information that is difficult to know using guidebooks alone. In this study, we used data from a tourist review site, but with our method it seems that it would be possible to generate a paragraph vector from various text data and tag information attached to extracted spots by conventional methods.

There has been research on relative retrieval and analogy retrieval, such as finding content corresponding to one domain in another domain. Nakajima and Tanaka [14] proposed a method of searching by using the relative relationship between two domains by mapping the space of one domain onto the other domain. In addition, Kato et al. [15] focused on two target keywords and a keyword representing their relationship. They proposed a method of automatically extracting the relationship keyword from a Web search index and a method of finding other corresponding keywords using the relationship keyword. These studies are similar to our research in that they are methods for discovering equivalent tourist spots in other locations and categories. In this research, we used a paragraph vector to carry it out and focused on the functional feature.

5 Conclusion

We have proposed a tourist spot search method using similarity of function based on distributed representations of user reviews. We generated a paragraph vector for tourist spots using user reviews on a tourist review site, and we generate a paragraph vector for location/category tags based on the paragraph vector for the spot. In this way, we made it possible to calculate spots and locations/categories. We assessed the characteristics of the generated paragraph vector for a spot by preliminary experiments. We can say that the generated paragraph vector for a spot represents the spot's functional feature; at the same time, however, it includes location features and category features. Therefore, we experimented with a method to generate a paragraph vector for location/category that appropriately subtracts location features and category features from the paragraph vector for a spot. The results confirmed the effectiveness of the proposed method using the average vector for the corresponding spot as the paragraph vector for the location/category tags.

As our future work, we need to develop a suitable method to amplify the generated paragraph vector for location/category tags because the contribution of the generated paragraph vector for location/category tags to the query vector is small. In addition, we will extend the paragraph vector generating method to treat not only location/category tags but any keyword.

Acknowledgements This work was supported by MEXT KAKENHI Grant Number JP15K16091

References

1. One of the most famous travel website around the world. www.tripadvisor.com
2. One of the largest travel website in Japan. www.jalan.net
3. T. Yoshida, D. Kitayama, S. Nakajima, K. Sumiya, A tourist spot search method using similarity of function based on distributed representations of user reviews, in *Lecture Notes in Engineering and Computer Science: Proceedings of The International MultiConference of Engineers and Computer Scientists 2017, 15–17 March, 2017, Hong Kong*, pp. 473–478
4. Q.V. Le, T. Mikolov, Distributed representations of sentences and documents," in *Proceedings of the 31th International Conference on Machine Learning, ICML 2014* (2014), pp. 1188–1196
5. T. Mikolov, I. Sutskever, K. Chen, G.S. Corrado, J. Dean, Distributed representations of words and phrases and their compositionality. Adv. Neural Inf. Process. Syst. **26**, 3111–3119 (2013)
6. Implementation of the algorithm proposed by Mikolov et al. https://code.google.com/archive/p/word2vec
7. Z.S. Harris, Distributional structure. Word **10**(2–3), 146–162 (1954)
8. R. Řehůřek, P. Sojka, Software Framework for Topic Modelling with Large Corpora, in *Proceedings of the LREC 2010 Workshop on New Challenges for NLP Frameworks* (2010), pp. 45–50. http://is.muni.cz/publication/884893/en
9. Gensim's API Reference. www.radimrehurek.com/gensim/models/doc2vec.html
10. D.J. Crandall, L. Backstrom, D. Huttenlocher, J. Kleinberg, Mapping the world's photos, in *Proceedings of the 18th International Conference on World Wide Web* (2009), pp. 761–770
11. M. Hirota, M. Shirai, H. Ishikawa, S. Yokoyama, "Detecting relations of hotspots using geo-tagged photographs in social media sites," in *Proceedings of Workshop on Managing and Mining Enriched Geo-Spatial Data* (2014), pp. 7:1–7:6

12. K. Oku, F. Hattori, Mapping geotagged tweets to tourist spots considering activity region of spot, in *Tourism Informatics*, vol. 90 (Springer, Heidelberg, 2015), pp. 15–30
13. K. Fujii, H. Nanba, T. Takezawa, A. Ishino, Enriching travel guidebooks with travel blog entries and archives of answered questions, in *Information and Communication Technologies in Tourism, Proceedings of the International Conference in Bilbao, Spain, 2–5 February 2016*, ed. by A. Inversini, R. Schegg, vol. 2016 (Springer International Publishing, 2016) pp. 157–171
14. S. Nakajima, K. Tanaka, Relative queries and the relative cluster-mapping method. in *Proceedings of the Database Systems for Advanced Applications: 9th International Conference, DASFAA 2004, Jeju Island, Korea, 17–19 March 2003*, ed. by Y. Lee, J. Li, K.-Y. Whang, D. Lee. (Springer, Heidelberg, 2004), pp. 843–856
15. M.P. Kato, H. Ohshima, S. Oyama, K. Tanaka, Query by analogical example: Relational search using web search engine indices, in *Proceedings of the 18th ACM Conference on Information and Knowledge Management* (2009), pp. 27–36

Learning Method of Fuzzy Inference Systems for Secure Multiparty Computation

Hirofumi Miyajima, Noritaka Shigei, Hiromi Miyajima, Yohtaro Miyanishi, Shinji Kitagami and Norio Shiratori

Abstract Many studies on privacy preserving of machine learning and data mining for cloud computing have been done in various methods by use of randomization techniques, cryptographic algorithms, anonymization methods, etc. Data encryption is one of typical approaches. However, its system requires both encryption and decryption for requests of client or user, so its complexity of computation is very high. Therefore, studies on secure computation using shared or divided data are made to avoid secure risks being abused or leaked and to reduce computing cost. The secure multiparty computation (SMC) is one of these methods. So far, some studies have been done with SMC using divided data, but complex calculation processing such as machine learning has never proposed yet. In the previous paper, we proposed BP learning method for SMC on cloud computing system. In this paper, we propose learning method (Fuzzy modeling) of fuzzy inference system for SMC and prove the validity of it. Further, the performance of the proposed method is shown in numerical simulations.

H. Miyajima (✉)
Okayama University of Science, Okayama, Japan
e-mail: miya@mis.ous.ac.jp

N. Shigei · H. Miyajima
Kagoshima University, Kagoshima, Japan
e-mail: shigei@eee.kagoshima-u-ac.jp

H. Miyajima
e-mail: miya@eee.kagoshima-u-ac.jp

Y. Miyanishi
Information Systems Engineering and Management, Tokyo, Japan
e-mail: miyanisi@jade.dti.ne.jp

S. Kitagami
Waseda University Graduate School of Global Information and Telecommunication Studies
(GITS), Tokyo, Japan
e-mail: kitagami.shinji@meltec.co.jp

N. Shiratori
Chuo University, Tokyo, Japan
e-mail: norio@shiratori.riec.tohoku.ac.jp

© Springer Nature Singapore Pte Ltd. 2018
S.-I. Ao et al. (eds.), *Transactions on Engineering Technologies*,
https://doi.org/10.1007/978-981-10-7488-2_17

Keywords Cloud computing · Control problem · Data division · Fuzzy
modeling · Privacy preserving learning method · Secure multiparty computation

1 Introduction

Privacy preserving machine learning and data mining on the cloud system can be
achieved in various methods by use of randomization techniques, cryptographic
algorithms, anonymization methods, etc. [1–7]. However, data encryption system
requires both encryption and decryption for requests of client or user, so its complexity is very high. The problem is the trade-off between the security and the complexity.
As one of these studies, secure multiparty computation (SMC) has been introduced
[8–10]. The purpose of SMC is to allow servers (parties) to carry out distributed
computing tasks in secure way. Most of the works in SMC are developed on applying the model of SMC on different data distributions such as vertically, horizontally
and arbitrarily partitioned data. They are the methods that each server performs its
processing for the subset of data [10–14]. As the another study of them, SMC systems dividing data itself to each server attract attention, and some studies with them
have been done. A simple method to divide data was proposed and they were applied
to some problems [13, 14]. In the previous paper, we proposed BP learning for neural networks and clustering methods for SMC [15–17]. In this paper, we propose
learning method (fuzzy modeling) of fuzzy inference systems for SMC and show the
effectiveness of them in numerical simulations. The difference between BP learning
and fuzzy modeling is as follows [18, 19]: In BP learning input and output relation
for learning data is represented as weights between layers of neural network, but it
is difficult to know the interpretability of input and output learning data. In fuzzy
modeling, input and output relation for learning data is represented as fuzzy rules
for fuzzy inference system while privacy preserving and it is comparatively easy to
understand the interpretability between input and output learning data. In this paper,
we will show this relation using a control problem. In Sect. 2, we describe the idea
for sharing data securely. Further, we also introduce the conventional learning algorithm for fuzzy inference systems. In Sect. 3, we describe our proposed learning
algorithm for SMC. In Sect. 4, some simulation results are presented to demonstrate
the effectiveness of the proposed method.

2 Preliminary

2.1 System Configuration of Cloud System and Related
Works

Let us consider a system for SMC of cloud computing (See Fig. 1). The system is
composed of a client and m servers (parties) [15]. The client sends data to each server
and each server memorizes them. If the client requires data processing, each server

performs one's computation and sends each result to client. The client computes the final result using them. If the result is not obtained by one processing, data processing between client and servers is iterated until the final result is obtained. The problem is how data are shared and the computation for each server is carried out.

Let us consider about conventional works related with SMC. Three partitioned representation of data such as horizontally, vertically and any partitioned methods for SMC are known [10, 11, 20]. Let us explain about them using an example of Table 1. Let $m = 2$. In Table 1, a and b are original data (marks) and ID is student identifier. The purpose of computation is to get the average of them.

First, let us explain about horizontally partitioned method (HPM) using Table 1. All the dataset are shared into two servers, Servers 1 and 2 as follows:

Server 1: dataset for ID = 1, 2

Server 2: dataset for ID = 3, 4.

In this case, two averages with subsets A and B for Server 1 are computed as $(90 + 60)/2$ and $(55 + 82)/2$, respectively. Likewise, two averages with subsets A and B for Server 2 are $(70 + 40)/2$ and $(30 + 70)/2$, respectively. As a result, two averages for subsets A and B are 65.0 and 59.25, respectively. These are obtained as the sums of average of $a^{(1)}$ and $a^{(2)}$, and of $b^{(1)}$ and $b^{(2)}$, where $a = a^{(1)} + a^{(2)}$ and $b = b^{(1)} + b^{(2)}$. Each server cannot know half of the dataset, so privacy preserving holds.

Likewise, vertically partitioned method (VPM) is introduced. In this case, dataset for subjects A and B are assigned to Servers 1 and 2, respectively.

At third, let us consider about any partitioned method for SMC. All the dataset are partitioned into two servers, mixed horizontally and vertically partitioned data.

These methods need a large number of servers to keep security, but the proposed method using secure divided data seems to keep them by small number of data.

2.2   The Representation of Secure Divided Data

Let us explain data representation for the proposed method using Fig. 1 [13, 14]. Let a and b be two positive integers. First, two integers a and b are divided into m real numbers. Let $a = a^{(1)} + \cdots + a^{(m)}$ and $b = b^{(1)} + \cdots + b^{(m)}$ as the addition form and $a = A^{(1)} \cdots A^{(m)}$ and $b = B^{(1)} \cdots B^{(m)}$ as the multiplication form. Then the following results hold:

(1) $a + b = (a^{(1)} + b^{(1)}) + \cdots + (a^{(m)} + b^{(m)})$

(2) $a - b = (a^{(1)} - b^{(1)}) + \cdots + (a^{(m)} - b^{(m)})$

(3) $ab = (A^{(1)}B^{(1)}) \cdots (A^{(m)}B^{(m)})$

(4) $a/b = (A^{(1)}/B^{(1)}) \cdots (A^{(m)}/B^{(m)})$

That is, four basic operations of arithmetic (addition, subtraction, multiplication, and division) hold as integration of the result computed independently by each server. In this paper, $a^{(k)}$ and $A^{(k)}$ for $1 \leq k < m - 1$ are selected randomly in [0, 1] and [−1, 1], respectively. Further, $a^{(m)}$ and $A^{(m)}$ are selected as $a - \sum_{k=1}^{m-1} a^{(k)}$ and $a/\prod_{k=1}^{m-1} A^{(k)}$, respectively.

Fig. 1 The representation
of secure divided data

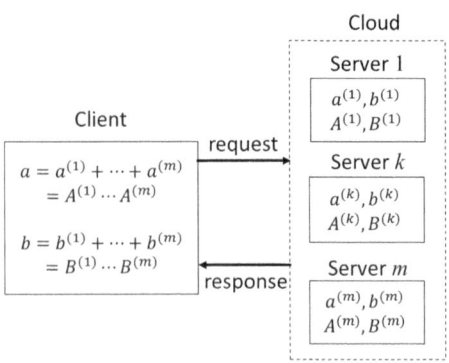

Let us show an example of divided data using Table 1 as follows [14]:

$a = a^{(1)} + a^{(2)} : a^{(1)} = a(r_1/10)$ and $a^{(2)} = a(1 - r_1/10)$,

$b = b^{(1)} + b^{(2)} : b^{(1)} = b(r_1/10)$ and $b^{(2)} = b(1 - r_2/10)$,

$a = A^{(1)}A^{(2)} : A^{(1)} = \sqrt{a}(r_1/10)$ and $A^{(2)} = \sqrt{a}(10/r_1)$,

$b = B^{(1)}B^{(2)} : B^{(1)} = \sqrt{b}(r_2/10)$ and $B^{(2)} = \sqrt{b}(10/r_2)$,

where r_1 and r_2 are real random numbers for $-9 \leq r_1 \leq 9$ and $0.2 \leq r_2 \leq 9$, $r_1 \neq 1$, $r_2 \neq 1$, respectively. For example, $a^{(1)}$ and $a^{(2)}$ for ID = 1 are computed as $a^{(1)} = 90 \times (2/10) = 18$ and $a^{(2)} = 90 \times (1 - 2/10) = 72$ and data $A^{(1)}$ and $A^{(2)}$ for ID = 1 are computed as $A^{(1)} = \sqrt{90} \times (4/10) = 3.79$ and $A^{(2)} = \sqrt{90} \times (10/4) = 23.71$, respectively. Note that Server 1 has all the data in column-wise of $a^{(1)}$, $b^{(1)}$, $A^{(1)}$ and $B^{(1)}$ for each ID and Server 2 has all the data in column-wise of $a^{(2)}$, $b^{(2)}$, $A^{(2)}$ and $B^{(2)}$ for each ID as shown in Table.1.

Remark that each data for server is randomized and the method does not need to use decrypted data.

2.3 The Conventional Fuzzy Inference System

The conventional fuzzy inference system is described [18, 19]. Let $Z_j = \{1, \ldots, j\}$ for the positive integer j. Let R be the set of real numbers. Let $\mathbf{x} = (x_1, \ldots, x_N)$ and y^* be input and output data, respectively, where $x_j \in R$ for $j \in Z_N$ and $y^* \in R$. Then the rule of fuzzy inference model is expressed as

$$R_i : \text{if } x_1 \text{ is } M_{i1} \text{ and } \cdots \text{ and } x_j \text{ is } M_{ij} \text{ and } \cdots \text{ and } x_N \text{ is } M_{iN}$$
$$\text{then } y \text{ is } c_i, \qquad (1)$$

where $i \in Z_n$ is a rule number, $j \in Z_N$ is a variable number, M_{ij} is a membership function of the antecedent part, and c_i is a real number of the consequent part.

A membership value of the antecedent part μ_i for input $\mathbf{x}$ is expressed as

Table 1 Data on Server 1 and Server 2

ID	Subject A	Subject B	Addition form					Multiplication form				
	a	b	r_1	a		b		r_2	A		B	
				$a^{(1)}$	$a^{(2)}$	$b^{(1)}$	$b^{(2)}$		$A^{(1)}$	$A^{(2)}$	$B^{(1)}$	$B^{(2)}$
1	90	55	2	18	72	11	44	4	3.79	23.71	2.97	18.54
2	60	82	−3	−18	78	−24.6	106.6	5	3.87	15.49	4.53	18.11
3	70	30	5	35	35	15	15	0.4	0.33	209.17	0.22	136.93
4	40	70	−6	−24	64	−42	112	3	1.90	21.08	2.51	27.89
Average	65	59.25		2.75	62.25	−10.15	69.4					

$$\mu_i = \prod_{j=1}^{N} M_{ij}(x_j). \tag{2}$$

If Gaussian membership function is used, then M_{ij} is expressed as follow

$$M_{ij} = \exp\left(-\frac{1}{2}\left(\frac{x_j - a_{ij}}{b_{ij}}\right)^2\right). \tag{3}$$

where a_{ij} and b_{ij} are the center and the width values of M_{ij}, respectively.
The output y^* of fuzzy inference is calculated by the following equation:

$$y^* = \frac{\sum_{i=1}^{n} \mu_i \cdot c_i}{\sum_{i=1}^{n} \mu_i}. \tag{4}$$

2.4 The Conventional Learning Algorithm for Fuzzy Inference Systems

In order to construct the effective model for fuzzy modeling, the conventional learning method is introduced. The objective function E is defined to evaluate the inference error between the desirable output y^r and the inference output y^*. In this section, we describe the conventional learning algorithm. Let $D = \{(x_1^p, \ldots, x_N^p, y_p^r)|p \in Z_P\}$ be the set of learning data. The objective of learning is to minimize the following mean square error (MSE):

$$E = \frac{1}{P}\sum_{p=1}^{P}(y_p^* - y_p^r)^2. \tag{5}$$

where y_p^* is the inference output for the p-th input x^p. In order to minimize the objective function E, each parameter $\alpha \in \{c_i, a_{ij}, b_{ij}\}$ is updated based on the descent method as follows [18, 19]:

$$\alpha(t+1) = \alpha(t) - K_\alpha \frac{\partial E}{\partial \alpha} \tag{6}$$

where t is iteration time and K_α is a constant. When Gaussian membership function for $i \in Z_n$ and $j \in Z_N$ are used, the following relation holds.

$$\frac{\partial E}{\partial c_i} = \frac{\mu_i}{\sum_{i=1}^{n} \mu_i} \cdot (y - y^*) \tag{7}$$

$$\frac{\partial E}{\partial a_{ij}} = \frac{\mu_j}{\sum_{i=1}^{n} \mu_i} \cdot (y - y^*) \cdot (c_i - y) \cdot \frac{x_j - a_{ij}}{b_{ij}^2} \tag{8}$$

$$\frac{\partial E}{\partial b_{ij}} = \frac{\mu_i}{\sum_{i=1}^{n} \mu_i} \cdot (y - y^*) \cdot (c_i - y) \cdot \frac{(x_j - a_{ij})^2}{b_{ij}^3}$$

Then, the conventional learning algorithm is shown as follows [18, 19]:

Learning Algorithm A

Step A1: The threshold θ of inference error and the maximum number of learning time T_{max} are given. The initial assignment of fuzzy rules is set to equally intervals. Let n be the number of rules and $n = d^N$ for an integer d, where d^N means the number of divided input domains. Let $t = 1$.

Step A2: The parameters c_i, a_{ij} and b_{ij} are set to the initial values.

Step A3: Let $p = 1$.

Step A4: A data $(x_1^p, \ldots, x_N^p, y_p^r) \in D$ is selected randomly.

Step A5: From Eqs. (2) and (4), μ_i and y^* are computed.

Step A6: Parameters c_i, a_{ij} and b_{ij} are updated by Eqs. (7), (8) and (9).

Step A7: If $p = P$ then go to Step A8 and if $p < P$ then go to Step A4 with $p \leftarrow p + 1$.

Step A8: Let $E(t)$ be inference error at step t calculated by Eq. (5). If $E(t) > \theta$ and $t < T_{max}$ then go to Step A3 with $t \leftarrow t + 1$ else if $E(t) \leq \theta$ and $t \leq T_{max}$ then the algorithm terminates.

Step A9: If $t > T_{max}$ and $E(t) > \theta$ then go to Step A2 with $n = d^N$ as $d \leftarrow d + 1$ and $t = 1$.

3 Secure Multiparty Computation

Let us consider a system composed of a client and m servers (See Fig. 1). In learning on cloud system, learning data and parameters are divided to each server in addition or multiplication form. Each server updates divided parameters and sends the computation result to the client. The client can get new parameters by adding or multiplying the results of m servers. The process is iterated until the error (difference) between the inference and the desired output becomes sufficiently small. The problem is how parameters on the client are updated using the set of learning data divided on each server. The divided representation of learning data $\{(x^l, d(x^l)) | l \in Z_P\}$ and parameters are given as follows:

$$x^l = \left(x_1^l, \ldots, x_j^l, \ldots, x_N^l \right) \tag{9}$$

for $l \in Z_p$ and

$$x_j^l = \sum_{k=1}^{m} (x_j^l)^k \tag{10}$$

for $j \in Z_N$,

$$d(x^l) = \sum_{k=1}^{m} (d(x^l))^k, \tag{11}$$

where $d(x^l)$ is the desirable output for the l-th input x^l and m is the number of servers. Note that Eqs. (10) and (11) are in addition form for divided data.

In this case, Eqs. (7)–(9) are renewed as follows:

$$\triangle c_i^k = \frac{\mu_i}{\sum_{i=1}^{n} \mu_i} \cdot (y - y^*) \tag{12}$$

$$(c_i^k)(t+1) = (c_i^k)(t) + K\triangle c_i^k(t) \tag{13}$$

$$\triangle a_{ij}^k = \frac{\mu_j}{\sum_{i=1}^{n} \mu_i} \cdot (y - y^*) \cdot (c_i - y) \cdot \frac{x_j - a_{ij}}{b_{ij}^2} \tag{14}$$

$$(a_{ij}^k)(t+1) = (a_{ij}^k)(t) + K\triangle a_{ij}^k(t) \tag{15}$$

$$\triangle b_{ij}^k = \frac{\mu_i}{\sum_{i=1}^{n} \mu_i} \cdot (y - y^*) \cdot (c_i - y) \cdot \frac{(x_j - a_{ij})^2}{b_{ij}^3} / b_{ij}^k \tag{16}$$

$$(b_{ij}^k)(t+1) = (b_{ij}^k)(t) + K\triangle b_{ij}^k(t) \tag{17}$$

where each of parameters $\{a_{ij}, b_{ij}, c_i | i \in Z_n, j \in Z_N\}$ is represented as $a_{ij} = \sum_{k=1}^{m} a_{ij}^k$, $b_{ij} = \Pi_{k=1}^{m} b_{ij}^k$ and $c_i = \sum_{k=1}^{m} c_i^k$ using data representation in Sect. 2.2.

Equation (17) means that each server can update the parameter by dividing by b_{ij}^k for the conventional method.

From these results, learning of the fizzy inference system is shown in Table 2. The validity of the algorithm is proved from the results of Eq. (12)–(17). Let us explain the relation between Algorithm A and Table 2.

In Step1, learning data is selected randomly. In Steps 2 and 3, $x_j - a_{ij}$ is computed using the result of each server. In Steps 4 and 5, $(x_j - a_{ij})/b_{ij}$ and Eq. (3) are computed. In Step 6, the denominator of Eq. (4) and the numerator of Eq. (4) for each server are computed. In Step 7, Eq. (4) is computed and each divided result of it is sent to each server. In Steps 8 and 9, $\triangle c_i^k$, $\triangle a_{ij}^k$ and $\triangle b_{ij}^k$ for Eqs. (12), (14), and (16) are computed, respectively, and sent them to each server. In Step10, c_i, a_{ij} and b_{ij} are updated. In Step 11, if the error E is sufficient small, or the maximum learning time is attained then the algorithm terminates else go to Step 1.

Table 2 Learning process of Fuzzy Inference Learning for SMC

	Client	k-th Server
Initial condition	The parameter $\{a_{ij}, b_{ij}, c_i\}$ is selected randomly, and send $a_{ij}^k (a_{ij} = \sum_{k=1}^m a_{ij}^k)$, $b_{ij}^k (b_{ij} = \Pi_{k=1}^m b_{ij}^k)$ and $c_i^k (c_i = \sum_{k=1}^m c_i^k)$ to each server for $i \in Z_n, j \in Z_N$ and $k \in Z_m$. Set $t = 1$.	$\{(x^l)^k, (d(x^l))^k \mid l \in Z_L\}$
Step 1	A number l is selected randomly	
Step 2		Compute $(x_j^l)^k - a_{ij}^k$ for $i \in Z_n$ and $j \in Z_N$ and send it to Client
Step 3	Compute $dist_{ij} = \sum_{k=1}^m \left((x_j^l)^k - a_{ij}^k\right)$ and send $dist_{ij}^k (dist_{ij} = \Pi_{k=1}^m dist_{ij}^k)$ to each server	
Step 4		Compute $p_{ij}^k = dist_{ij}^k / b_{ij}^k$ and $q_{ij}^k = dist_{ij}^k / (b_{ij}^k)^2$ and send them to Client
Step 5	Compute $\mu_i = \exp\left(-(\Pi_{k=1}^m p_{ij}^k)^2\right)$ and send it to each server	
Step 6	Compute $S = \sum_{i=1}^r \mu_i$	Compute $\sum_{i=1}^n \mu_i c_i^k$ and send it to Client
Step 7	Compute output $y = \sum_{k=1}^m (\sum_{i=1}^n \mu_i c_i^k)/S$ and send $y^k (y = \sum_{k=1}^m y^k)$ to each server	
Step 8		Compute $s^k = c_i^k - y^k$ and $\triangle^k = (d(x^l))^k - y^k$ and send them to Client
Step 9	Compute $\alpha_{ij} = \frac{(\sum_{k=1}^m \triangle^k)(\sum_{k=1}^m s^k)(\Pi_{k=1}^m q_{ij}^k)\mu_i}{S}$, $\beta_{ij} = \frac{(\sum_{k=1}^m \triangle^k)(\sum_{k=1}^m s^k)(\Pi_{k=1}^k (\sigma_{ij}^k)^2)\mu_i}{S}$ and $\gamma_i = \frac{(\sum_{k=1}^m \triangle^k)\mu_i}{S}$ and send them to each server	
Step 10		Update $a_{ij}^k \leftarrow a_{ij}^k + K_a \alpha_{ij}, b_{ij}^k \leftarrow b_{ij}^k + K_b \beta_{ij}/b_{ij}^k$ and $c_i^k \leftarrow c_i^k + K_c \gamma_i$
Step 11	If $t \neq T_{max}$ or $E > \theta$ then go to Step 1 with $t \leftarrow t + 1$ else the algorithm terminates	

4   Numerical Simulations

In this section, numerical simulations of two-category classification and control problem for conventional and proposed methods are performed. The conventional method means the method without dividing data and the proposed method is ones with $m = 3$ and $m = 10$.

4.1   Two-Category Classification

In this section, two-category classification problems with spheres are performed. In the classification problems, points on $[0, 1] \times [0, 1] \times [0, 1]$ are classified into two classes: class 0 and class 1. The class boundaries are given as spheres centered at $(0.5, 0.5, 0.5)$. For Sphere, the inside of sphere with a radius of 0.3 is associated with class 1 and the outside with class 0. For Double-Sphere with radiuses of 0.2 and 0.4, the area between Spheres 1 and 2 is associated with class 1 and the other area with class 0 (See Fig. 2a). For triple-Sphere with radiuses of 0.1, 0.2 and 0.4, the inside of Sphere 1 and the area between Sphere 2 and Sphere 3 is associated with class 1 and the other area with class 0 (See Fig. 2b). The desired output y_p^r is set as follows: if x^p belongs to class 0, then $y_p^r = 0.0$. Otherwise $y_p^r = 1.0$. The numbers of learning and test data selected randomly are 512 and 6400, respectively. The simulation conditions are $K_c = 0.001$, $K_b = 0.001$, $K_w = 0.01$ and $T_{max} = 50000$. The results on the rate of misclassification are shown in Fig. 3, where #Para means the number of parameters. The result of simulation is the average from ten trials. Table 3 shows that the results for the conventional and proposed methods are almost the same ones.

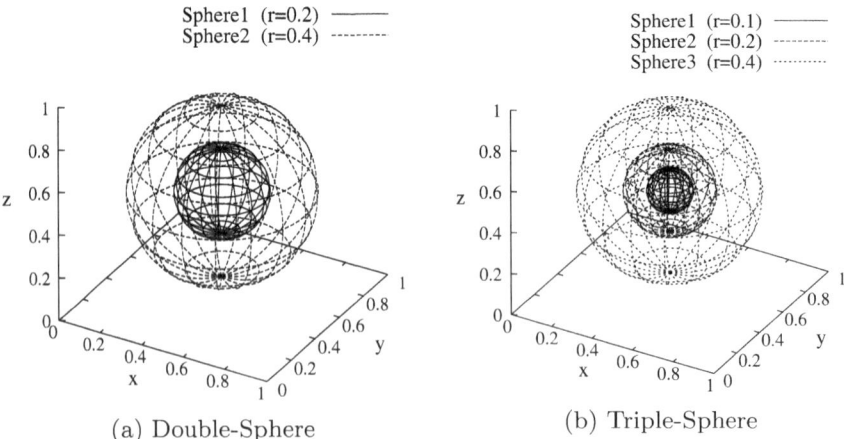

(a) Double-Sphere   (b) Triple-Sphere

Fig. 2 Two-category classification problems

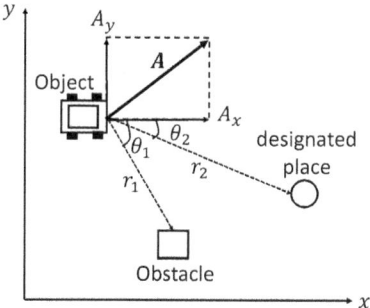

Fig. 3 Simulation on obstacle avoidance and arriving at the designated place, where A_x is constant and A_y is adjusted

Table 3 Simulation result for two-category classification problem

		Sphere	Double-Sphere	Triple-Sphere
Conventional	learning (%)	1.2	2.6	3.8
	test (%)	2.8	7.5	8.5
	#Para	189	189	189
Proposed $m = 3$	learning (%)	0.1	2.1	2.6
	test (%)	1.8	7.1	7.0
	#Para	567	567	567
Proposed $m = 10$	learning (%)	1.3	5.8	6.4
	test (%)	2.9	8.7	8.6
	#Para	945	945	945

4.2 Obstacle Avoidance and Arriving at Designated Point

In order to show the interpretability of the proposed model for SMC, let us perform simulation of control problem [21]. The problem is how the object avoids the obstacle and arrives at the designated place. As shown in Fig. 3, the distance r_1 and the angle θ_1 between object and obstacle and the distance r_2 and the angle θ_2 between object and the designated place are selected as input variables, where θ_1 and θ_2 are normalized.

Fuzzy inference rules for the conventional and proposed methods are constructed from learning data of $P = 400$ points shown in Fig. 4. An obstacle is placed at $(0.5, 0.5)$ and a designated place is placed at $(1.0, 0.5)$. The number of rules for each method is 81 and the number of attributes is 3. The object moves with the vector A at each step, where A_x of A is constant and A_y of A is determined by learning. Four paths shown in Fig. 4 are used as learning data with 400 points. Learning for the conventional and proposed methods is successful and three scenarios are performed as test simulations (Table 4).

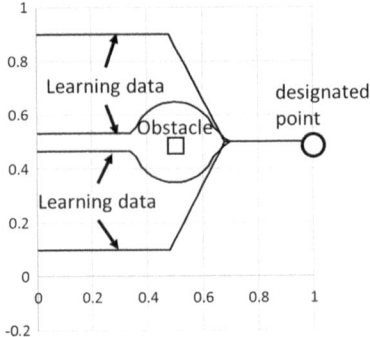

Fig. 4 Learning data to avoid obstacle and arrive at the designated place (1.0, 0.5)

Table 4 Initial condition for simulation of obstacle avoidance

	Conventional	Proposed
T_{max}	5000	1000
K_c	0.001	0.001
K_b	0.001	0.001
K_w	0.05	0.05
d	3	3
Initial a_{ij}	Equal intervals	
Initial b_{ij}	$\frac{1}{2(d-1)} \times$ (the domain of input)	
Initial c_i	0.0	

(1) Scenario 1 is simulation for obstacle avoidance and arriving at the designated place when the mobile object starts from various places (See Fig. 5). Figure 5 shows the results of moves of object for starting places at $(0.1, 0)$, $(0.2, 0)$,..., $(0.8, 0)$, $(0.9, 0)$ after learning. Simulations are successful for all cases. Figure 5 shows the result of conventional and proposed methods for $m = 10$.

(2) Scenario 2 is simulation for the case where the mobile object avoids obstacle placed at different place and arrives at the different designated place. Simulations with obstacle placed at the place $(0.4, 0.4)$ and arriving at the designated place $(1, 0.6)$ are performed for all methods. The results are successful as shown in Fig. 6. Figure 6 shows the result of conventional and proposed methods for $m = 10$.

(3) Scenario 3 is simulation for the case where obstacle moves with the fixed speed. Simulations with obstacle moving with the speed $(0.01, 0.02)$ from the place $(0.3, 1.0)$ to the place $(0.8, 0.0)$ and object arriving at the place $(1, 0.6)$ are performed. Simulations are successful for all cases. Figure 7 shows the results of simulations for conventional and proposed methods for $m = 10$. Since the object does not collide in the obstacle at $t = 30$, thereafter it does not collide in the obstacle.

Let us consider interpretability of fuzzy rules obtained by the proposed method. Let us consider fuzzy rules constructed for the proposed method of $m = 10$ by

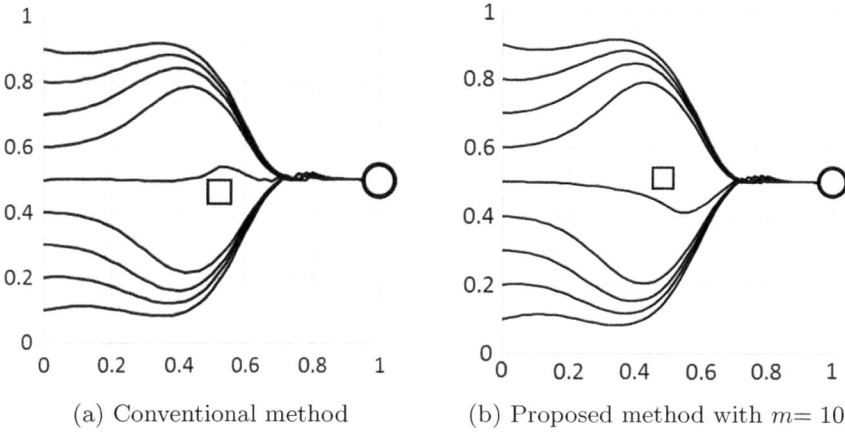

(a) Conventional method   (b) Proposed method with $m=10$

Fig. 5 Simulation for obstacle avoidance and arriving at the designated place starting from various places after learning for the proposed method with $m = 10$

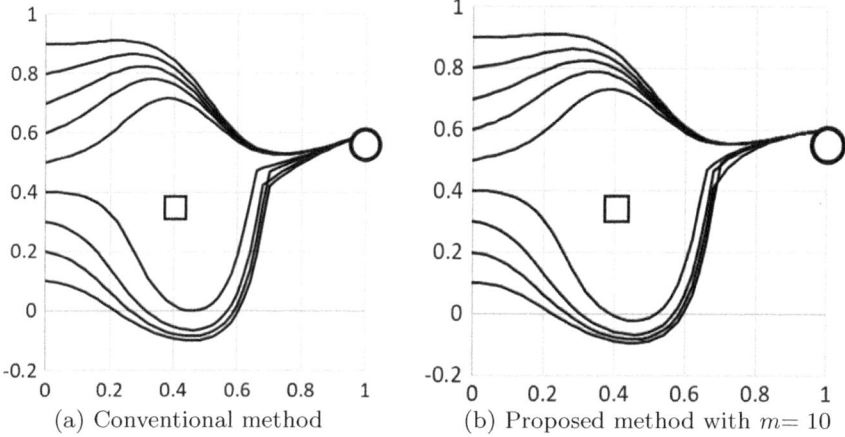

(a) Conventional method   (b) Proposed method with $m=10$

Fig. 6 Simulation for obstacle placed at the different place $(0.4, 0.4)$ and arriving at the different place $(1.0, 0.6)$

learning. Assume that three attributes are short, middle and long for r_1 and r_2, minus, central and plus for θ_1 and θ_2 and left, center and right for the direction of A_y, respectively. Then, main fuzzy rules for the proposed method are constructed as shown in Table 5. From Table 5, we can get the rules: "If the object approaches to the obstacle, move in the direction away from the object." and "If the object approach to the goal, then move toward to the goal". For example, Rule 1 means that if the object is near the obstacle (r_1 is short.), the object is far from the designated place (r_2 is long.), the object is above the obstacle (θ_1 is plus in Fig. 3.), the object is above the designated

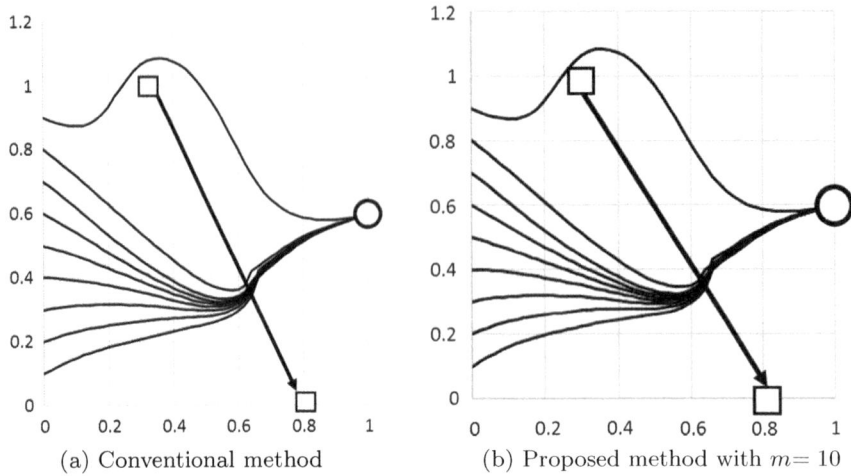

(a) Conventional method   (b) Proposed method with $m = 10$

Fig. 7 Simulation for moving obstacle avoidance with fixed speed and the different designated place $(1.0, 0.6)$

Table 5 Main fuzzy rules obtained for the proposed method with $m = 10$

	r_1	r_2	θ_1	θ_2	A_y
Rule 1	short	long	plus	plus	left
Rule 2			minus	minus	right
Rule 3	middle	middle	plus	plus	right
Rule 4			minus	minus	left

place (θ_2 is plus.) then the object moves in the direction away from the obstacle (A_y is left(plus) in Fig. 3). The result shows that obtained fuzzy rules are near our intuition.

Lastly, let us explain the features of the proposed method for SMC and the meaning of simulations.

(1) There are many forms to divide each item of (learning) data. The addition and multiplication forms, used in this paper, are the most standard ones. In the addition form, the whole calculation can be replaced by partial calculation. On the other hand, in the multiplication form, it does not necessarily hold as shown in Chap. 3. Both forms in fuzzy modeling are used.

(2) It is impossible to get the solution to the problem only using parameters of each server, and only the client having all the information on parameters can obtain the approximate solution. Especially in fuzzy modeling, it is possible to perform complicated calculation processing on the server, and the client can integrate and use the result.

(3) As mentioned in this paper, each server can not know complete information of data and parameters, but the client can know all information, that is, fuzzy modeling is performed while privacy preserving.

5   Conclusion

In this paper, we proposed a learning method (fuzzy modeling) of fuzzy inference system for SMC and proved the validity of it. Further, the performance of the proposed method was shown in numerical simulations. The advantage of fuzzy modeling compared to other learning methods is that input and output relation for learning data is represented as fuzzy rules and it is comparatively easy to understand the interpretability of input and output learning data. The idea of our study is to perform interpretable fuzzy modeling as "privacy preserving fuzzy modeling = divided learning data + parallel algorithm". That is, we performed to find the representation of divided data and to construct parallel algorithm. In the future work, we will consider improved methods to reduce the computation of client and develop AUI (Application User Interface) for the client.

References

1. C.C. Aggarwal, P.S. Yu, *Privacy-Preserving Data Mining: Models and Algorithms* (Springer, 2009). ISBN 978-0-387-70991-8
2. S. Subashini, V. Kavitha, A survey on security issues in service delivery models of cloud computing. J. Netw. Comput. Appl. **34**, 1–11 (2011)
3. C. Gentry, Fully homomorphic encryption using ideal lattices, in *STOC* (2009), pp. 169–178
4. A. Shamir, How to share a secret. Commun. ACM **22**(11), 612–613 (1979)
5. J. Yuan, S. Yu, Privacy preserving back-propagation neural network learning made practical with cloud computing. IEEE Trans. Parallel Distrib. Syst. **25**(1), 212–221 (2013)
6. A. Beimel, Secret-sharing schemes: a survey, in *Proceedings of the Third International Conference on Coding and Cryptology (IWCC 11)* (2011)
7. HElib, *An Implementation of homomorphic encryption*. https://github.com/shaih/HElib
8. R. Canetti et al., Adaptively secure multi-party computation. STOC **96**, 639–648 (1996)
9. A. Ben-David et al., Fair play MP: a system for secure multi-party computation, in *ACM CCS' 08* (2008)
10. S.S. Rathna, T. Karthikeyan, Survey on recent algorithms for privacy preserving data mining. Int. J. Comput. Sci. Inf. Technol. **6**(2), 1835–1840 (2015)
11. F.O. Catak, Secure multi-party computation based privacy preserving extreme learning machine algorithm over vertically distributed data, in *ICONIP 2015*, vol. 9490 (LNCS, 2015) pp. 337–345
12. T. Chen, S. Zhong, Privacy-preserving back propagation neural network learning. IEEE Trans. NN **20**(10), 1554–1564 (2009)
13. K. Chida, et al., A lightweight three-party secure function evaluation with error detection and its experimental result. IPSJ J. **52**(9), 2674–2685 (2011) (in Japanese)
14. Y. Miyanishi, A. Kanaoka, F. Sato, X. Han, S. Kitagami, Y.Urano, N. Shiratori, New methods to ensure security to increase user's sense of safety in cloud services, in *Proceedings of The 14th IEEE International Conference on Scalable Computing and Communications (ScalCom-2014)*, Bali, Dec 2014 pp. 859–865
15. H. Miyajima, N. Shigei, H. Miyajima, Y. Miyanishi, S. Kitagami, N. Shiratori, A proposal of back propagation learning for secure multi-party computation methods, in *Proceedings of The International MultiConference of Engineers and Computer Scientists 2016*, Hong Kong, 16–18 Mar 2016, Lecture Notes in Engineering and Computer Science, pp. 381–386, http://www.iaeng.org/publication/LNECS/

16. H. Miyajima, N. Shigei, H. Miyajima, Y. Miyanishi, S. Kitagami, N. Shiratori, New privacy preserving clustering methods for secure multiparty computation. Artif. Intell. Res. **6**(1) (2017)
17. H. Miyajima, N. Shigei, H. Miyajima, Privacy preserving fuzzy modeling for secure multiparty computation, in *Proceedings of The International MultiConference of Engineers and Computer Scientists 2017, Hong Kong*, 15–17 Mar 2017, Lecture Notes in Engineering and Computer Science, pp. 451-456, http://www.iaeng.org/publication/LNECS/
18. M.M. Gupta, L. Jin, N. Homma, *Static and Dynamic Neural Networks*. IEEE Press (2003)
19. H. Miyajima, N. Shigei, H. Miyajima, SIRMs fuzzy inference model with linear transformation of input variables and universal approximation, in *Advances in Computational Intelligence: Proceedings of the 13th International Work-Conference on Artificial Neural Networks*, vol. 9094 (LNCS, 2015), pp. 561–575
20. N. Schlitter, A protocol for privacy preserving neural network learning on horizontal partitioned data Privacy, in Statistics in Database(PSD) (2008)
21. H. Miyajima, N. Shigei, H. Miyajima, On the capability of a fuzzy inference system with improved interpretability, *Proceedings of The International MultiConference of Engineers and Computer Scientists 2015, Hong Kong*, 18–20 Mar 2015, Lecture Notes in Engineering and Computer Science, pp. 51–56, http://www.iaeng.org/publication/LNECS/

Solving the Fagnano's Problem via a Dynamic Geometry Approach

Yiu-Kwong Man

Abstract The Fagnano's problem is a famous historical problem in plane geometry, which involves finding an inscribed triangle with minimal perimeter in a given acute triangle. We discuss how to solve this problem via a dynamic geometry approach and derive a simple formula for finding the perimeter of the orthic triangle, which is the solution of the Fagnano's problem. Some illustrative examples are included.

Keywords Billiard trajectory · Dynamic geometry approach · Fagnano's problem · GeoGebra · Heron's theorem · Minimal perimeter
Orthic triangle

1 Introduction

In 1775, the Italian mathematician Giovanni Francesco Fagnano (1715–1797) used calculus to solve the following optimization problem in plane geometry [2, 3, 5, 13].

Given an acute triangle △ABC, determine the inscribed triangle with minimal perimeter in △ABC.

The solution of the problem is the orthic triangle, which is the triangle formed by joining the feet of the altitudes of △ABC (see Fig. 1). It is also called altitude triangle in some literature (see [2]).

Besides using calculus, some alternative methods can be used to solve the problem. Readers may refer to [2, 3, 9, 10, 15, 17, 18] for the details. In this chapter, we will describe how to solve the Fagnano's problem via a simple dynamic geometry approach. We will also describe a simple formula for finding the perimeter of the orthic triangle. We expect this approach will be found useful for

Y.-K. Man (✉)
Department of Mathematics and Information Technology,
The Education University of Hong Kong, 10 Lo Ping Road, Tai Po,
New Territories, Hong Kong
e-mail: ykman@eduhk.hk

© Springer Nature Singapore Pte Ltd. 2018
S.-I. Ao et al. (eds.), *Transactions on Engineering Technologies*,
https://doi.org/10.1007/978-981-10-7488-2_18

243

Fig. 1 ΔDEF is the orthic
triangle of ΔABC

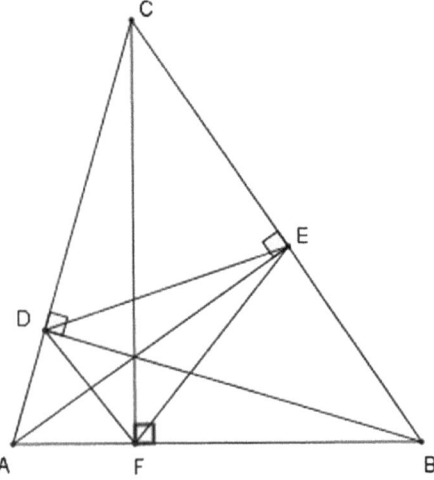

reference by school teachers and lecturers who are interested to introduce this
interesting and challenging geometry problem to their own students.

2 Properties of the Orthic Triangle

Here are some basic properties of orthic triangle and the proofs can be found in
[2, 3, 16].

Theorem 1 *The altitudes CF, AE and BD bisect the angles ∠DFE, ∠FED and
∠EDF, respectively.*

Theorem 2 *There are some equal angles in Fig. 1, namely:*

1. ∠ACB = ∠AFD = ∠BFE
2. ∠ABC = ∠ADF = ∠CDE
3. ∠CAB = ∠BEF = ∠CED
4. ∠CAE = ∠CFD = ∠CFE = ∠CBD
5. ∠BCF = ∠BDE = ∠BDF = ∠BAE
6. ∠ABD = ∠AEF = ∠AED = ∠ACF

3 How to Solve the Fagnano's Problem

First, let us introduce a very useful technique for solving a shortest path problem
below [3, 6].

   *Given two points A, B on the same side of a fixed line. How to determine a point,
say O, on the line such that the path A-O-B is the shortest possible?*

Suppose the fixed line is denoted by ℓ. By using it as an axis of reflection, we can determine the mirror image of B, say B_1. Then, we can determine the position of the point O, which is the intersection of AB_1 and ℓ (see Fig. 2). Now, $AO + OB = AO + OB_1 = AB_1$. Since the latter is a line segment, so the path A-O-B is the shortest possible. In fact, if we choose an arbitrary point P ($\neq$ O) on ℓ, the path A-P-B will be longer than A-O-B because $AP + PB = AP + PB_1 > AB_1$, due to the *Triangular Inequality* (see Fig. 3). With the use of dynamic geometry software, say GeoGebra, we can easily complete the constructions shown in Figs. 2 and 3. Also, we can easily deduce that $\angle\alpha = \angle\beta$, $\angle\beta = \angle\gamma$ and hence $\angle\alpha = \angle\gamma$. Thus, the results can be summarized in the theorem below, which is called the Heron's Theorem by some authors (see [2, 11]).

Theorem 3 *Given two points A, B on the same side of a fixed line ℓ. The shortest path joining A, B and meeting ℓ is a broken line whose parts make equal angles with ℓ.*

Fig. 2 Finding the shortest path by the reflection technique

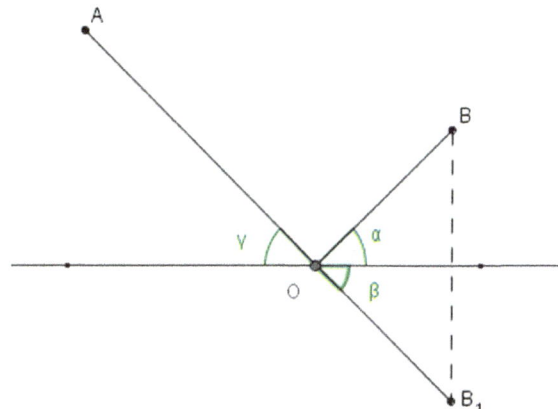

Fig. 3 The path A-P-B is longer than A-O-B

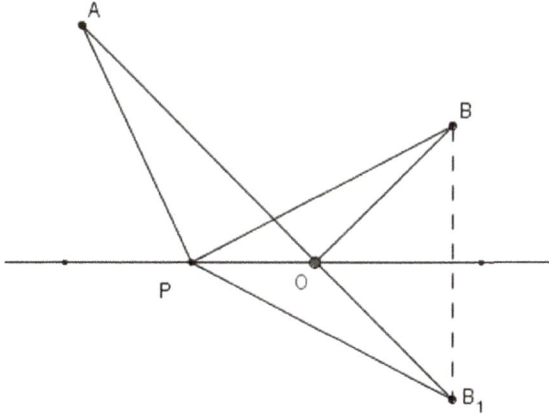

Now, we can describe how to solve the Fagnano's problem via the use of GeoGebra. First, an acute triangle $\triangle ABC$ with altitudes AE, BD, CF and a point P on AB are drawn (see Fig. 4).

By using BC and AC as the axes of reflection, we can use the GeoGebra tools to find the mirror images of P, say Q and R, respectively. Then, QR is drawn to meet AC and BC at the points M, N respectively (see Fig. 5).

It is obvious that $\angle CMN = \angle BMP$ and $\angle CNM = \angle ANP$. However, $\angle APN$ may not equal to $\angle BPM$ in general. Anyway, we can drag the point P slowly along AB and observe the angle measures of $\angle APN$ and $\angle BPM$ until they are equal (see Fig. 6). By doing so, we can deduce that $\angle APN = \angle BPM$, $\angle CMN = \angle BMP$ and $\angle CNM = \angle ANP$ occur when P, M, N coincide with the points F, E and D, respectively. By Theorem 3, the shortest paths from N to M via P, P to N via M and

Fig. 4 An acute triangle $\triangle ABC$ with three drawn altitudes

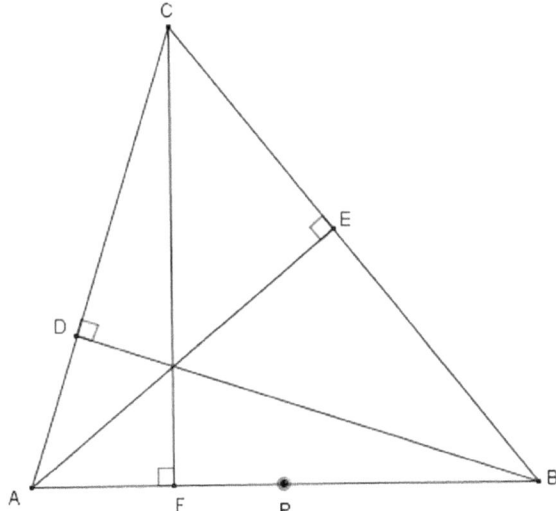

Fig. 5 $\triangle PMN$ is an inscribed triangle of $\triangle ABC$

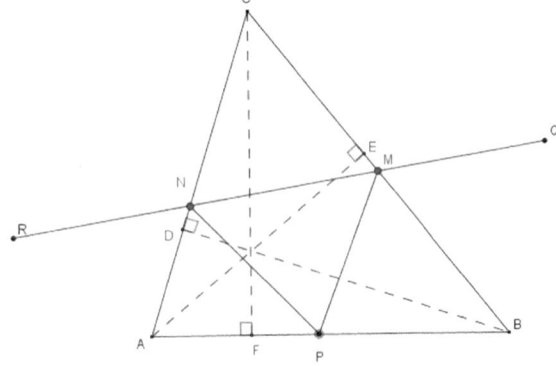

Fig. 6 The orthic triangle has the minimal perimeter

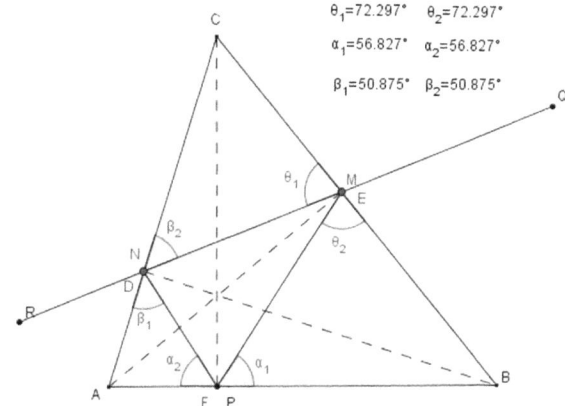

M to P via N, are N-P-M, P-M-N and M-P-N, respectively. It implies that the inscribed triangle of $\triangle ABC$ with minimal perimeter is the orthic triangle $\triangle DEF$.

Let us summarize the result as follows.

Theorem 4 *Given an acute triangle $\triangle ABC$, the inscribed triangle with the minimal perimeter is the orthic triangle of $\triangle ABC$.*

If $\triangle ABC$ is regarded as a triangular billiard table with smooth boundary and P as a billiard such that it always moves in a straight line at a constant speed in the interior of $\triangle ABC$ and the angle of incidence is equal to the angle of reflection whenever P hits the boundary of the table, then the trajectory of P is the orthic triangle. In fact, it is called the Fagnano billiard trajectory or a closed 3-periodic billiard trajectory by some researchers. Readers may refer to [1, 17] and the references therein for more details and related results.

4 The Perimeter of the Orthic Triangle

Let $BC = a$, $AC = b$, $AB = c$, $\angle CAB = \theta_1$, $\angle ACB = \theta_2$ and $\angle ABC = \theta_3$. We now derive a formula for finding the perimeter of the orthic triangle $\triangle DEF$ in the acute triangle $\triangle ABC$ (see Fig. 7).

By Theorem 1, $\theta_1 = \alpha_1 = \alpha_2$, $\theta_2 = \beta_1 = \beta_2$, $\theta_3 = \gamma_1 = \gamma_2$. Since $\triangle ADF \sim \triangle ABC$, we have $DF/BC = AF/AC$.

So,

$$DF = a \times \frac{b \sin(\pi/2 - \theta_1)}{b} = a \cos \theta_1$$

Also, $\triangle BFE \sim \triangle BCA$, we have $FE/AC = BE/AB$.

Fig. 7 Some equal angles in
$\triangle ABC$

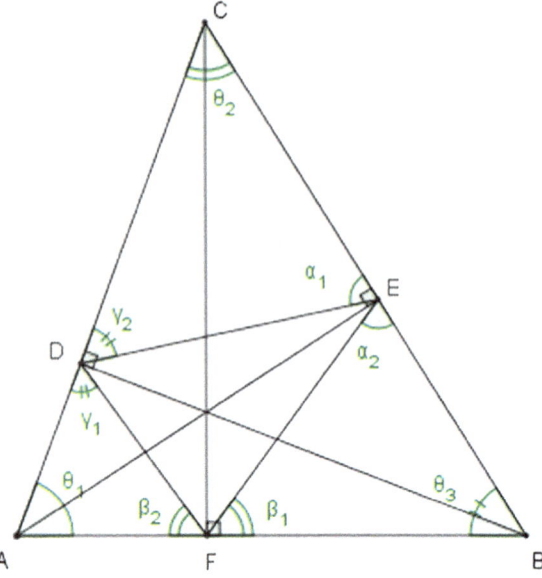

So,

$$EF = b \times \frac{c\ \sin(\pi/2 - \theta_3)}{c} = b\ \cos\ \theta_3$$

Similarly, $\triangle CED \sim \triangle CAB$, we have $DE/AB = CE/AC$.
Hence,

$$DE = c \times \frac{b\ \sin(\pi/2 - \theta_2)}{b} = c\ \cos\ \theta_2$$

Theorem 5 *The side lengths of the orthic triangle $\triangle DEF$ in the acute triangle $\triangle ABC$ are equal to $a \cos \theta_1$, $b \cos \theta_3$ and $c \cos \theta_2$, respectively.*

Theorem 6 *The perimeter of the orthic triangle $\triangle DEF$ in the acute triangle $\triangle ABC$, say $\delta_{\triangle DEF}$, can be found by using the formula:*

$$\delta_{\triangle DEF} = a\ \cos\ \theta_1 + b\ \cos\ \theta_3 + c\ \cos\ \theta_2.$$

By using cosine law and polynomial factorizations, we can rewrite the formula as follows:

$$\delta_{\triangle DEF} = \frac{(a+b+c)(a+b-c)(a+c-b)(b+c-a)}{2abc}.$$

Example. Suppose $\triangle ABC$ is an equilateral triangle with unit side length. Determine the perimeters of the sequence of successive orthic triangles in $\triangle ABC$.

Solution. Let $\triangle_i$ denotes the i-th orthic triangle in $\triangle ABC$. By Theorem 6, we have

$$\delta_{\triangle_1} = 3 \cos \frac{\pi}{3} = \frac{3}{2}$$

$$\delta_{\triangle_2} = 3 \times \frac{1}{2} \cos \frac{\pi}{3} = \frac{3}{4}$$

$$\delta_{\triangle_3} = 3 \times \frac{1}{4} \cos \frac{\pi}{3} = \frac{3}{8}$$

In general, we have: $\delta_{\triangle_i} = 3 \times \frac{1}{2^{i-1}} \cos \frac{\pi}{3} = \frac{3}{2^i}$.

Example. Prove that the perimeter of the orthic triangle $\triangle DEF$ in the acute triangle $\triangle ABC$ is less than twice the length of any altitude of $\triangle ABC$.

Solution. Consider the Fig. 8, where $\triangle DEF$ denotes the orthic triangle of $\triangle ABC$. By using BC, AC as the axes of reflection, we can find the mirror images G and H of F, respectively. According to the property of reflection, DF = DG, FE = EH, GC = HC = FC = altitude of $\triangle ABC$, so DF + DE + FE = GD + DE + EH = GH < GC + HC = 2 × FC due to the *Triangular Inequality*. Therefore, the perimeter of $\triangle DEF$ is less than twice the length of the altitude FC. By similar arguments, we can prove that the perimeter of $\triangle DEF$ is less than twice the length of the altitude BD or AE. It completes the proof.

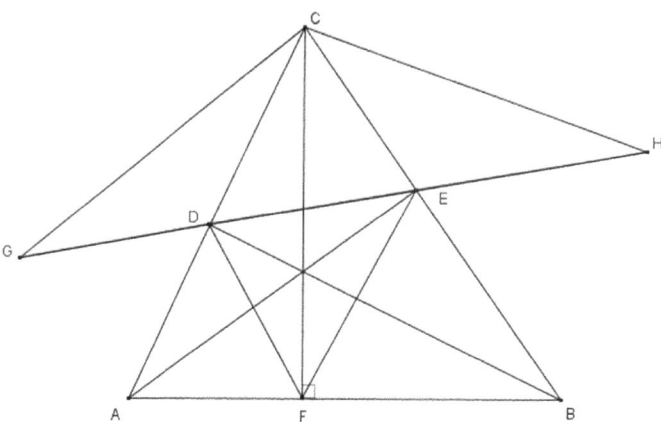

Fig. 8 The perimeter of $\triangle DEF$ is equal to GH

Fig. 9 An open periodic
billiard orbit in a right triangle

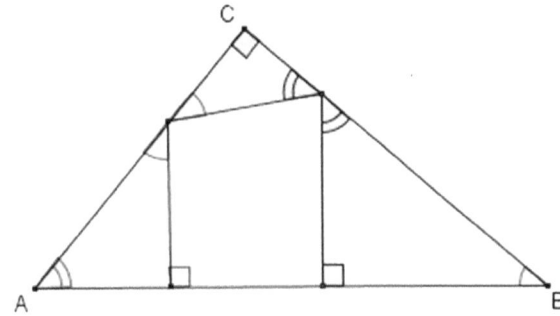

5 Final Remarks

We have discussed how to use a simple dynamic geometry approach to solve the
famous Fagnano's problem and provided a simple formula for finding the perimeter
of the orthic triangle of a given acute triangle. It is worth to point out that some
recent researches on periodic billiard trajectories in polygons originated from the
Fagnano's problem, where the solution, namely the orthic triangle, is often called a
closed 3-periodic billiard trajectory or orbit. If we replace the acute triangle by a
right triangle, then the orthic triangle will degenerate to an altitude of the right
triangle concerned. However, periodic billiard orbit does exist in right triangles,
although the orbits are open rather than closed. Figure 9 illustrates such an
example. In fact, there are still many open problems in the research areas of periodic
billiard orbits in triangles or convex n-sided (n > 3) polygons. Readers may refer to
[4, 7, 8, 13–15] for some of the open problems and recent results in these interesting
and fascinating research areas.

Acknowledgements This article is a revised version of the paper presented orally by the author at
the International Multi-Conference of Engineers and Computer Scientists (IMECS 2017) held on
15–17 March 2017 at Hong Kong [12].

References

1. N. Alkoumi, F. Schlenk, Shortest closed billiard orbits on convex tables. Manuscripta Math.
   147, 365–380 (2015)
2. R. Courant, H. Robbins, *What is mathematics? An Elementary Approach to Ideas and
   Methods*, 2nd edn. (Oxford University Press, Oxford, 1996)
3. H.S.M. Coxeter, S.L. Greitzer, *Geometry Revisited* (AMS, Washington, DC, 1967)
4. V. Cyr, A number theoretic question arising in the geometry of plane curves and in billiard
   dynamics. Proc. Am. Math. Soc. **140**(9), 3035–3040 (2012)
5. H. Dorrie, *100 Great Problems of Elementary Mathematics: Their History and Solutions*
   (Dover, New York, 1963)
6. D. Gay, *Geometry by Discovery* (Wiley, New York, 1998)
7. D. Grieser, S. Maronna, Hearing the shape of a triangle. Not. AMS **60**(11), 1440–1447 (2013)

8. E. Gutkin, Billiard dynamics: an updated survey with the emphasis on open problems. Chaos: Interdiscip. J. Nonlinear Sci. **22**(2) (2012). http://dx.doi.org/10.1063/1.4729307
9. N.M. Ha, Another proof of Fagnano's inequality. Forum Geometricorum **4**, 199–201 (2004)
10. F. Holland, Another verification of Fagnano's theorem. Forum Geometricorum **7**, 207–210 (2007)
11. R.A. Johnson, *Advanced Euclidean Geometry* (Dover, New York, 2007)
12. Y.K. Man, A dynamic geometry approach to the Fagnano's problem, in *Proceeding of International Multi-Conference of Engineers and Computer Scientists 2017, IMECS 2017*, Lecture Notes in Engineering and Computer Science, 15–17 Mar 2017, Hong Kong, pp. 117–120
13. P.J. Nahin, *When Least is Best* (Princeton University Press, NJ, 2004)
14. J.R. Noche, Periodic billiard paths in triangles, in *Proceeding of Bicol Mathematics Conference*, Ateneo de Naga University, 4 Feb 2012, pp. 35–39
15. A. Ostermann, G. Wanner, *Geometry by Its History* (Springer, Berlin, 2012)
16. A.S. Posamentier, *Advanced Euclidean Geometry* (Wiley, NJ, 2002)
17. S. Tabachnikov, *Geometry and Billiards* (AMS, Providence, RI, 2005)
18. P. Todd, Using force to crack some geometry chestnuts, in *Proceeding of 20th Asian Technology Conference in Mathematics*, Leshan, China, 2015, pp. 189–197

An Innovative Approach to Video Based Monitoring System for Independent Living Elderly People

Thi Thi Zin, Pyke Tin and Hiromitsu Hama

Abstract In these days the population of elderly people grows faster and faster and most of them are rather preferred independent living at their homes. Thus a new and better approaches are necessary for improving the life quality of the elderly with the help of modern technology. In this chapter we shall propose a video based monitoring system to analyze the daily activities of elderly people with independent living at their homes. This approach combines data provided by the video cameras with data provided by the multiple environmental data based on the type of activity. Only normal activity or behavior data are used to train the stochastic model. Then decisions are made based on the variations from the model results to detect the abnormal behaviors. Some experimental results are shown to confirm the validity of proposed method in this paper.

Keywords Elderly daily activities · Event based model · Human interactions with environments · Layered Markov model · Stochastic model Unusual behavior · Video based monitoring system

1 Introduction

Due to the increasing number of elderly people worldwide with independent living and they have their own problems because of aging such as memory disorders physical illness, the elderly people need to receive sufficient care to improve the

T. T. Zin (✉)
Faculty of Engineering, University of Miyazaki, Miyazaki, Japan
e-mail: thithi@cc.miyazaki-u.ac.jp

P. Tin
Center for International Relations, University of Miyazaki, Miyazaki, Japan
e-mail: pyketin11@gmail.com

H. Hama
Osaka City University, Osaka, Japan
e-mail: hama@ado.osaka-cu.ac

© Springer Nature Singapore Pte Ltd. 2018   253
S.-I. Ao et al. (eds.), *Transactions on Engineering Technologies*,
https://doi.org/10.1007/978-981-10-7488-2_19

quality of life so that they can extend the period of their independent living. According to statistics provided by the World Health Organization, Worldwide, the number of persons over 60 years is growing faster than any other age group. It was estimated that by the year 2050, the number of elderly population will grow up to two billion. This projection will make in that time; the number of aging people will be much larger than that of young ones under the age of 14 years for the first time in human history [1]. This trend will lead to the number of people requiring care will grow accordingly. Thus a system permitting elderly to live safely at home is more than needed. Most of health care and medical professional believe that one of the best ways to detect emerging physical and mental health problems, before it becomes critical—particularly for the elderly—is analyzing the human behavior and looking for changes in the activities of daily living. The general nature daily activities involve sleeping and eating, waking up and making up meals, showering and towering, cleaning and laundering, turning on and turning off TV, dressing and getting toilet, opening and closing doors, taking medicines and making snacks and so on [2]. Therefore, we need an automatic recognition technique for these activities to establish an automatic health monitoring system to alarm the persons concerned [3]. This makes how an activity monitoring system is important in the future health applications. In these days, advances in technology have an amazing potential to provide equipment turning the home environment smart with a network of cameras through which each and every activity can be observed and analyzed. By using those observations along with stochastic behavior modeling a video based activity monitoring system can assist elderly people for safety independent living at their homes. This kind of system by using information about human key postures will be beneficial to all of us especially for recognizing unusual activities of elderly people among their daily activities Thus in this paper we will focus on recognizing activities that elderly people are able to do on daily basis for example ability of elderly person to reach and open a kitchen cupboard or ability of taking medicine regularly. The recognition of these activities can help the medical experts (gerontologists) to evaluate the degree of frailty of elderly. The changes in their behavior patterns are necessary to be thoroughly investigated. It is also worthwhile to focus on detecting critical situations of elderly (e.g. feeling faint, falling down), which can indicate the presence of health disorders (physical and/or mental). The detection of these critical situations can enable early assistance for elderly people.

Another area of research topics concerning with assisting technology for elderly and patient is video based system for detecting and recognizing accident leading actions such as falling from bed or falling in the bath rooms. The falling event among others is the biggest problem for elderly people because it may cause a lot of side effects such as delaying recovery, brain injury or even may lead to losing life. According to the statistics provided in World Health Organization (WHO), 28–35% of elderly people over 70 years old have some experience associated with falls in nearly every year [4, 5]. In the case that nobody is around an elderly living alone, then his/her fall cannot be known. This can cause serious results such as life-threatening or even result in death. To avoid this situation, services to detect the fall and notify it to observers his/her family, relatives, supporters or rescuers, are required immediately.

All in all, in order to realize the ideas described in the above we believe that a video based monitoring system composed of simple techniques will be more beneficial to both users and society. Thus in this chapter we propose an innovative approach to the visual elderly people activity monitoring system for elderly people living alone at their homes. The proposed system aims to have a general structure for detecting and recognizing various types of activities for different applications and different environments. More over another objective of the visual monitoring system for elderly people daily activity analysis is to improve the quality of elderly life with the aid of modern technology and prolong the period of independent living in their own home. Due to aging some common problems arise among the elderly people such as memory disorder, action obstacles and etc.

So, we organize the chapter as follows. In Sect. 2 some related works are presented followed by the overview of proposed method in Sect. 3. Some experimental results of the proposed method are shown in Sect. 4. Finally, conclusion and future works are described in Sect. 5.

2 Some Related Works

An automatic video based monitoring of elderly daily activities has been drawn much attention from the image processing researchers in particular. Over the last a few years back, a lot of efforts has been put into developing and employing a variety of cameras and sensors to monitoring activities at homes, including activity detection, tracking and recognition for elder people [2, 6–8]. Some other researchers utilize wearable sensors to detect human posture and using accelerometers to detect some types of activities such as sitting, standing, walking, lying down and bending for examples [9, 10]. But employing the wearable sensors is not included in our proposed not to put mental and physical pressures on the elderly.

Thus, an innovative video based monitoring system is more suitable for elderly people daily activity observation and analysis while they are living alone at homes so that necessary precautions can be taken [9, 11]. Using the observed data, the system will learn the activities and behaviors of elderly people who involved the system. From the learned models, the system will detect any unusual or suspicious behaviors or activities and by this time the system will send an alarm signal to the persons in charge of the elders so that helps can be given immediately.

In general, the system consists of two sets of functions namely, the basic set and the interactive set. The function of the basic set includes the basic tasks of everyday life such as eating, bathing, walking, toileting and dressing. The interactive set includes the activities that people do in interactive manners such as phone use, watch TV, talk visitors etc. [12, 13].

There has been a variety of methods to tackle the problems of elderly activity monitoring system. Among them, Markovian methods including Hidden Markov Models (HMM), Bayesian approaches and Machine Learning Techniques including Neural Network and Deep Learning algorithms are some to name a few [14–17]. In HMM approach the types of activities or behaviors are considered as Hidden States to be discovered and the corresponding features are behaved as observations along with emission probabilities. Combining the transition probabilities of Hidden States, the emission probabilities with the initial probabilities, the HMM is trained for known activities. Then the trained HMM is employed to detect and recognize the daily activities of elderly people. Hoe ever the Bayesian approach utilizes the posterior probability for choosing the states to merge and for the stopping criterion. Sometimes, a variable-length Markov models can be used to efficient Bayesian to analyze the motion detector data, to learn the behavior of the elders.

Recently, the machine learning techniques based on the convolutional neural networks are used for human activities classification [18–21]. General speaking, the input video sequences are processed through convolution layer by using a certain kernel, then the resultants pass through Rectified Linear stage and maximum pooling stage. This type of layers can be repeated with different kernels and continued until we stop by fully connected layer. In the final fully connected layer we can employ SVM or Soft max classifiers to attain final outputs.

In this paper we will employ different approach which is composed of multiple layers similar to deep learning but modified by using concept of Markov Chain and HMM concepts.

3   Proposed Elderly Activity Monitoring System

The general architecture of the proposed video based activity monitoring system which is an extended version of our conference paper [22], is made up as a Multi-Layered Markov Model consisting five Interactive Layers. Layer I is the root or input layer and the last Layer V is the output layer. The middle three layers, Layer II to Layer IV are environment layer, activity layer and objects layer respectively. The overview of this architecture is described in Fig. 1. In order to realize the proposed system, we define a terminology called Mode as the set of all possible actions and interactions acted by the elderly people. Thus Mode may contain behaviors of individuals, their interactions against the system or environments. The degree of personal dependency will be also included in the Mode. We also define another mode to consider the restraints and difficulties for elderly persons making them obstacles in performing their daily activities. Thus the task of monitoring system is to use an efficient model to recognize the mode of the elderly person so that they can perform towards the system such as: Off, On, Alarm, Warning. The majority of previous works were based on the Markov model as a model for the recognition of old people activities focused only on a particular event of eating dinner [20, 21].

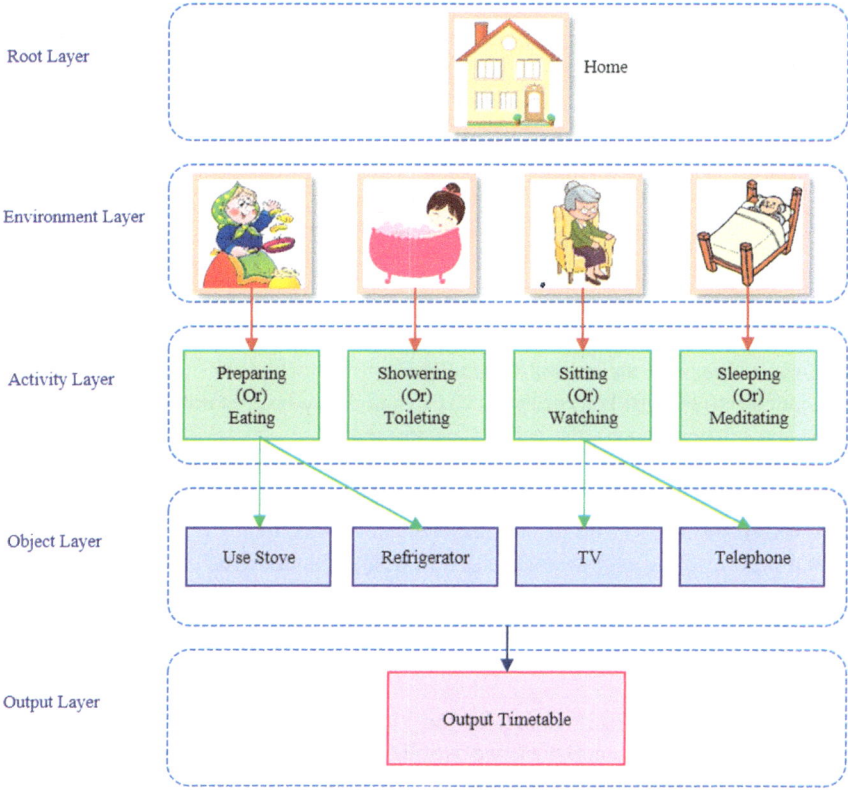

Fig. 1 Overview of the proposed system

In the present proposed system each particular activity is taken place in its corresponding environment. For example, "the person cannot prepare a meal in the bedroom", "the person cannot sleep in the bathroom", etc. Also each activity is composed of a series of sub-activities before the main activity is performed. In the activity of meal preparation, it may include sub-activity of opening and closing refrigerator or turning on and turning off the corresponding accessories such as toaster, oven and stove etc. It is true for all activities performed by the elderly people. For simplification of recognition process, we tailor each activity to the location where it is performed. Here in this chapter, we assume only a common home architecture that consists of: Bedroom, Bathroom (including toilets), Kitchen and Living room. Also we assume the home is furnished with essential utilities as described in Table 1. Then a Hidden Markov Model and Markov Chain Model are employed for detection, recognition and classification each activity based on the room location in which the person is involved.

In order to compute a Hidden Markov Model, we first define the relationship between layers. For this model, we started at Home Layer as Layer I. The Layer II

Table 1 Description of
home functions

	Activities	Utilities
Kitchen	Preparing meals Eating, Drinking Washing dishes	Stove Refrigerator Coffee maker
Bath room	Taking shower, Bath Toileting	Bath tub, Soap Toilet
Living room	Watching TV Using telephone Meeting with visitor	TV, CD Player Telephone Chair table

represents a mode of all possible places in the home. Layer III is concerned with the utilities or objects in the home. All activities and actions performed by the elder are embedded into the last layer, Layer IV. Each of every layer is directly related to the other layer depending on the activity performed.

For example, if the person opens the refrigerator the following action will be the most likely to have a drink or to make a meal preparation. In similar fashion, if the person turn-on the TV, he or she may be likely watching TV. Consequently, we cannot pass from layer n to layer $n + 2$. These links allow us to make a complete mode starting with the kind of place where the action is realized by the resident.

Concerning the "Layer III", each activity can have from 1 to n object(s) with n is the number of objects used in this activity. The link between each object in the same "Layer IV" is the <and>.

Example: the person at home, in the kitchen, preparing meal, use stove and refrigerator. By using an HMM, we are able to detect the abnormal activities by discovering the hidden states.

3.1 Technical Details for Hidden Markov Model

In order to formulate the daily activity of elderly care problem as Hidden Markov Model, we first divide a day into 48-time interval having 30 min length of each time interval.

Let $S_j(t)$ be the state that the elderly person is in room j (j = 1, 2, 3, 4) at time t for t = 1, 2, …, 48.

By using visual monitoring system, we can estimate the following transition probabilities, initial probabilities and emission probabilities.

Transition Probability: $p_{ij}(t) = \Pr\{S_j(t+1)|S_i(t)\}$
Transition Probability Matrix: $A(t) = [p_{ij}(t)]$
Initial Probability Vector: $\pi = \{\pi_1, \pi_2, \pi_3, \pi_4\}$
Emission Probability Matrix: $B = \{b_i(k)\}$

where, $b_i(k) = \Pr\{O_t = k \mid S_i(t)\}$ in which O_t is the observation of the system at time t.

We then establish a Hidden Markov Model $\lambda = (\pi, A, B)$ and employ a forward algorithm to find the state sequence.

$$S = \{S_1(t), S_2(t), S_3(t), S_4(t)\}, \quad \text{for } t = 1, 2, \ldots, 48.$$

From this state sequence we can detect the abnormal behavior of elderly people by observing the time duration at which the person spent at one location more necessary (this can be done by using predefined threshold).

For example, we consider a piece of three rooms where each room is equipped with a visual monitoring system for presence, motion, etc. The information provided by each subsystem allows to know about the activities and the presence of the person. In our simulation, each room is linked to specific activity. The simulation period is fixed a duration say 5 or 6 h. We simulate these activities in a random way. Each mode has a specific set of visual cameras regarding each room. In our case, we simulate these modes during 6 h: from 6:00 to 12:00 A.M. Each room contains a specific mode which contains from 1 to n activities. Example: at 6:00 A.M. the person wakes up in the bedroom, takes toileting at 6:15 in the bathroom. In the kitchen, elderly prepares his breakfast at 8:00. Afterwards, he washes the dishes then goes to the living room to watch TV at 9:30. Later on, the elderly has entered the kitchen to prepare his meal at 12:00 then he takes his medication in the bedroom at 12:45.

This example contains three modes: The kitchen mode: preparing breakfast at 8:00; Wash these dishes at 8.30; preparing meal at 11:00. The bedroom mode: wakes up at 6:00; Take medication at 12.45. The living room scenario: watch TV at 9.30. We then simulate the model to investigate the daily activities of elderly people living independently.

3.2   Use of Time Dependent Markov Chain Model

We have already noted that each main activity consists of a series of sub-activities to pass through. For example, for activity of preparation and having breakfast, the sub-activities may include opening and closing refrigerator, use of toaster, making coffee, turning on and off water taps, sitting and eating so on.

Let us M_i denote for $i = 1, 2, \ldots, K$ as ith main activity where K is the total number of daily activities performed by the elders.

Let $U_j(i)$ be jth sub-activity in jth main activity for $j = 1, 2, \ldots, L$, where L stands for the number of sub-activities to complete the main activity i.

In order to perform activity analysis we will assume that both processes $\{M_i\}$ and $\{U_j(i)\}$ time dependent Markov Chain processes.

Therefore, we can estimate the transition probabilities for both processes by using collected data from the monitoring system. We then define m_{ij} as the transition probability between M_i and M_j. Similarly, $u_{jk}(i)$ be defined as the transition probability between $U_j(i)$ and $U_k(i)$. In matrix forms we will write

$$M = m_{ji} \text{ and } U_i = u_{jk}(i).$$

Let T_i for $i = 1, ..., K$ be a random variable representing time taken to complete the activity i. So that T_i can interpreted as the first passage time taken from starting sub-activity to the last sub-activity in ith main activity. Since there are L sub-activities to complete ith main activities, we have to calculate the expected value of T_i starting from sub-activity 1 to ending sub-activity L.

Let us denote the expected value of T_i starting from sub-activity j to ending sub-activity k by $u_{jk}(i)$. We then have from the fundamental properties of Markov Chain Theorem for first time probability, satisfies the following equation.

$$u_{jk}(i) = 1 + \sum_{l \neq k} u_{ji}(i)u_{lk}(i)$$

By solving this equation, we obtain the average passage time going one sub-activity to another sub-activity. If this time is more a threshold value, then the activity of the person is suspicious and to be taken care of from the system. Thus the decision is the action of elderly is abnormal if $u_{lk}(i) \geq Thresholdvalue$.

4   Experimental Simulation Results

Scenario 1
In this section, we shall illustrate the validity of the proposed models by showing some experimental simulation results. In order to do so, we first estimate the parameters involved in the models. They are (N, A, B, Π) where N is the number of state, A is the transition matrix, B is the emission matrix and Π is the initial matrix. The following matrices A, B and Π are row stochastic, which means that each element is a probability and the elements of each row sum to 1, that is, each row is a probability distribution. In this case, we implement our modes with the following matrices.

Transition matrix represents the probabilities of the transition between the states where each state represents a human mode.

A	Mode K	Mode L	Mode B
Mode K	0.15	0.715	0.1
Mode L	0.1	0.8	0.1
Mode B	0.1	0.11	0.8

Initial matrix represents the probabilities of the initial state.

	Mode K	Mode L	Mode B
Π	0.6	0.2	0.2

Emission matrix contains the emission probabilities, the probability to emit each observation for each state.

	Mode K	Mode L	Mode B
B	0.2	0.6	0.2

By using the synthetic data, we estimate the three Modes (Mode L, Mode K, and Mode B) according to the time. We can perform the estimation of the hidden state "in this context the state is the place/location of the person" for each room in (300 min) using the Mode L (Living-room scenario), K (Kitchen) and B (Bath-room). We obtain the transition matrix as

0.105263	0.561407	0.33333
0.105263	0.561407	0.33333
0.105263	0.561396	0.333341

The state vector after 300 min, we get

$$0.105263 \quad 0.561405 \quad 0.333332$$

These results show that the elderly people spent more time in the living room rather than the bed room. More real data will be needed to make accurate results.

Scenario 2
In this scenario, we consider an activity based analysis. Suppose the person is in the Kitchen mode to prepare and have the breakfast. We assume that the corresponding sub-activities are use of refrigerator, use of toaster, eating meal and washing dishes. We also assume that these sub-activities will be performed in series. From the collected information from the monitoring system, we can obtain the estimated parameters for the matrix $\mathbf{U}(i)$ in Sect. 3, here i stands for the main activity of having the breakfast.

	Sub/sub	1	2	3	4
$U(i)$	1	0.21	0.19	0.28	3.2
	2	0.2	0.27	0.28	0.25
	3	0.19	0.26	0.31	0.24
	4	0.16	0.26	0.36	0.22

By using the mean first passage time described in Sect. 3, we obtain the following simulated results. From this table we can see that the average passage time to enter a state

Table 2 Mean first passage times (minutes)

No	Starting state	Entering state	Mean passage time
1	1	4	3.56
2	1	3	3.35
3	1	2	4.33
4	1	1	5.28
5	2	4	3.85
6	2	3	3.37
7	2	2	3.99
8	2	1	5.31
9	3	4	3.90
10	3	3	3.24
11	3	2	4.02
12	3	1	5.41
13	4	4	3.98
14	4	3	3.10
15	4	2	4.03
16	4	1	5.53

Table 3 Mean passage time independent of starting state

No	Entering state	Mean passage time (min)
1	1	26.53
2	2	16.37
3	3	13.46
4	4	14.29

independent of starting state is given in Table 3. This can interpret as follows: The time taken for preparation and having breakfast is (26.53 + 16.37 + 13.46 = 56.36) nearly one hour a big longer than usual time of 30 min. Therefore, the activity is unusual or suspicious so that necessary measure should be taken (Table 2).

5 Conclusion and Future Work

In this work, we have proposed an innovative video based monitoring system for analyzing daily activities of elderly people who are living alone at homes. By using Multi-Layered Stochastic and time dependent Markov Chain models, we are able to develop methods for activity and location based systems. The simulation results reveal that the proposed model is efficient for activities recognition with an observation error rate that is not very large compared to our hidden states. In the next steps of this work, we will explore the enrichment of our approach by investigating the learning-based systems (such as neural networks with a new learning algorithm) in order to recognize the events accurately. More over real life scenarios will be taken for further research.

Acknowledgements This work is partially supported by the Grant of Telecommunication Advanced Foundation.

References

1. WHO Library Cataloguing-in-Publication Data WHO global report on falls prevention in older age 2007
2. B. Boulay, F. Bremond, M. Thonnat, Applying 3D human model in a posture recognition system. Pattern Recogn. Lett. **27**(15), 1788–1796 (2006)
3. A.A. vanzi, F. Bremond, C. Tornieri, M. Thonnat, Advances in intelligent vision system: methods and applications. Design and assessment of an intelligent activity monitoring platform. EURASIP J. Appl. Signal Process. Spec. Issue **60**, 870–880 (2015)
4. T. Sumiya, Y. Matsubara, M. Nakano, M. Sugaya, A mobile robot for fall detection for elderly-care, in *19th International Conference in Knowledge Based and Intelligent Information and Engineering Systems*, 1 Jan 2015, vol. 60, pp. 870–880
5. R. Kaur, P.D. Kaur, Review on fall detection techniques based on elder people. Int. J. Adv. Res. Comput. Sci. **3**(8) (2017)
6. T. Moeslund, A. Hilton, V. Kruger, A survey of advances in vision based human motion capture and analysis. Comput. Vis. Image Understand. (CVIU) **104**(2), 90–126 (2006)
7. F.Z. Bremond, M. Thonnat, A. Anfosso, E. Pascual, P. Mallea, V. Mailland, O. Guerrin, A computer system to monitor older adults at home: preliminary results. Gerontechnol. J. **8**(3), 129–139 (2009)
8. N. Zouba, B. Boulay, F. Bremond, M. Thonnat, Monitoring activities of daily living (ADLs) of elderly based on 3D key human postures. Int. Cogn. Vis. Workshop **5329**, 37–50 (2008)
9. D. Bruckner, B. Sallans, Behavior learning via state chains from motion detector sensors, *Presented at Bio-Inspired Models of Network, Information and Computing Systems*, 10 Dec 2007, pp. 176–183
10. E. Munguia-Tapia, S.S. Intille, K. Larson, Activity recognition in the home setting using simple and ubiquitous sensors. Proc. Pervasive **4**, 158–175 (2004)
11. G. Yin, D. Bruckner, Daily activity learning from motion detector data for ambient assisted living, *Presented at the 3rd International Conference on Human System Interaction*, 13 May 2010, pp. 89–94
12. T. Lemlouma, S. Laborie, P. Roose, Toward a context-aware and automatic evaluation of elderly dependency in smart home and cities, in *IEEE 14th International Symposium and Workshops on World of Wireless, Mobile and Multimedia Networks*, 2013, pp. 1–6
13. H. Msahli, T. Lemlouma, D. Magoni, Analysis of dependency evaluation models for e-health services, in *IEEE International Conference on Global Communication Conference*, 2014, pp. 2429–2435
14. Z. Liouane, T. Lemlouma, P. Roose, F. Weis, H. Messaoud, A Markovian-based approach for daily living activities recognition. arXiv:1603.03251, 10 Mar 2016
15. T. Duong, H. Bui, D. Phung, S. Venkatesh, Activity recognition and abnormality detection with the switching hidden semi Markov model, in *Proceedings of the IEEE Computer Society Conference on Computer Vision and Pattern Recognition (CVPR)*, 2005, vol. 1, pp. 838–845
16. N.M. Oliver, B. Rosario, A.P. Pentland, A Bayesian computer vision system for modelling human interactions. IEEE Trans. Pattern Anal. Mach. Intell. **22**(8), 831–843 (2000)

17. T.T. Zin, P. Tin, H. Hama, An innovative deep machine for human behavior analysis, in *Proceedings of 12th International Conference on Innovative Computing, Information and Control (ICICIC2017)*, Kurume, Japan, 28–30 Aug 2017
18. W.E. Hahn, S. Lewkowitz, D.C. Lacombe, J.E. Barenholtz, Deep learning human actions from video via sparse filtering and locally competitive algorithms. Multimedia Tools Appl. **74** (22), 10097–10110 (2015)
19. M. Baccouche, F. Mamalet, C. Wolf, C. Garcia, A. Baskurt, Sequential deep learning for human action recognition, *International Workshop on Human Behavior Understanding 2011*, 16 Nov 2011 (Springer, Berlin, Heidelberg), pp. 29–39
20. Z. Zhang, T. Tan, K. Huang, An extended grammar system for learning and recognizing complex visual events. IEEE Trans. Pattern Anal. Mach. Intell. **33**(2), 240–255 (2011)
21. Z. Liouane, T. Lemlouma, P. Roose, F. Weis, H. Messaoud, A Markovian-based approach for daily living activities recognition, in *Proceedings of the 5th International Conference on Sensor Networks*, 2016, vol. 1, pp. 214–219
22. T.T. Zin, P. Tin, H. Hama, Visual monitoring system for elderly people daily living activity analysis, in *Proceedings of The International MultiConference of Engineers and Computer Scientists 2017*. Lecture Notes in Engineering and Computer Science, 15–17 Mar 2017, Hong Kong, pp. 140–142

Color Blindness Image Segmentation Using Rho-Theta Space

Yung-Sheng Chen, Long-Yun Li and Chao-Yan Zhou

Abstract Segmentation of color information in RGB space is considered as the detection of clouds in rho-theta space. The conversion between RGB space and rho-theta space is first derived. Then the peak detection in the cloud-like rho-theta image is developed for color plane segmentation. The color blindness images are used for illustrations and experiments. Results confirm the feasibility of the proposed method. In addition, the segmentation of pattern and background for a color blindness image is also further demonstrated by means of the spatial distance computation among segmented color planes as well as the traditional K-means algorithm.

Keywords Cloud image · Color blindness image · Image segmentation
K-means · Peak detection · RGB · Rho-theta space

1 Introduction

Color image segmentation is of great importance in the field of image processing and pattern recognition. It is known that the color visual perception from human eyes is primarily reflected by red (R), green (G), and blue (B) color components. Thus the constructed color space for color information processing is usually named as RGB space. In order to facilitate the specific applications, several color space conversion methods have been proposed. For example, CMYK (Cyan-Magenta-Yellow-Black) space is frequently used in color printer; HSI (Hue-Saturation-Intensity) space is often adopted for the investigation of human visual phenomena; Lab (L for lightness,

Y.-S. Chen (✉)
Department of Electrical Engineering, Yuan Ze University, Taoyuan, Taiwan, ROC
e-mail: eeyschen@saturn.yzu.edu.tw

L.-Y. Li · C.-Y. Zhou
School of Physics and Telecommunication Engineering, South China Normal University,
Guangzhou, People's Republic of China
e-mail: 416785643@qq.com

C.-Y. Zhou
e-mail: 412503135@qq.com

© Springer Nature Singapore Pte Ltd. 2018   265
S.-I. Ao et al. (eds.), *Transactions on Engineering Technologies*,
https://doi.org/10.1007/978-981-10-7488-2_20

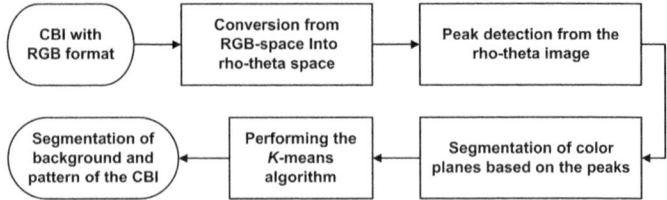

Fig. 1 Flowchart of the proposed approach

a and b for color-opponent dimensions) can be regarded as a device-independent color space; and YUV space is used in the traditional video display. Jin and Li proposed a switching vector median filter based on the Lab color space converted from the RGB space [1]. Lee et al. investigated a robust color space conversion between RGB and HSI for date maturity evaluation [2]. Mukherjee et al. used YUV colors space for image demosaicing [3].

Several well-known methods have been developed for performing the image segmentation such as thresholding based on histogram, clustering, region growing, edge detection, blurring, etc., which can also be extended and applied for color image segmentation. Underwood and Aggarwal projected the three-dimensional color space into two-dimensional plane and analyzed the color characteristics of detected tree outlines for reporting the degree of infestation present [4]. By means of the competitive learning technique, Uchiyama and Arbib presented a color image segmentation method which can efficiently divide the color space into clusters [5]. By combining region growing and region merging processes, Tremeau and Borel proposed a color segmentation algorithm [6]. Chen and Hsu adopted self-organizing feature map for performing color blindness image segmentation [7] and further developed an active-and-passive approach for understanding the figure in the color blindness image (CBI) [8].

Even there has a much progress in the field of color image segmentation during the past two decades, it still has a room for exploring the color space conversion on this topic. Since the CBI is often used for the investigation of human visual perception [8–11], in this study the CBI is adopted for investigating the characteristics between RGB space and a newly defined rho-theta (or ρ-θ) space. Based on this new space, the segmentation of color information can be readily demonstrated. The proposed approach can be simply depicted in Fig. 1. The adopted CBI is acquired in RGB-space, which is a three dimensional space. Let white color point [255, 255, 255] be an anchor point, any color point $[r, g, b]$ can be represented by two parameters ρ and θ (which will be defined in next section). Thus for a given CBI image, our approach includes: (1) conversion from RGB-space into ρ-θ space, (2) peak detection from the ρ-θ image, (3) segmentation of the color planes based on the peaks in the ρ-θ image, (4) performing the K-means clustering algorithm, and finally segmentation of background and pattern of the CBI. The partial work of this study has been presented in the IMECS 2017 [12].

The reset of this paper is organized as follows. Based on the proposed approach, Sect. 2 first investigates some observations of RGB space, then introduces the ρ-θ space, shows how to find the peaks in the ρ-θ image and segment the corresponding color planes based on the detected peaks. Section 3 presents the segmentation of background and pattern of a CBI based on the K-means algorithm. Experimental results and discussions are given in Sect. 4. The conclusion of this paper is finally drawn in Sect. 5.

2 Segmentation of Color Planes

2.1 Observation of a Color Image in RGB Space

Figure 2a–c show three CBIs, which are from the well-known Ishihara test plates and usually adopted for the study of color perception as mentioned previously. From our visual inspection, it mainly consists of the size-varied color dots with orange, brown, purple, green, and cyan colors. When it is used for the inspection of human color blindness, e.g., dichromats, the major colors will be focused on. In this case, the majority of color components are of red, whereas the minority is of green. For a normal vision, the figure "5" being composed of greenish dots can be perceived successfully, in which the majority of reddish dots is usually regarded as background. Note here that the white color is not used for perception and can be regarded as a reference.

Let a color pixel p including red, green, and blue components be denoted as $p[r, g, b]$. The color image in Fig. 2a, for illustration, can be easily converted into the well-known RGB space and shown in Fig. 2d. In this space, it is obvious that the color dots are mainly grouped into five color lines (i.e., purple, brown, orange, green, and cyan color lines) and separated from one to another. The concept of "color line" can be found in [13]. Therefore it is our goal in this study to develop a feasible method for the color segmentation method based on this observation.

2.2 ρ-θ Space

Figure 3 depicts the RGB space, where $o[0, 0, 0]$ represents the origin point or black point. However, as mentioned before, the white color can be regarded as a reference point, namely $w[255, 255, 255]$, in RGB space. This phenomenon can be found in Fig. 2d, where the distribution of each grouped color line diffuses from the reference point to the space. Therefore, a rho-theta parameterization like scheme can be applied for this study. Consider a pixel $p[r, g, b]$ in RGB space, a vector can be constructed from reference point $w[255, 255, 255]$ to it. In this study, two parameters ρ and θ representing the included angle of $\overrightarrow{wp}$ and B-axis and that of $\overrightarrow{wp}$ and R-

(a)   **(b)**   **(c)**

w[255, 255, 255]

(d)

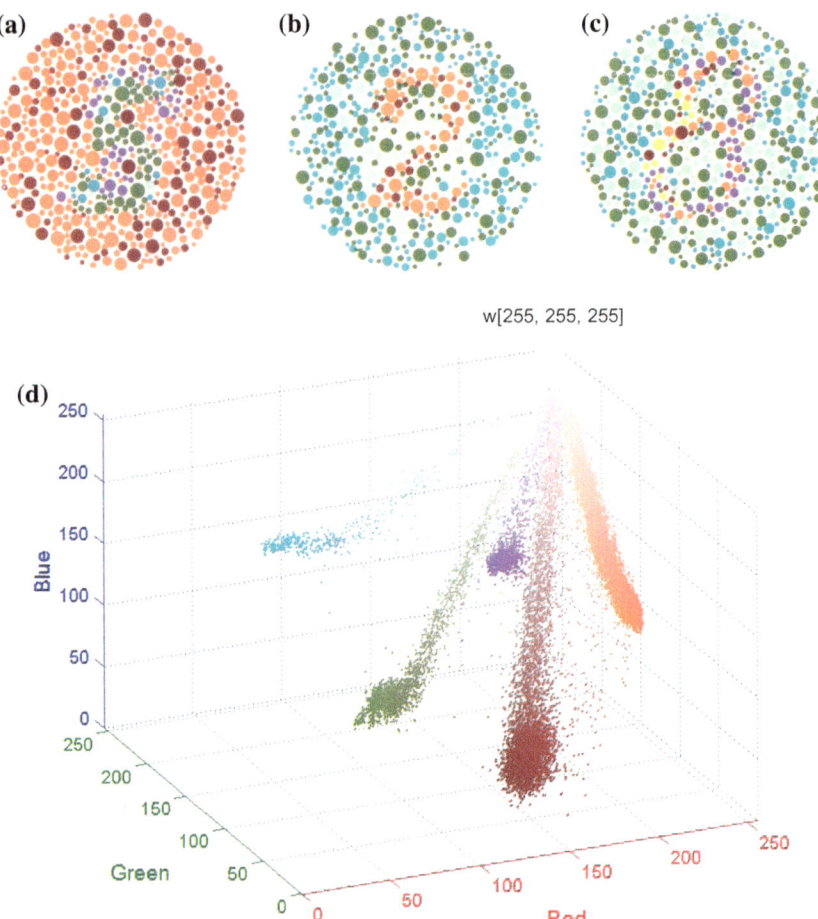

Fig. 2 **a–c** Three color blindness images from the ishihara test plates, and **d** the color distribution in RGB space for the CBI in (**a**), where five major color groups are easily observed

axis, respectively are used enough for further transformation. Ideally, each grouped color line distribution should be more gathered up in the new ρ-θ space. Based on this concept, the color segmentation could be readily performed in the ρ-θ space. According to the relationships in Fig. 3, the three components r, g, b of the pixel p can be represented as follows.

$$b = 255 - |\overrightarrow{wp}| \cos \rho \qquad (1)$$

$$r = 255 - |\overrightarrow{wp}| \cos \theta \qquad (2)$$

Fig. 3 Illustration of transforming a color pixel from RGB space into ρ-θ space

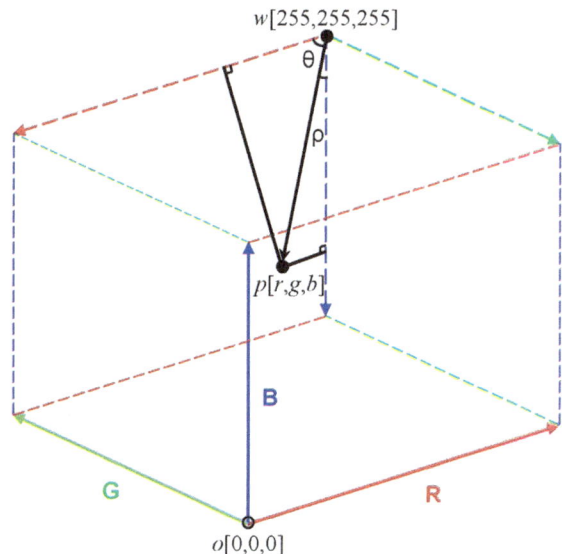

where

$$g = 255 - |\overrightarrow{wp}| \sqrt{1 - (\cos \rho)^2 - (\cos \theta)^2} \tag{3}$$

Thus we have

$$\rho = \cos^{-1}\left(\frac{255 - b}{|\overrightarrow{wp}|}\right) \tag{4}$$

and

$$\theta = \cos^{-1}\left(\frac{255 - r}{|\overrightarrow{wp}|}\right) \tag{5}$$

to represent the so-called ρ-θ space.

For the sake of performing the color segmentation, the ρ-θ space is designed as a two-dimensional array like the generalized Hough transform [14] uses. That is, if one case of ρ and θ occurs, it will be increased one in the memory location of (ρ, θ). The accumulated amount of (ρ, θ) indicates the number of those color pixels having ρ and θ values, which are treated as the same category as one color line in Fig. 2d displays. In order to make the color information to be more apparent for segmentation, all the pixels p having $|\overrightarrow{wp}| < 30$ (regarded as a white pixel) are ignored and will not be accumulated in the (ρ, θ) array, where the content at location (ρ, θ) is denoted as $C_{\rho\theta}$. The following procedure is next used for yielding the ρ-θ image. The threshold $\text{TH}_1 = 2$ is selected experimentally in this study.

Fig. 4 Cloud-like ρ-θ image transformed from the RGB information in Fig. 2d by the proposed method

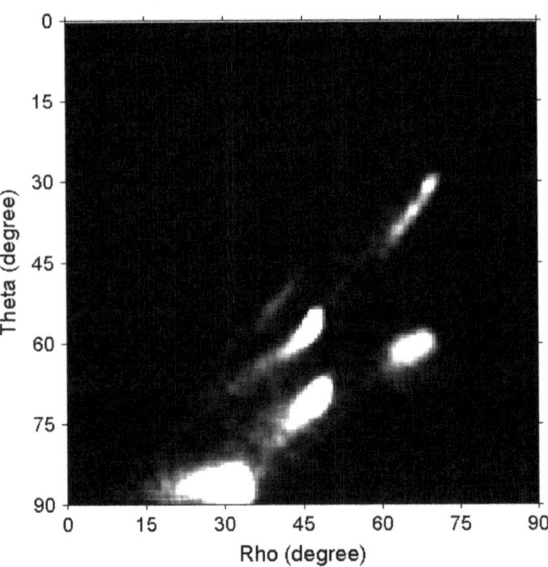

1. For each memory location (ρ, θ), do steps 2–3.
2. Compute ρ_p, and θ_p for all color pixels p (not white).
3. If $|\rho - \rho_p| <$ TH$_1$ and $|\theta - \theta_p| <$ TH$_1$, then $C_{\rho\theta} \leftarrow C_{\rho\theta} + 1$.

Note here that the range of ρ and θ is within $[0°, 90°]$ and $C_{\rho\theta}$ is normalized with $\widehat{C}_{\rho\theta} = C_{\rho\theta}/\overline{C}$, where $\overline{C}$ is the mean of all none-zero $(C_{\rho\theta})$s. Along this manipulation for Fig. 2d, a cloud-like ρ-θ image can be obtained as shown in Fig. 4, where we can find five major groups (or clouds) containing brightness area as the five color lines indicated previously. In addition, one small-area group with less brightness is also shown. This cue is very helpful for the color image segmentation.

2.3 Find the Peaks in ρ-θ Image

According to the property of ρ-θ image, the segmentation of the groups can be transformed into finding the respective peaks. There are two steps, namely local maxima detection and small-peak removal, in this process, where a $(2s+1) \times (2s+1)$ sliding window is used. We adopted $s = 3$ for our experiments. In the step of local maximal detection, a peak is labelled if its value is the local maximum within the corresponding local area. Furthermore, there exist many unwanted small peaks which shall be ignored. In this study, only the peak having $\widehat{C}_{\rho\theta} > (2\text{TH}_1)^2/\overline{C}$ will be remained.

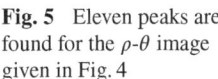

Fig. 5 Eleven peaks are found for the ρ-θ image given in Fig. 4

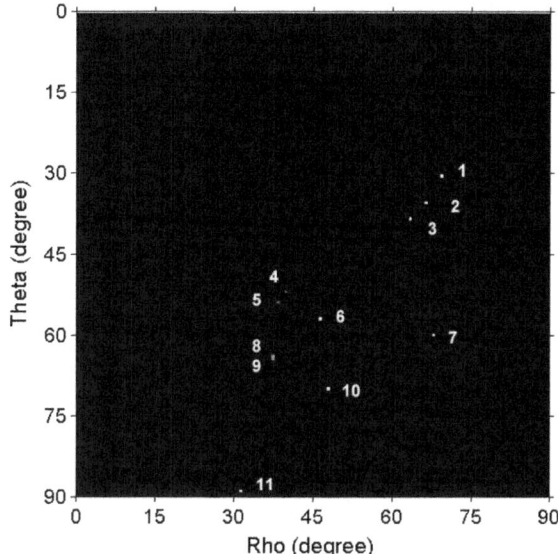

After performing such a process on the ρ-θ image given in Fig. 4, eleven peaks are detected as shown in Fig. 5.

By observing the images in Figs. 4 and 5, one cloud may include several peaks and one peak should have a local maximum within its cloud. In addition, the trend of pixel value changing is decreased gradually from the peak to the outer. Based on this property, we can use the relationship between any two peaks P and Q in the peak-image of Fig. 5 to represent whether they belong to the same group or not. Let v_P and v_Q be the value of peak P and Q respectively, $\overline{PQ}$ denote the line segment, and $S_{\overline{PQ}}$ be the set of all pixel values within the segment in the ρ-θ image of Fig. 4. In addition, note here that the value of "dark area" in Fig. 4 may not be exactly zero. Therefore a threshold $\mathrm{TH}_{PQ} = \min(v_P, v_Q)/2.4$ is used in this study. Within any line segment $\overline{PQ}(P \neq Q)$ we say peak P is not related to peak Q if there exists one pixel $\in S_{\overline{PQ}}$ whose value is less than TH_{PQ}. Otherwise, P and Q are related. As an illustration given in Fig. 6, TH_{PQ} represents the threshold for the two peaks P and Q. Since there are some peaks $\in S_{\overline{PQ}}$ whose values are less than TH_{PQ}, peaks P and Q are mutually independent ones. Contrarily, all the peaks $\in S_{\overline{QR}}$ whose values are greater than TH_{QR} for the two peaks Q and R, thus they are satisfied to the equivalent relationship and belong to the same cloud.

After performing such an equivalent relationship process for Figs. 4 and 5, an equivalent relationship table can be obtained as given in Table 1. Here peaks 1–3 are regarded as the same cloud; peaks 4 and 5 the same cloud; and peaks 6, 8 and 9 the same cloud. Others (peaks 7, 10 and 11) are independent clouds. If several peaks are related, the peak having the maximal value can be used to represent the

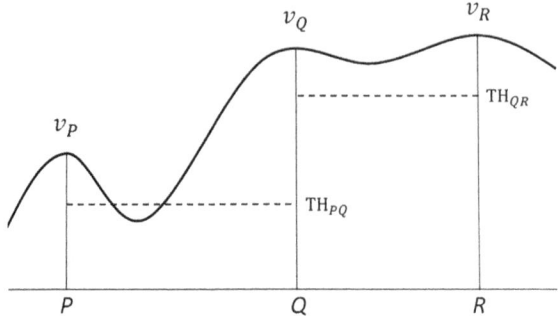

Fig. 6 Illustration of finding the equivalent relationship between two peaks, where peak Q and R are related, but peak P and Q (also R) are not

Table 1 Equivalent relationship table for the peaks in Fig. 5. here '1' and '0' denote with relationship or not for two peaks

Peak	1	2	3	4	5	6	7	8	9	10	11
1	–	1	1	0	0	0	0	0	0	0	0
2	1	–	1	0	0	0	0	0	0	0	0
3	1	1	–	0	0	0	0	0	0	0	0
4	0	0	0	–	1	0	0	0	0	0	0
5	0	0	0	1	–	0	0	0	0	0	0
6	0	0	0	0	0	–	0	1	1	0	0
7	0	0	0	0	0	0	–	0	0	0	0
8	0	0	0	0	0	1	0	–	1	0	0
9	0	0	0	0	0	1	0	1	–	0	0
10	0	0	0	0	0	0	0	0	0	–	0
11	0	0	0	0	0	0	0	0	0	0	–

newly grouped peak. Figure 7 shows the six peaks finally obtained for the current illustration.

2.4 Color Planes

According to the final six peaks shown in Fig. 7, we have six coordinates, $(\rho_i, \theta_i), i = 1, 2, \ldots, 6$, in the ρ-θ image. Since each RGB pixel has its (ρ, θ) based on Eqs. (4) and (5), the pixel classification can be performed based on the distance between (ρ, θ) and (ρ_i, θ_i), i.e., $d(\rho, \theta; \rho_i, \theta_i)$.

If $d(\rho, \theta; \rho_k, \theta_k) = \min_{\forall i} d(\rho, \theta; \rho_i, \theta_i)$, then the pixel with (ρ, θ) is assigned to the class k. In the current illustration, there are six classes, and thus six color planes are obtained as shown in Fig. 8. By observing these color planes in detail, there possibly

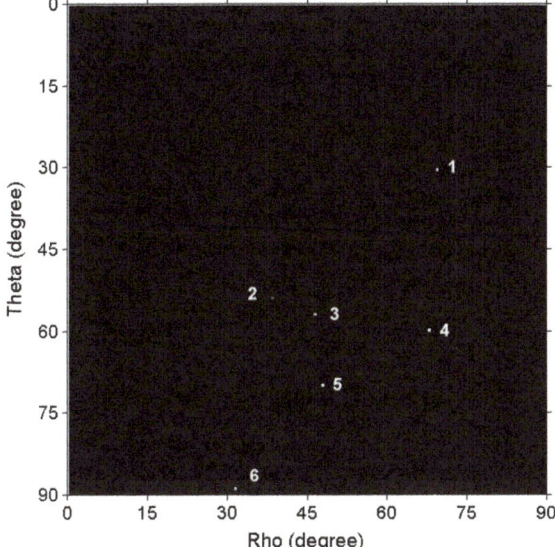

Fig. 7 Six peaks are obtained finally for the cloud-like ρ-θ image given in Fig. 4

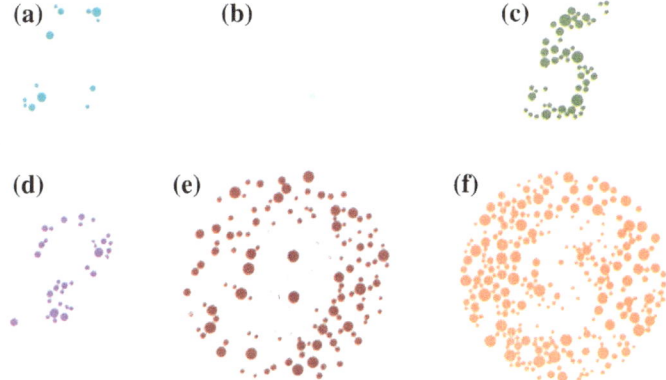

Fig. 8 Six color planes with color pixel assignment

exist some tiny noisy pixels (see Fig. 8e for example) which can be easily removed by means of the median filtering. Accordingly the final segmentation of color planes (namely 1, 2, ..., 6 respectively) from the given CBI can be displayed in Fig. 9.

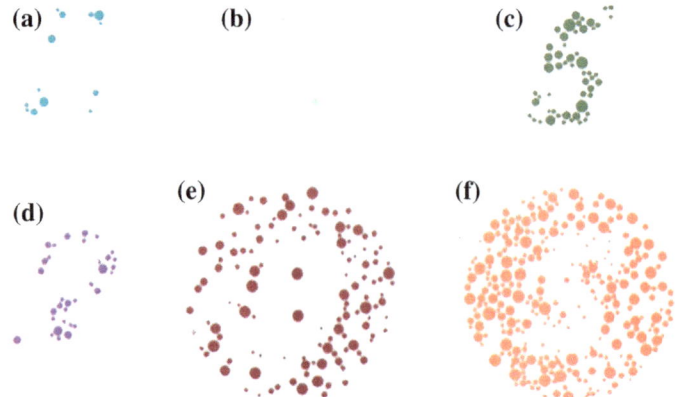

Fig. 9 Six color planes (namely 1–6 respectively) after median filtering

3 Segmentation of Background and Pattern

In the case study of CBI, the CBI is usually divided into two classes, i.e., background and pattern, for further computer vision application [8, 11]. Even the main goal of this study has been achieved with rho-theta parameterization for color image segmentation, in this section a useful process of classifying background and pattern based on the found color planes will also be demonstrated for the further application. It mainly includes two steps: spatial distance computation among color planes [7] and classification using K-means [15], which are simply reviewed as follows.

The color plane can be represented as a binary image $\mathbf{I}$, in which let the non-white pixel be 1-pixel, otherwise 0-pixel. Consider any two color planes $\mathbf{I}_a$ and $\mathbf{I}_b$, let the distance between a 1-pixel $(i, j) \in \mathbf{I}_a$ (or $I_a(i, j) = 1$) and a 1-pixel $(u, v) \in \mathbf{I}_b$ (or $I_b(u, v) = 1$) be expressed by Euclidean one

$$d(i, j; u, v) = \sqrt{(i - u)^2 + (j - v)^2} \tag{6}$$

Then the distance between a 1-pixel $(i, j) \in \mathbf{I}_a$ and the color plane $\mathbf{I}_b$ may be expressed by

$$d(i, j; \mathbf{I}_b) = \min_{\forall I_b(u,v)=1} d(i, j; u, v) \tag{7}$$

Thus the distance between color plane $\mathbf{I}_a$ and $\mathbf{I}_b$ can be defined as

$$d(\mathbf{I}_a, \mathbf{I}_b) = \frac{1}{n(\mathbf{I}_a)} \sum_{\forall I_a(i,j)=1} d(i, j; \mathbf{I}_b) \tag{8}$$

Table 2 Symmetrical distance matrix obtained by the spatial distance computation for the six color planes shown in Fig. 9

Color plane	1	2	3	4	5	6
1	0.00	10.21	7.50	10.22	21.11	18.88
2	10.21	0.00	8.29	14.81	25.18	23.38
3	7.50	8.29	0.00	8.40	19.96	17.69
4	10.22	14.81	8.40	0.00	19.53	16.47
5	21.11	25.18	19.96	19.53	0.00	4.20
6	18.88	23.38	17.69	16.47	4.20	0.00

where $n(\mathbf{I}_a)$ and $n(\mathbf{I}_b)$ denote the number of non-zero pixels of color planes $\mathbf{I}_a$ and $\mathbf{I}_b$, respectively. Because the $\mathbf{I}_a$ and $\mathbf{I}_b$ are different; the $n(\mathbf{I}_a)$ and $n(\mathbf{I}_b)$ are also different. Therefore, the average distance $\bar{d}_{ab}$ of $d(\mathbf{I}_a, \mathbf{I}_b)$ and $d(\mathbf{I}_b, \mathbf{I}_a)$ are usually adopted in such an application [7]. That is,

$$\bar{d}_{ab} = \frac{d(\mathbf{I}_a, \mathbf{I}_b) + d(\mathbf{I}_b, \mathbf{I}_a)}{2} \tag{9}$$

Note here that $\bar{d}_{ab} = \bar{d}_{ba}$, and the obtained spatial distance matrix will be of symmetry.

By performing the spatial distance computation for the six color planes shown in Fig. 9, a symmetrical distance matrix can be obtained as given in Table 2. It means that six 6-element distance vectors have been obtained. With the found distance vectors, it can be fed into the K-means algorithm to perform the clustering. Since only background and pattern will be classified in the current consideration, K is set to be 2. Note that in this study the background is also composed of size-varied color dots but does not show any pattern information. The K-means algorithm adopted here can be briefly summarized as follows.

Fig. 10 Final segmentation of **a** background and **b** pattern for the CBI "5" given in Fig. 2a

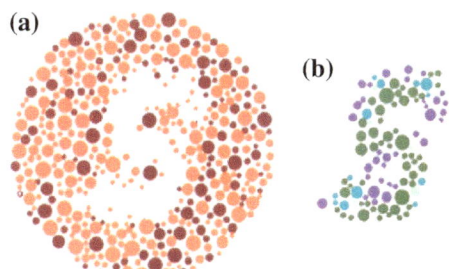

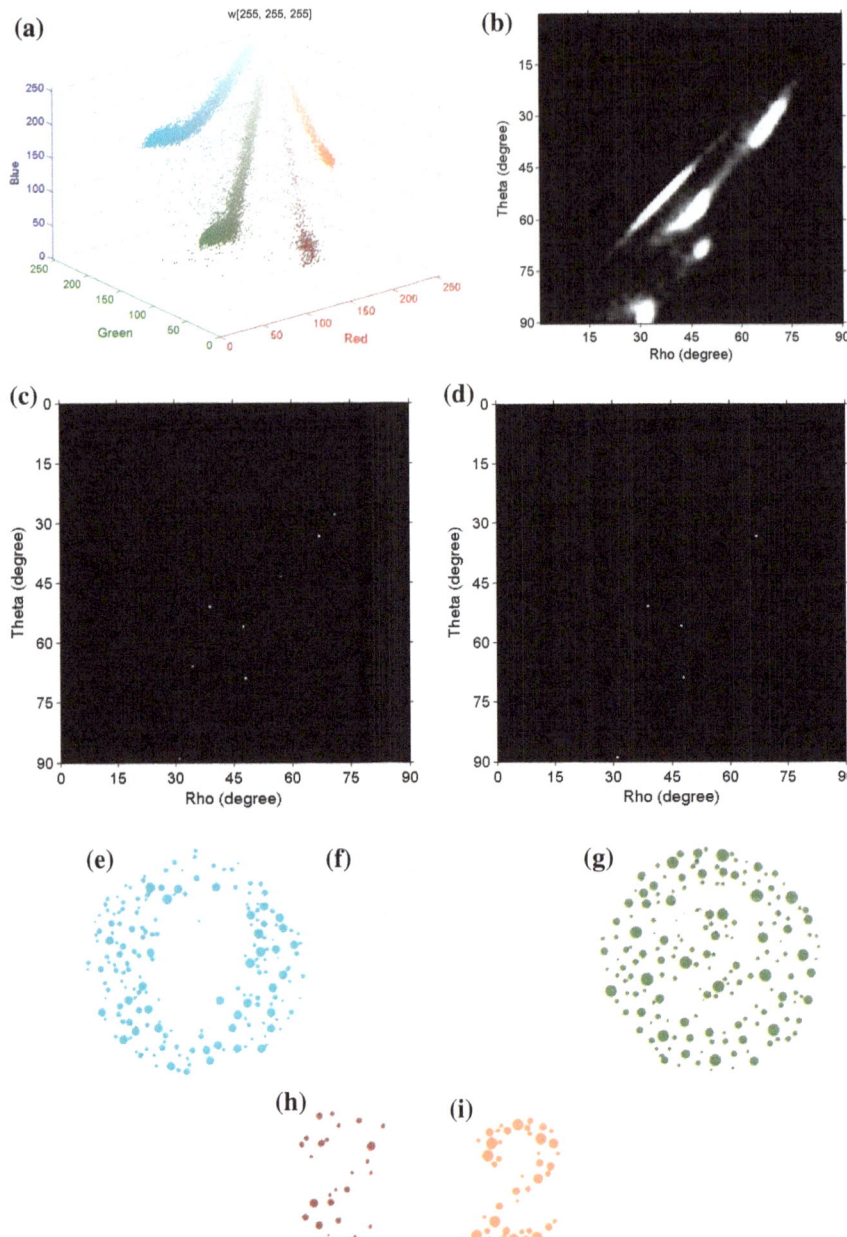

Fig. 11 Intermediate results for CBI given in Fig. 2b. **a** Color distribution in RGB-space. **b** Cloud-like ρ-θ image. **c** Initially detected peaks. **d** Final peaks. **e**–**i** Segmented color planes

Fig. 12 Segmentation of **a** background and **b** pattern for the CBI "2" given in Fig. 2b

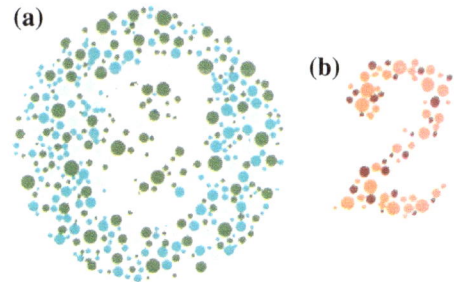

1. Input n m-element distance vectors. In current illustration, $n = 6$ and $m = 6$.
2. Select K data from the input data set as the cluster centers. For two clusters of background and pattern, $K = 2$. They are the initial cluster centers, cc_b and cc_p.
3. Compute the distances for each input data to the cc_b and cc_p, and perform the clustering assignment based on the distances.
4. After clustering assignment, recompute the cluster centers, $\hat{cc}_b$ and $\hat{cc}_p$.
5. If there are changes between the new and old cluster centers, update $cc_b \leftarrow \hat{cc}_b$ and $cc_p \leftarrow \hat{cc}_p$, and go to step (3).
6. Output the final cluster centers.

After the K-means computation, the color planes 1–4 are assigned to be the same class; whereas the color planes 5 and 6 the same class. Moreover since the number of background pixels is usually greater than that of pattern pixels, the background and pattern of the CBI "5" are readily obtained as shown in Fig. 10a, b, respectively.

4 Experimental Results and Discussion

Each CBI used in this study is of 196×196 pixels. The algorithm is implemented with MATLAB R2013a. Except for the results of the current illustrations for the CBI "5" given in Fig. 2a, the related results of the other two CBIs given in Fig. 2b, c are also displayed respectively in this section. For the CBI "2" given in Fig. 2b, its color distribution in RGB-space is shown in Fig. 11a; the found cloud-like ρ-θ image is shown in Fig. 11b; and there are eight peaks initially detected as shown in Fig. 11c. After the equivalent relationship table construction, the number of peaks are reduced to five as given in Fig. 11d, and thus there are five corresponding color planes segmented as shown in Fig. 11e–i, where the median filtering have been applied for. The reader can inspect the relationship among the color distribution in RGB-space

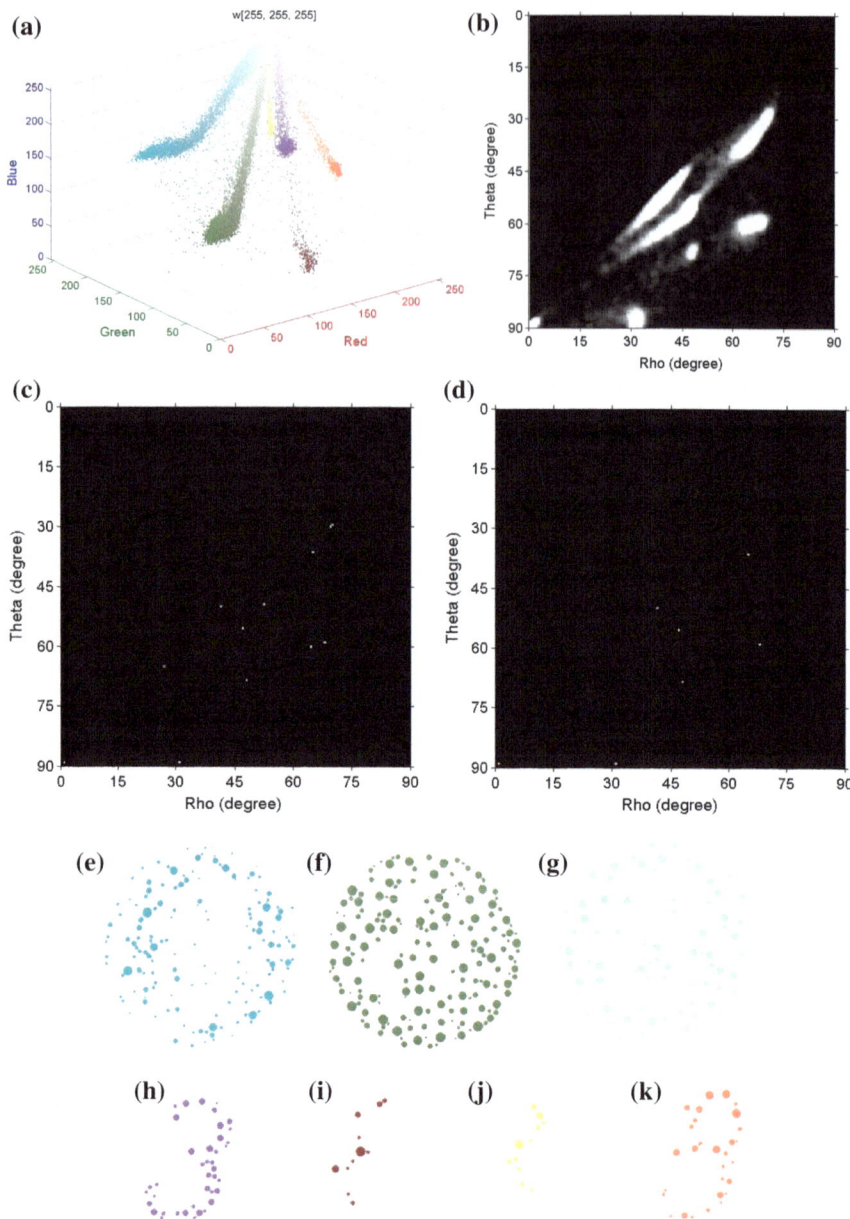

Fig. 13 Intermediate results for CBI given in Fig. 2c. **a** Color distribution in RGB-space. **b** Cloud-like ρ-θ image. **c** Initially detected peaks. **d** Final peaks. **e–k** Segmented color planes

Fig. 14 Segmentation of **a** background and **b** pattern for the CBI "8" given in Fig. 2c

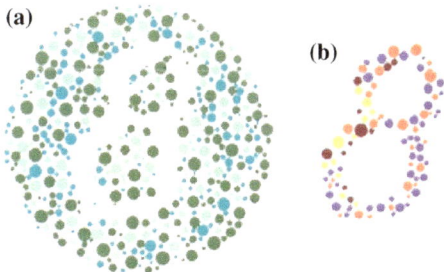

(Fig. 11a), the found peaks (Fig. 11d), and the segmented color planes (Fig. 11e–i). In addition, it is also easily observed that the color planes (Fig. 11e–g) belong to the background, whereas those (Fig. 11h, i) belong to the pattern. After applying the K-means algorithm to these color planes, the final background and pattern can be obtained as displayed in Fig. 12. Figures 13 and 14 display the corresponding results for the CBI "8" given in Fig. 2c. As a result, the feasibility of the proposed method has been confirmed by these experiments.

5 Conclusion

So far we have presented a rho-theta parameterization approach for color blindness image segmentation. The rho-theta space presents the cloud information for each clustered color distribution, which can benefit the peak detection and thus the segmentation of color planes. With the assistance of spatial distance computation among color planes and the K-means algorithm, the segmentation of pattern and background of a CBI can also be obtained. Our experiments have confirmed the feasibility of the proposed approach. As a future work, based on the concept from color distribution to the cloud in rho-theta space, some manipulation schemes on the rho-theta space could be developed possibly for solving the wanted topics (like the identification of colored object and finding the relationship among colored objects in a color image) on the traditional RGB color space.

Acknowledgements This work was supported in part by the Ministry of Science and Technology, Taiwan, Republic of China, under the grant number MOST103-2221-E-155-040 and MOST105-2221-E-155-063.

References

1. L. Jin, D. Li, A switching vector median filter based on the CIELAB color space for color image restoration. Sig. Proc. **87**, 1345–1354 (2007)

2. D.J. Lee, J.K. Archibald, Y.C. Chang, C.R. Greco, Robust color space conversion and color distribution analysis techniques for date maturity evaluation. J. Food Eng. **88**, 364–372 (2008)
3. J. Mukherjee, M.K. Lang, S.K. Mitra, Demosaicing of images obtained from single-chip imaging sensors in YUV color space. Pattern Recogn. Lett. **26**, 985–997 (2005)
4. S.A. Underwood, J.K. Aggarwal, Interactive computer analysis of aerial color infrared photographs. Comput. Graph. Image Process. **6**, 1–24 (1977)
5. T. Uchiyama, M.A. Arbib, Color image segmentation using competitive learning. IEEE Trans. Pattern Anal. Mach. Intell. **16**, 1197–1206 (1994)
6. A. Tremeau, N. Borel, A region growing and merging algorithm to color segmentation. Pattern Recogn. **30**, 1191–1203 (1997)
7. Y.S. Chen, Y.C. Hsu, Image segmentation of a color-blindness plate. Appl. Opt. **33**, 6818–6822 (1994)
8. Y.S. Chen, Y.C. Hsu, Computer vision on a colour blindness plate. Image Vis. Comput. **13**, 463–478 (1995)
9. C.E. Martin, J.G. Keller, S.K. Rogers, M. Kabrisky, Color blindness and a color human visual system model. IEEE Trans. Syst. Man Cyber. **30**, 494–500 (2000)
10. T. Wachtler, U. Dohrmann, R. Hertel, Modeling color percepts of dichromats. Vis. Res. **44**, 2843–2855 (2004)
11. Y.S. Chen, C.Y. Zhou, L.Y. Li, Perceiving stroke information from color-blindness images, in *Proceedings of the IEEE International Conference on Systems, Man, and Cybernetics*, Budapest, Hungary (2016), pp. 70–73
12. Y.S. Chen, L.Y. Li, C.Y. Zhou, Rho-theta parameterization for color blindness image segmentation, in *Proceedings of the International MultiConference of Engineers and Computer Scientists, Lecture Notes in Engineering and Computer Science*, Hong Kong (2017), pp. 405–409
13. I. Omer, M. Werman, Color lines: image specific color representation, in *Proceedings of IEEE Computer Society Conference on Computer Vision and Pattern Recognition*, Washington, DC, USA (2004), pp. 946–953
14. R.O. Duda, P.E. Hart, Use of the hough transformation to detect lines and curves in pictures. Comm. ACM **15**, 11–15 (1972)
15. J. MacQueen, Some methods for classification and analysis of multivariate observations, in *Proceedings of the Fifth Berkeley Symposium on Mathematical Statistics and Probability*. University of California Press, USA (1967), pp. 281–297

Estimation of Optimal Global Threshold Based on Statistical Change-Point Detection and a New Thresholding Performance Criteria

Rohit Kamal Chatterjee and Avijit Kar

Abstract Global thresholding of gray-level image is a method of selecting an optimal gray-value that partition it into two mutually exclusive regions called background and foreground (object). The aim of this paper is to interpret the problem of optimal global threshold estimation in the terminology of statistical change-point detection (CPD). An important advantage of this approach is that it does not assume any prior statistical distribution of background or object classes. Further, this method is less influenced by the presence of outliers due to our judicious derivation of a robust criterion function depending on Kullback-Leibler (KL) divergence measure. Experimental results manifest the efficacy of proposed algorithm compared to other popular methods available for global image thresholding. In this paper, we also propose a new criterion for performance evaluation of thresholding algorithms. This performance criterion does not depend on any ground truth image. This performance criterion is used to compare the results of proposed thresholding algorithm with most cited global thresholding method found in the literature.

Keywords Change-point detection · Independence of ground truth image
Global image thresholding · Kullback-Leibler divergence · Robust statistical
measure · Thresholding performance criteria

R. K. Chatterjee (✉)
Department of Computer Science & Engineering, Birla Institute of Technology,
Mesra, Ranchi, India
e-mail: rkchatterjee@bitmesra.ac.in

A. Kar
Department of Computer Science & Engineering, Jadavpur University, Kolkata, India
e-mail: avijit_kar@cse.jdvu.ac.in

© Springer Nature Singapore Pte Ltd. 2018
S.-I. Ao et al. (eds.), *Transactions on Engineering Technologies*,
https://doi.org/10.1007/978-981-10-7488-2_21

1 Introduction

A gray-valued digital image is a discrete two-dimensional signal $I: \mathbb{Z} \times \mathbb{Z} \to L$, with $L = \{l_i \in \mathbb{R} \text{ and } i = 1, 2, ..., M\}$ is the set of M gray-levels. The problem of global thresholding is to estimate an optimal gray-value t_0 which segments the image into two meaningful sets, viz. background $B = \{b_b(x, y) = 1 \mid I(x, y) < t_0\}$ and foreground $F = \{b_f(x, y) = 1 \mid I(x, y) \geq t_0\}$ or the opposite. The function $I(x, y)$ can take any random gray-value $l_i \in L$; so, the sampling distribution of gray levels becomes an important deciding factor for t_0. Deciding a threshold in an empirical manner is possible by trial, but in most cases, it becomes a tedious job and is prone to errors. Automating the process of deciding an optimal threshold, therefore, is extremely important for low-level segmentation or even final segmentation of object and background.

In general, automatic thresholding algorithms are of two types, viz. global and local methods. Global methods estimate a single threshold for the entire image; local methods find an adaptive threshold for each pixel depending on the characteristics of its neighborhood. Local methods are useful when statistical features of the background and foreground classes are non-stationary. On the other hand, global methods are used if the image is considered as a mixture of two or more statistical distributions. In this paper, we address the global thresholding methods guided by the image histogram.

Most of the classical global thresholding methods try to estimate the threshold (t_0) iteratively by optimizing a criterion function [1]. Ridler and Calvard's [2] algorithm is a classic example of early iterative thresholding methods. Some other methods attempt to estimate optimal t_0 depending on histogram shape [3, 4], image attribute such as topology [6] or some clustering techniques [5, 31]. Comprehensive surveys discussing various aspects of thresholding methods can be found in the references [1, 10, 11].

Two most popular and often cited algorithms are proposed by Otsu [7] and Kittler-Illingworth [8]. Otsu's algorithm proposes the between-class variance as a severability measure and the threshold that maximizes this measure is considered as the optimal threshold. Otsu's method performs well when the foreground and the background distributions are approximately normal with nearly equal within-class variances [9]. Minimum error threshold selection algorithm proposed by Kittler and Illingworth assumes that foreground and background gray values are normally distributed and the threshold is selected by minimizing some measure of the potential error involved in the process of classification.

An entropy-based method is initially proposed by Kapur et al. [13]. This method determines the optimal threshold by maximizing the overall entropy subject to inequality constraints derived from measures of uniformity and the shape of image regions. Some advanced entropy based thresholding techniques were advocated by Abutaleb [14] and Brink [15], which take into account the spatial structure by considering 2-D entropy of the original image. Afterwards, Sahoo et al. [16] uses Renyi entropy and later, Sahoo and Arora [17] propose using Tsallis-Havrda-Charvat entropy as an improvement in determining an optimal threshold.

In recent times, numerous fuzzy logic based global thresholding techniques have been explored; some of these are listed in references [18–21]. The underlying idea of these techniques is to reduce the ambiguity of the threshold by using different fuzzy measures. More recently, Bazi et al. [22] have suggested a parametric method that assumes Generalized Gaussian (GG) distribution for the two classes as an alternative to Gaussian distribution used in most classical techniques. Despite its impressive results, the algorithm is quite complex and takes a large amount of time compared to other classical methods. Wang et al. [23] have suggested a nonparametric algorithm using Parzen window technique to estimate the spatial probability distribution of gray-levels.

Many of these classical and recent schemes perform remarkably well for images with matching underlying assumptions but fail to yield desired results otherwise. Some of the explicit or implicit reasons for their failure could be: (i) assumption of some standard distribution (e.g. Gaussian) for both the classes, in reality though, foreground and background classes have skewed and asymmetric distributions, (ii) use of non-robust measures for computing criterion functions which get influenced by outliers. Further, the effectiveness of these algorithms greatly decreases when the areas under the two classes are highly unbalanced. Some of the methods depend on user-specified constant (e.g. Renyi or Tsallis entropy-based methods), greatly compromising their performance without its appropriate value.

We propose in this paper a novel algorithm addressing these drawbacks. This algorithm uses a well-known statistical technique called *change-point detection (CPD)*. For the last few decades, researchers in statistics and control theory have been attracted by the problem of detecting abrupt changes in the statistical behavior of an observed signal or time series [24, 25]. They are collectively called statistical change-point detection.

The general technique of change-point detection considers an observed sequence of *independent* random variables $\{Y_k\}_{1 \leq x \leq n}$ with a probability density function (pdf) $p_\theta(y)$ depending on a *scalar* parameter θ. Basically, it is assumed in CPD that θ takes values θ_0 and θ_1 ($\neq \theta_0$) before and after the unknown change time t_0. The problem is then to estimate this change in the parameter and the change time t_0. A very important property of the log-likelihood ratio, $\Lambda(y) = log\left((p_{\theta_1}(y))/(p_{\theta_0}(y))\right)$, becomes a tool for reaching this goal. Let $E_{\theta_0}(\Lambda)$ and $E_{\theta_1}(\Lambda)$ denote the expectations of Λ with respect to two distributions p_{θ_0} and p_{θ_1}; then it can be shown that $E_{\theta_0}(\Lambda) < 0$ and $E_{\theta_1}(\Lambda) > 0$. In other words, a change in the parameter θ is reflected as a change in the sign of the expectation of the log-likelihood ratio. This statistical property can be used to detect the change in θ [24]. Given the *Kullback-Leibler (KL)* divergence [29] $K(p_{\theta_1}||p_{\theta_0}) = E_{\theta_1}(\Lambda)$, difference between the two mean values is

$$E_{\theta_1}(\Lambda) - E_{\theta_0}(\Lambda) = K(p_{\theta_1}||p_{\theta_0}) + K(p_{\theta_0}||p_{\theta_1}) \tag{1}$$

From this, we infer that the detection of a change can also be made with the help of the KL divergence before and after the change. This concept is used in this paper for deciding the threshold in an image histogram.

Section 4 of this paper proposes a new performance criterion for the evaluation of thresholding algorithms. No ground truth image is required for its evaluation. It depends on the structural difference between the shapes of background and foreground. We use this performance criterion to compare different thresholding algorithms including ours.

Rest of the paper is organized as follows: Sect. 2 provides a short introduction to the problem of statistical change-point detection, Sect. 3 formulates and derives the global thresholding as a change-point detection problem, Sect. 4 describes our proposal for thresholding performance criteria, Sect. 5 presents the experimental results and compares the results with various cited global thresholding algorithms, and finally Sect. 6 summarizes main ideas in this paper.

2 Overview of the Change-Point Detection (CPD) Problem

Change-point detection (CPD) problem can be classified into two broad categories: real-time or online change-point detection, which targets applications where the instantaneous response is desired such as robot control; on the other hand, retrospective or offline change-point detection is used when longer reaction periods are allowed e.g. image processing problems [26]. The latter technique is likely to give robust and accurate detection since the entire sample distribution is accessible. Since the image is available to us and the corresponding empirical distribution (histogram) can be built, our discussion will concentrate on offline or retrospective change point detection and simply use the phrase "change point detection (CPD)" throughout this chapter. Further, we will assume that there is only one change point throughout the given observations $\{y_k\}_{1 \leq k \leq n}$. When required, this assumption can easily be relaxed and extended to multiple change point detection that can be applied in multi-level threshold detection problems.

2.1 Problem Statement

When taking an offline point of view, it is convenient to introduce the following null hypothesis (H_0) about the observations y_1, y_2 ..., y_n with corresponding probability distribution functions (cdf) F_1, F_2, ..., F_n, belong to a common parametric family $F(\theta)$, where $\theta \in R^p$, p > 0, then the change point problem is to test the null hypothesis about the population parameter θ_j, j = 1, 2, ..., n:

$$H_0 : \boldsymbol{\theta}_j = \boldsymbol{\theta} \quad for \ 1 \leq j \leq n$$

versus an alternative hypothesis

$$
\begin{aligned}
H_1 : \boldsymbol{\theta}_j &= \boldsymbol{\theta} \quad for \ 1 \leq j \leq k - 1 \\
&= \boldsymbol{\theta}' \ for \ k \leq j \leq n
\end{aligned}
\tag{2}
$$

where $\boldsymbol{\theta} \neq \boldsymbol{\theta}'$ with unknown time of change k. These hypotheses together determine if any change point exists in the process and estimating the *time of change $t_0 = k$.* The likelihood ratio corresponding to the hypotheses H_0 and H_1 is given by

$$
\Lambda_1^n(k) = \frac{\prod_{j=1}^{k-1} p_{\theta_0}(y_j) \times \prod_{j=k}^{n} p_{\theta_1}(y_j)}{\prod_{j=1}^{n} p_{\theta_0}(y_j)}
\tag{3}
$$

where p_{θ_0} is the probability density when H_0 is true and p_{θ_1} when H_1 is true. In this situation, the standard statistical approach is to use maximum likelihood estimation (MLE) for detection of unknown time of change t_0. Therefore, we consider the following statistic

$$
t_0 = \arg \max_{1 \leq k \leq n} \Lambda_1^n(k)
\tag{4}
$$

2.2 Offline Estimation of Change-Point

We consider a sequence of observations $\{y_j\}_{1 \leq j \leq n}$ and the same hypotheses as in the previous subsection. Also, we assume the existence of a change point (typically, in the present case, we assume parameters θ_0 and θ_1 are known from observations) and the problem is to estimate the time of change. Therefore, considering Eqs. (3) and (4) and the fact that $\prod_{j=1}^{n} p_{\theta_0}(y_j)$ is a constant for given data, the corresponding MLE estimate is

$$
\widehat{t}_0 = \arg \max_{1 \leq k \leq n} \left(\ln \left(\prod_{j=1}^{k-1} p_{\theta_0}(y_j) \times \prod_{j=k}^{n} p_{\theta_1}(y_j) \right) \right)
\tag{5}
$$

where $\hat{t}_0$ is maximum log-likelihood estimate of t_0. Rewriting Eq. (5) as

$$\hat{t}_0 = \arg\max_{1 \leq k \leq n} \left[\ln\left(\frac{\prod_{j=k}^{n} p_{\theta_1}(y_j)}{\prod_{j=k}^{n} p_{\theta_0}(y_j)} \right) + \ln\left(\prod_{j=1}^{n} p_{\theta_0}(y_j) \right) \right] \tag{6}$$

As $\ln\left(\prod_{j=1}^{n} p_{\theta_0}(y_j) \right)$ remains a constant for given observation, estimation of $\hat{t}_0$ is simplified as

$$\hat{t}_0 = \arg\max_{1 \leq k \leq n} \sum_{j=k}^{n} \ln\left(\frac{p_{\theta_0}(y_j)}{p_{\theta_0}(y_j)} \right) \tag{7}$$

Therefore, the MLE of the change time $\mathbf{t_0}$ is the value which maximizes the sum of log-likelihood ratio corresponding to all k possible value given by Eq. (7).

3 Global Thresholding: A Change-Point Detection Formulation

3.1 Assumptions

Let us assume a one-dimensional observation *of discrete gray-values* $Y_1, ..., Y_N$, which are independent and identically distributed (i.i.d.) coming from the family of probability distributions $\{F_\theta\}_{\theta \in \Theta}$ defined on the measurable space (χ, β_χ) where β_χ is the σ-field of Borel subsets $\mathbf{A} \subset \chi$. The parameter space Θ is assumed as an open convex subset of $\mathbb{R}^p$. We consider a finite population Π of all gray-level images with each sample point (pixel) could be classified into M categories or classes $L = \{l_1, ..., l_M\}$ of gray values.

3.2 Change-Point Detection Formulation [33]

Since we are mainly interested in discrete gray-level data, we consider the *multinomial distribution* model. Let $\wp = \{E_i\}$, $i = 1, ..., M$ be a partition of χ. The formula $Pr_\theta(E_i) = p_i(\theta)$, $i = 1, ..., M$, defines a discrete statistical model with a probability of the l_i^{th} gray-level. Further, we assume $\{Y_1, ..., Y_n\}$ to be a random sample from the population described by the random variable $\mathbf{Y}$, representing the gray-level of a pixel. And let $N_i = \sum_{j=1}^{n} I_{E_i}(Y_j)$, where $\mathbf{I_E}$ is the index function. Then we can approximate $p_i(\theta) \approx N_i/n$, $i = 1, ..., M$. Estimating θ by maximum likelihood method consists of maximizing the joint probability distribution for fixed $n_1, \ldots, n_M$,

$$Pr_\theta(N_1 = n_1, N_2 = n_2, \ldots, N_M = n_M) = \frac{n!}{n_1! n_2! \ldots n_M!} (p_1(\theta))^{n_1} (p_2(\theta))^{n_2} \ldots (p_M(\theta))^{n_M} \tag{8}$$

or equivalently maximizing the log-likelihood function

$$\Lambda(\theta) = ln \left[\frac{n!}{n_1! n_2! \ldots n_M!} (p_1(\theta))^{n_1} (p_2(\theta))^{n_2} \ldots (p_M(\theta))^{n_M} \right] \tag{9}$$

Therefore, referring to Eq. (5), problem of estimating the threshold by MLE can be stated as

$$\hat{\tau}_0 = arg \max_{1 \leq j \leq M} ln \left(\frac{n!}{n_1! n_2! \ldots n_M!} \prod_{i=1}^{j-1} (p_i(\theta_0))^{n_i} \prod_{i=j}^{M} (p_i(\theta_1))^{n_i} \right) \tag{10}$$

where unknown parameter $\theta = \theta_0$ before the change and $\theta = \theta_1$ after the change. Now, Eq. (10) can be expanded as

$$\hat{\tau}_0 = arg \max_{1 \leq j \leq M} \left[ln \left(\frac{n!}{n_1! n_2! \ldots n_M!} \right) + ln \left(\frac{\prod_{i=j}^{M} (p_i(\theta_1))^{n_i}}{\prod_{i=j}^{M} (p_i(\theta_0))^{n_i}} \right) + ln \left(\prod_{i=1}^{M} (p_i(\theta_0))^{n_i} \right) \right] \tag{11}$$

The first term within the bracket on the right side of Eq. (11) is a constant and the last term is independent of j, i.e. it cannot influence the MLE. So, eliminating these terms from Eq. (11) and simplifying we get

$$\hat{\tau}_0 = arg \max_{1 \leq j \leq M} \sum_{i=j}^{M} ln \left(\frac{p_i(\theta_1)}{p_i(\theta_0)} \right)^{n_i} \tag{12}$$

$$= arg \max_{1 \leq j \leq M} n \sum_{i=j}^{M} \left(\frac{n_i}{n} \right) ln \left(\frac{p_i(\theta_1)}{p_i(\theta_0)} \right) \tag{13}$$

$$\approx arg \max_{1 \leq j \leq M} n \sum_{i=j}^{M} p_i(\theta_1) ln \left(\frac{p_i(\theta_1)}{p_i(\theta_0)} \right) \tag{14}$$

where we assume $p_i(\theta) \approx n_i/n$. Expression in (14) under the summation is the *Kullback-Leibler (KL)* divergence between the density $p(\theta_1)$ and $p(\theta_0)$; therefore Eq. (14) can be written as

$$\hat{t}_0 = arg \max_{1 \le j \le M} nK_j^M(p(\boldsymbol{\theta}_1)\|p(\boldsymbol{\theta}_0)) \tag{15}$$

Now, since total sum $\sum_{i=1}^{M} p_i(\boldsymbol{\theta}_1) ln\left(\frac{p_i(\boldsymbol{\theta}_1)}{p_i(\boldsymbol{\theta}_0)}\right)$ is independent of j i.e. a constant for a given observation (image)—the Eq. (15) can be equivalently written as

$$\hat{t}_0 = arg \min_{1 \le j \le M} nK_1^j(p(\boldsymbol{\theta}_1)\|p(\boldsymbol{\theta}_0)) \tag{16}$$

Hence, Eq. (16) provides maximum likelihood estimation of the threshold t_0. Derivation of above formula (16) can be stated as a proposition:

Proposition 1 *In a mixture of distributions, the maximum likelihood estimate of change-point is found by minimizing the Kullback-Leibler divergence of the probability mass across successive thresholds.*

In spite of this striking property, KL divergence is not a 'metric' since it is not symmetric. An alternative symmetric formula by "averaging" the two KL divergences is given as [28]

$$D(p_{\boldsymbol{\theta}_1}\|p_{\boldsymbol{\theta}_0}) = \frac{1}{2}\left(K(p_{\boldsymbol{\theta}_1}\|p_{\boldsymbol{\theta}_0}) + K(p_{\boldsymbol{\theta}_0}\|p_{\boldsymbol{\theta}_1})\right) \tag{17}$$

An attractive property of KL divergence is its robustness i.e. KL divergence is little influenced even when one component of mixture distribution is considerably skewed. A proof of robustness can be found for generalized divergence measures in [27, 28].

This method can be easily extended to find multiple thresholds for several mixture distributions by identifying multiple change-points simultaneously.

3.3   Implementation

Let us consider an image $I:\mathbb{Z} \times \mathbb{Z} \rightarrow L$, whose pixels assume M gray-values in the set $L = \{l_1, l_2, \ldots, l_M\}$. The empirical distribution of the image can be represented by a normalized histogram $p(l_i) = n_i/N$, where n_i is the number of pixels in ith gray-level and N is the total number pixels in the image.

Now, suppose we are grouping the pixels into two classes **B** and **F** (background and object) by thresholding at the level k. Histogram of gray-levels can be found for the classes **B** and **F**; let us denote them as $p_B(l_i)$ and $p_F(l_i)$. Following statistics are calculated for the level k

$$\mathbf{K}^B(k) = \sum_{i=1}^{k-1} p_F(\mathrm{I}_i) \ln\left(\frac{p_F(\mathrm{I}_i)}{p_B(\mathrm{I}_i)}\right) \tag{18}$$

$$\mathbf{K}^F(k) = \sum_{i=k}^{M} p_B(l_i) \ln\left(\frac{p_B(l_i)}{p_F(l_i)}\right) \tag{19}$$

and finally taking the average of (18) and (19) we get symmetric CPD measure

$$\mathbf{CPD(k)} = \frac{1}{2}\left(\mathbf{K^F}(k) + \mathbf{K}^B(k)\right) \tag{20}$$

The minimum value of *CPD(k)* for all values of *k* in the range [1, …, M] gives an optimal estimate of threshold t_0.

4 Thresholding Performance Criteria

The objective of the global thresholding algorithm is to decide a gray-value that divides an image into two binary images generally called background and foreground (object). Most of the histogram-based thresholding algorithms try to devise a criterion function which produces a threshold to separate the shapes and patterns of the foreground and background as much as possible. A good thresholding algorithm, therefore, can be judged by how well it splits apart the patterns present in the background and foreground binary images, i.e. how much dissimilarity exists between the foreground and the background. Since the background and foreground images are binary images dissimilarity between them can be measured by any binary distance measures. Based on this observation, we propose a threshold evaluation criterion, which tries to find the dissimilarity between the patterns and shapes in foreground and background.

A number of binary similarity and distance measures have been proposed, a comprehensive survey of them can be found in Choi et al. [30, 34]. Suppose that the thresholded foreground and background binary images containing N pixels are represented by a binary vector $x_i \in \beta^N$, where $\beta = \{0, 1\}$ and $i = 1, 2$. For $x_1, x_2 \in \beta^N$, binary scalar product is defined as $x_1^T x_2 = \sum_{k=1}^{N} x_{1k} x_{2k}$ and $(1 - x_i)$ is the complementary vector of x_i. This allows us to define the following counters

$a = x_1^T x_2$—the number of pixels in which both agree to one,
$b = x_1^T (1 - x_2)$—the number of pixels in which x_1 has 1, but x_2 has 0,
$c = (1 - x_1)^T x_2$—the number of pixels in which x_1 has 0, but x_2 has 1,
$d = (1 - x_1)^T (1 - x_2)$—the number of pixels in which both x_1 and x_2 has 0.

For various definitions of dissimilarity measures, a **2 × 2** contingency table is considered for each pair of images as presented in Table 1.

Hence, from the above definitions, *b* + *c* denote the total count where foreground and background pixels differ (Hamming distance) and *a* + *d* is the total count where they agree. In order to extract the shapes and patterns present in the

Table 1 Binary contingency table

Foreground	Background	
	1	0
1	a	b
0	c	d

foreground (**F**) and background (**B**) images, we use *binary morphological gradient*. The binary Morphological gradient is the difference between the *eroded* and *dilated* images. In this chapter, a simple binary distance measure known as Normalised Manhattan distance (D_{NM}) is used (21)

$$D_{NM}(F_g, B_g) = \frac{b+c}{a+b+c+d} \qquad (21)$$

where F_g and B_g denote Binary Morphological gradients of foreground (**F**) and background (**B**) respectively. The range of this distance measure is the interval *[0, 1]*. It is expected that well-segmented image will have D_{NM} close to **1**, while in the worst case $D_{NM} = 0$. The advantage of this algorithm is that it does not require any ground truth image.

5   Experimental Results and Discussion

To validate the applicability of proposed Change-Point Detection (CPD) thresholding algorithm, we provide experimental results and compare the results with existing algorithms. The first row of Fig. 1 shows test images that are labeled from

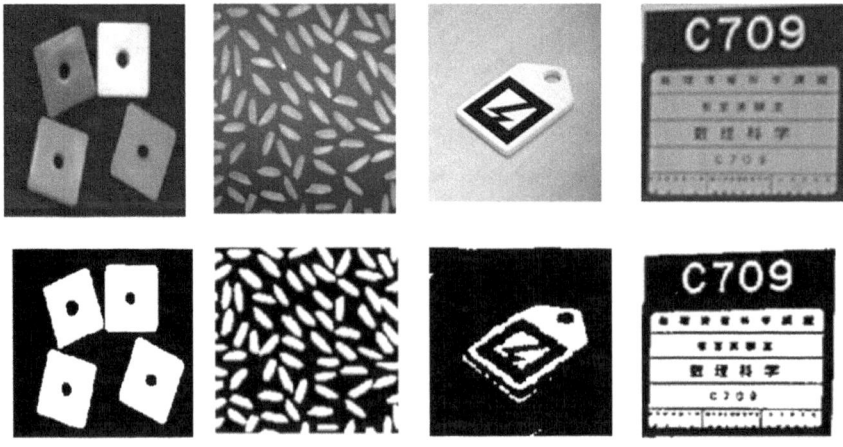

Fig. 1   Row-1: Test images: Dice, Rice, Object, and Number plate; **Row-2**: Corresponding ground truth images

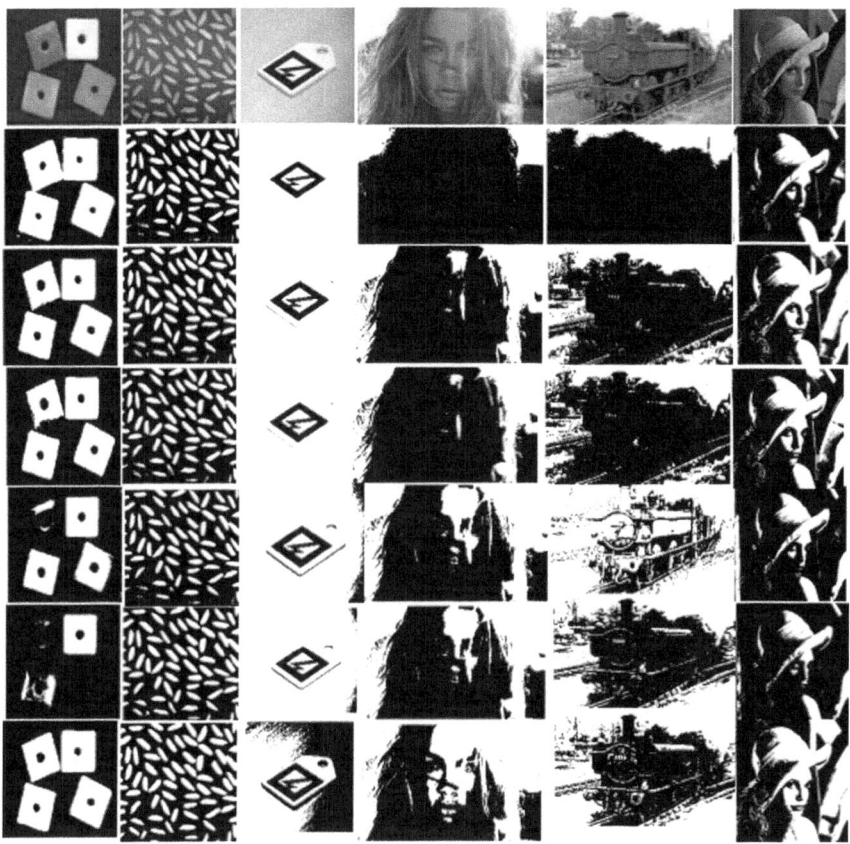

Fig. 2 A results of thresholding algorithms on tested images: **Row-1**: Original test images (left to right): Dice; Rice; Object; Denise; Train and Lena; **Row-2**: Kittler; **Row-3**: Otsu; **Row-4**: Kurita; **Row-5**: Sahoo; **Row-6**: Entropy: **Row-7**: CPD threshold

left to right as *Dice, Rice, Object, Number plate* respectively and the second row shows corresponding ground truth images. We have also selected few more test images shown in the first row of Fig. 2 with names *Denise, Train,* and *Lena* to visually compare the results. These images have deliberately been so selected that the difference of areas between foreground and background is hugely disproportionate. This gives us an opportunity to test the robustness of CPD algorithm. To compare the results, we selected five most popular thresholding algorithms, namely, Kittler-Illingworth [8], Otsu [7], Kurita [12], Sahoo [16] and Entropy [17].

Table 2 SSIM of the test images: Dice, Rice, Object, and Number plate

	Dice		Rice	
Algorithm	Threshold	SSIM	Threshold	SSIM
Entropy	157	0.6705	118	0.9377
Kittler	55	0.9182	132	0.8172
Kurita	106	0.9560	126	0.8709
Otsu	103	0.9469	124	0.8880
Sahoo	134	0.7040	133	0.8082
CPD	96	0.9907	107	0.9748
	Object		Number plate	
Algorithm	Threshold	SSIM	Threshold	SSIM
Entropy	150	0.8523	149	0.7923
Kittler	83	0.5381	20	0.0381
Kurita	107	0.7752	103	0.6752
Otsu	110	0.7638	102	0.6511
Sahoo	154	0.9014	153	0.7314
CPD	190	0.9601	141	0.8953

For quantitative comparison of the performance of the algorithms, Structural Similarity Index (SSIM) [32] is computed for the output images with respect to the ground truth images. Table 2 shows optimal thresholds of five selected algorithms and the CPD algorithm with SSIM score.

In Fig. 2 second row onwards are shown the outputs of different thresholding algorithms. The last row displays the outputs of the proposed CPD based thresholding algorithm. Due to substantial skewness in the distributions of gray-levels in object or background, most of the classical algorithms confused foreground with background. But results in the last row clearly show that CPD works significantly better in all cases.

For example, in the Denise image and Train image, Kittler-Illingworth algorithm totally fails to distinguish the object from the background due to its assumption of Gaussian distribution for both the classes. Otsu's and Kurita's method yield almost same output due to their common assumptions.

Histograms of Denise and Train image marked with threshold locations of all the six algorithms above are shown in Fig. 3. The threshold locations show that CPD algorithm is very little influenced by the asymmetry of an object or background distributions.

Table 3 shows the performance of CPD algorithm with respect to the five above mentioned algorithms using our proposed performance criteria (D_{NM}). It is clear that CPD performs reasonably well with respect to other algorithms.

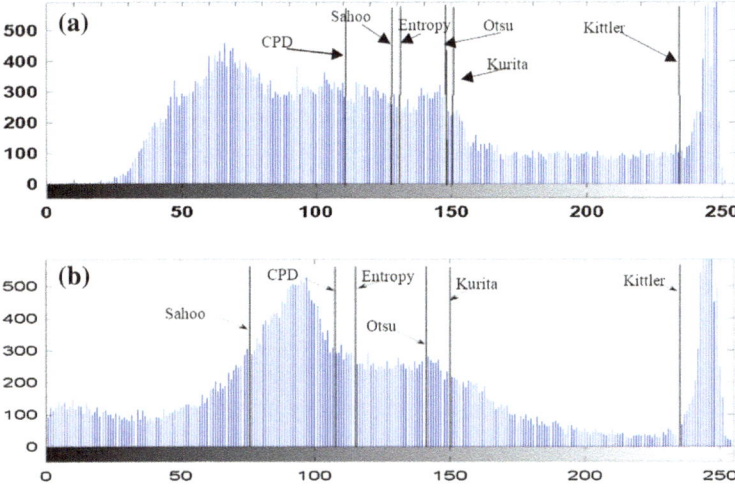

Fig. 3 Histogram of **a** Denise and **b** Train image with threshold locations of examined algorithms

Table 3 Comparison of test images using proposed threshold evaluation criterion (D_{NM}) for the images Dice, Rice, Object, Denise, Train, and Lena

	Dice		Rice	
Algorithm	Threshold	D_{NM}	Threshold	SSIM
Entropy	157	0.1046	118	0.2360
Kittler	55	0.1584	132	0.2210
Kurita	106	0.1574	126	0.2286
Otsu	103	0.1534	124	0.2306
Sahoo	134	0.1274	133	0.2190
CPD	96	0.2746	107	0.2458
	Object		**Denise**	
Algorithm	Threshold	D_{NM}	Threshold	D_{NM}
Entropy	150	0.1320	131	0.1768
Kittler	83	0.0558	235	0.0426
Kurita	107	0.0634	151	0.1612
Otsu	110	0.0652	147	0.1650
Sahoo	154	0.1550	128	0.1810
CPD	190	0.2944	111	0.2078
	Train		**Lena**	
Algorithm	Threshold	D_{NM}	Threshold	D_{NM}
Entropy	117	0.4186	118	0.1090
Kittler	234	0.0554	102	0.1278
Kurita	150	0.3226	89	0.1714
Otsu	142	0.3554	72	0.1874
Sahoo	80	0.4564	109	0.1120
CPD	108	0.4690	64	0.2056

6 Conclusion

In this paper we propose a novel global image thresholding algorithm based on the well-founded statistical technique of Change-Point detection (CPD), which depends on the symmetric version of Kullback-Leibler divergence measure.

The experimental results clearly show this algorithm is largely unaffected by disproportionate dispersal of object and background scene and also very little influenced by the skewness of mixture distribution of object and background classes compared to other well-known algorithms. Extension of this algorithm is easily possible for multilevel thresholding.

This paper also proposes a thresholding performance criterion using dissimilarity between foreground and background binary images. The advantage of this performance criterion is that it does not require any ground truth image.

References

1. M. Sezgin, B. Sankur, Survey over image thresholding techniques and quantitative performance evaluation. J. Electr. Imag. 13(1), 146–165 (2004)
2. T.W. Ridler, S. Calvard, Picture thresholding using an iterative selection method. IEEE Trans. Syst. Man Cybern. SMC-8, 630–632 (1978)
3. A. Rosenfeld, P. De la Torre, Histogram concavity analysis as an aid in threshold selection. IEEE Trans. Syst. Man Cybern. SMC-13, 231–235 (1993)
4. M.I. Sezan, A Peak detection algorithm and its application to histogram-based image data reduction. Graph. Models Image Process. 29, 47–59 (1985)
5. D.M. Tsai, A fast thresholding selection procedure for multimodal and unimodal histograms. Pattern Recogn. Lett. 16, 653–666 (1995)
6. A. Pikaz, A. Averbuch, Digital image thresholding based on topological stable state. Pattern Recogn. 29, 829–843 (1996)
7. N. Otsu, A threshold selection method from gray level histograms. IEEE Trans. Syst. Man Cybern. SMC-9, 62–66 (1979)
8. J. Kittler, J. Illingworth, Minimum error thresholding. Pattern Recogn. 19, 41–47 (1986)
9. J. Xue, D.M. Titterington, t-tests, F-tests and Otsu's Methods for image thresholding. IEEE Trans. Image Process. 20(8), 2392–2396 (2011)
10. N.R. Pal, S.K. Pal, A review on image segmentation techniques. Pattern Recogn. 26(9), 1277–1294 (1993)
11. P.K. Sahoo, S. Soltani, A.K.C. Wong, Y.C. Chen, A survey of thresholding techniques. Comp. Vis. Graph. Image Process. 41(2), 233–260 (1988)
12. T. Kurita, N. Otsu, N. Abdelmalek, Maximum likelihood thresholding based on population mixture models. Pattern Recogn. 25, 1231–1240 (1992)
13. J.N. Kapur, P.K. Sahoo, A.K.C. Wong, A new method for gray-level picture thresholding using the entropy of the histogram. Graph. Models Image Process. 29, 273–285 (1985)
14. A.S. Abutaleb, Automatic thresholding of gray-level pictures using two-dimensional entropy. Comp. Vis. Graph. Image Process. 47, 22–32 (1989)
15. A.D. Brink, Thresholding of digital images using two-dimensional entropies. Pattern Recogn. 25, 803–808 (1992)
16. P.K. Sahoo, C. Wilkins, J. Yeager, Threshold selection using Renyi's entropy. Pattern Recogn. 30, 71–84 (1997)

17. P.K. Sahoo, G. Arora, Image thresholding using two-dimensional Tsallis-Havrda-Charvat entropy. Pattern Recogn. Lett. 1999 **27**, 520–528 (2006)
18. H.D. Cheng, Y.H. Chen, Fuzzy partition of two-dimensional histogram and its application to thresholding. Pattern Recogn. **32**, 825–843 (1999)
19. C.A. Murthy, S.K. Pal, Fuzzy thresholding: a mathematical framework, bound functions and weighted moving average technique. Pattern Recogn. Lett. **11**, 197–206 (1990)
20. C.V. Jawahar, P.K. Biswas, A.K. Ray, Investigations on fuzzy thresholding based on fuzzy clustering. Pattern Recogn. **30**(10), 1605–1613 (1997)
21. H. Tizhoosh, Image thresholding using type II fuzzy sets. Pattern Recogn. **38**, 2363–2372 (2005)
22. Y. Bazi, L. Bruzzone, F. Melgani, Image thresholding based on the EM algorithm and the generalized Gaussian distribution. Pattern Recogn. **40**, 619–634 (2007)
23. S. Wang, F. Chung, F. Xiong, A novel image thresholding method based on Parzen window estimate. Pattern Recogn. **41**, 117–129 (2008)
24. H.V. Poor, O. Hadjiliadis, *Quickest Detection* (Cambridge University Press, New York, 2009)
25. B. E. Brodsky, B. S. Darkhovsky, Nonparametric methods in change point problems, in *Mathematics and Its Applications*, vol. 243 (Kluwer Academic Publishers, Dordrecht/ Boston/ London, 1993)
26. J. Chen, A.K. Gupta, Parametric statistical change point analysis, with applications to genetics, medicine, and finance, 2nd edn. (Birkhäuser, Boston, 2012)
27. L. Pardo, Statistical Inference Based on Divergence Measures (Chapman & Hall/CRC, 2006), p. 233
28. Y. Wang, Generalized information theory: a review and outlook. Inform. Tech. J. **10**(3), 461–469 (2011)
29. J. Lin, Divergence measures based on Shannon entropy. IEEE Trans. Inform. Theor. **37**(1), 145–151 (1991)
30. S. Choi, S. Cha, C.C. Tappert, A survey of binary similarity and distance. Meas. Syst. Cybern. Inf. **8**(1), 43–48 (2010)
31. F. Zhao, Y. Yang, W. Zhao, Adaptive clustering algorithm based on max-min distance and bayesian decision theory. IAENG Int. J. Comp. Sci. IJCS **44**(2), 24 May 2017
32. Z. Wang, A.C. Bovik, H.R. Sheikh, E.P. Simoncelli, Image quality assessment: from error visibility to structural similarity. IEEE Trans. Image Process. **13**(4), 600–612 (2004)
33. R.K. Chatterjee, A. Kar, Global image thresholding based on change-point detection, in *Proceedings of the International MultiConference of Engineers and Computer Scientists 2017*. Lecture Notes in Engineering and Computer Science (Hong Kong, 15–17 Mar 2017), pp. 434–438
34. E. Pekalska, R.P.W. Duin, The dissimilarity representation for pattern recognition-foundations and applications, in *Series in Machine Perception and Artificial Intelligence*, vol. 64, (World Scientific Publishing Co. Pte. Ltd., 2005), pp. 215–222

Automated Computer Vision System Based on Color Concentration and Level for Product Quality Inspection

Nor Nabilah Syazana Abdul Rahman, Norhashimah Mohd Saad, Abdul Rahim Abdullah and Farhan Abdul Wahab

Abstract Recently, the use of automated product quality inspection in industries is rapidly increasing. Quality is commonly related with product to satisfy the customer's desire and it is important to maintain it before sending to customers. This study presents a technique for product inspection using a computer vision approach. Soft drink beverages have been used as product that to be tested for quality inspection. The database is created to inspect the product based on color concentration and water level quality inspection. The system used Otsu' method for segmentation, histogram from combined red, green, blue (RGB) color model for features extraction, and quadratic distance classifier to classify the product based on color concentration. For water level, the coordinate of image is set to measure the range of water level. Internet Protocol (IP) camera is used while validate the performance of the system. The result shows that the proposed technique is 98% accurate using 246 samples.

Keywords Automatic visual inspection · Color classification
Internet protocol (IP) · Level analysis · Otsu' method · Quadratic distance
classifier · Red Green Blue (RGB) color

N. N. S. A. Rahman (✉) · N. Mohd Saad · A. R. Abdullah
Faculty of Electronic and Computer Engineering, Center for Robotics and Industrial
Automation, Universiti Teknikal Malaysia (UTeM), 76100 Durian Tunggal, Melaka,
Malaysia
e-mail: m021610013@student.utem.edu.my

N. Mohd Saad
e-mail: norhashimah@utem.edu.my

A. R. Abdullah
e-mail: abdulr@utem.edu.my

F. A. Wahab
Infineon Technologies Sdn. Bhd., Free Trade Zone, 75710 Batu Berendam, Melaka, Malaysia
e-mail: fabdulw@student.utem.edu.my

© Springer Nature Singapore Pte Ltd. 2018
S.-I. Ao et al. (eds.), *Transactions on Engineering Technologies*,
https://doi.org/10.1007/978-981-10-7488-2_22

1 Introduction

Nowadays, many industries have been upgraded from manual to automated visual inspection to inspect everything from pharmaceutical drugs to textile production. The human visual system is adapted to perform in a world of variety and change, the visual inspection process, on the other hand, requires observing the same type of image repeatedly to inspect the product [1]. Some reviews show that the accuracy of human visual inspection declines with dull, endlessly routine jobs [2]. Automated visual inspection is obviously the alternative to the human inspector. The demand for industrial automation and the general acceptance among manufacturers show that the automated systems will increase productivity and improve product quality [3, 4].

The ability of vision inspection to detect and prevent defective product packaging from being distributed to consumers is one of the good system. In recent years, retailers and consumers have become much less tolerant with poor packaging quality that can cause health risks or increased retailer costs because of manual inspection by humans [5].

Based on the manual inspection and quality of product issue, this paper is conducted to design the quality inspection for beverage product that can be used for automation process. In developing the quality inspection algorithm, new features will be constructed to carry out two processes of the beverage quality inspection. A quadratic distance classifier technique is proposed in this paper to classify a good or reject of beverage product, based on color concentration and level of water in a bottle. This technique is used to separate two or more classes of object by a quadratic surface. The main advantage of quadratic distance classifier is that the parameters of each class are estimated independently using samples of one class only [6]. It is also has high efficiency, and when the differences on images reach to some extent, it can get high accuracy.

This study presents an automatic visual inspection system for beverage product to develop a real-time system for product quality inspection using Matlab software. Besides, this paper proposed a design of algorithms to classify the color concentration and water level of beverage product. This paper also proposed a design GUI for color concentration and a water level of quality inspection that completing the system. The results show that the accuracy of the system with the proposed technique is better than manual inspection technique.

The remainders of this paper are shown in the followings. Section 2 is described and illustrated the method used to developed visual inspection system. Section 3 is the experimental result that has been analyzed. Section 4 is an overall conclusion about this paper.

2   Methodology

The color and water level quality inspection mainly employs a simple input device and software. Matlab software is chosen for the software development, as it provides a suitable GUI.

2.1   Image Acquisition

Image acquisition act as the action of retrieving image from the source. In the experimental setup which is shown in Fig. 2, the bottle is placed on the conveyer and IP camera 5.0 Megapixel is used to take the image as a sample. IP camera is connected to Wi-Fi wirelessly [7]. The position of the camera is set to 30 cm from the front of the bottle. The proper illumination is designed to provide diffuse illumination over the bottle surface.

The procedure below explained on how to integrate an IP camera to the software:

1. An Android application 'IP Webcam' is installed in smartphone.
2. Resolution of the image can be set in the smartphone.
3. Start the server in smartphone.
4. Note the 'url' will be shown at the bottom of the screen of the smartphone.
5. MATLAB is opened, then the 'url' shown in smartphone and the source code as Fig. 1 below is typed in the command window (Fig. 2).

This paper is used 246 samples, product which is based on colors for beverage products (tropical, strawberry, orange, root beer and grape). The captured image is analyzed using a Matlab software. There are three main steps to categorize the image: color segmentation, classification, and level.

Fig. 1 Source code for streaming live video from IP webcam

```
url = 'http://<ip address>/shot.jpg';
ss = imread(url);

fh = image(ss);
while(1)
    ss = imread(url);
    set(fh, 'CData', ss);
    drawnow;
end
```

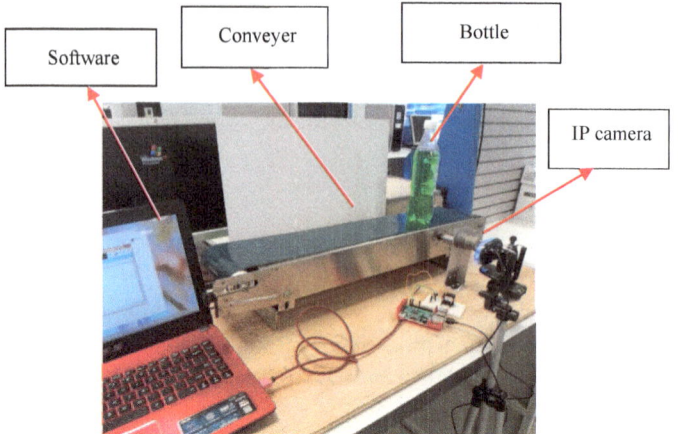

Fig. 2 Experimental setup

2.2 Color Process

Figure 3 shows the analysis framework for color beverages classification. The process is started by converting the input images of red, green, blue (RGB) color model for hue, saturation, value (HSV) representations. This is because HSV color space is similar to the way in which humans perceive color and less complex in terms of hue and saturation. Here, only S component is needed to do the segmentation and thresholding process. Otsu method and morphological operation are applied to segment the region of interest (ROI) area. Histogram analysis is performed to the ROI area. Lastly, the color saturation is classified by using the Quadratic distance classifier.

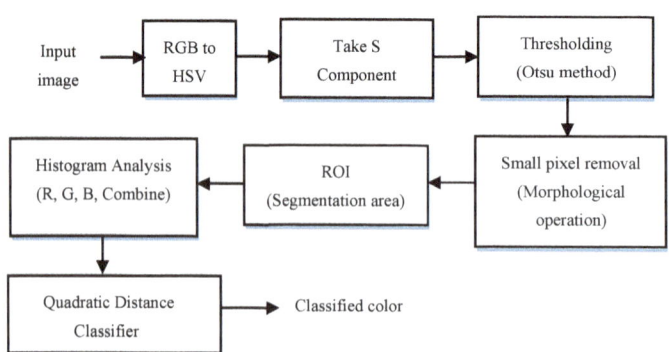

Fig. 3 Color classification process

2.3   Color Segmentation

The segmentation process is dividing an image into parts which are its shape area [8]. The color segmentation often used for indexing and management of data. It is used to separate an image into the certain division, which has similarity in the color information. Segmentation is a process that divides an image into its region or objects that have similar features or characteristics.

2.4   RGB to HSV

In RGB to HSV, the image is converted from RGB to HSV and saturation component. It is based on such intuitive color characteristics a tint, shade, and tone. The coordinate system is cylindrical, and the colors are defined inside a hexcone. The hue H is running from 0 to 360°. The saturation of S is the degree of strength or purity and it is from 0 to 1. Purity is how white is added to the color, so S = 1 makes the purest color. The brightness of V also ranges from 0 to 1, where 0 is the black.

2.5   Otsu' Method

Thresholding is a method of creating a binary image from the gray level image [9]. Otsu' method is utilized image thresholding or, the reduction of a gray level image to a binary image. Otsu' proposed a method that maximizes the separability of the resultant classes in gray levels utilizing a between-class variance function [10]. The description of the syntax are, level = graythresh(I) computes a global threshold (level) that can be used to convert an intensity image to a binary image with im2bw.

2.6   Color Classification

The image is classified by using the quadratic distance technique. From the feature extraction, the image is converted into a red, green and blue histogram to get the distribution data in the image. The image must be normalized to get the value from 0 to 1. The red, green and blue histogram is divided into 10 bins and then it is combined to become one RGB histogram which has 30 bins to view the distributed data from the image. Equation 1 shows the quadratic distance formula to classify color concentration.

$$Vthresh = \sqrt{\sum_{i=1}^{30} [P_1(i) - P_2(i)]^2}, i = 1, 2, 3, \ldots, 30 \tag{1}$$

where, $P_1(i)$ is reference image and $P_2(i)$ is the test image from the histogram bins. Figure 4 shows an example of the color process for the reference image.

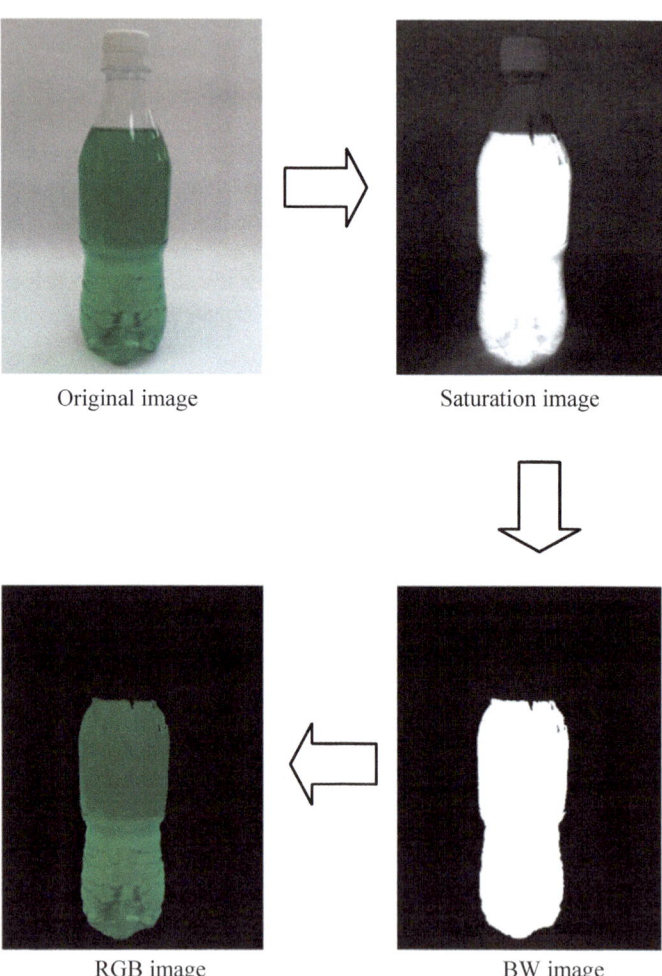

Fig. 4 Color process of reference image

2.7 Level Process

The level classification process is inspected based on three conditions which are level pass, level overfill and level underfill. The input image is processed by setting two points for range of level [11, 12]. The observation of this experiment shows

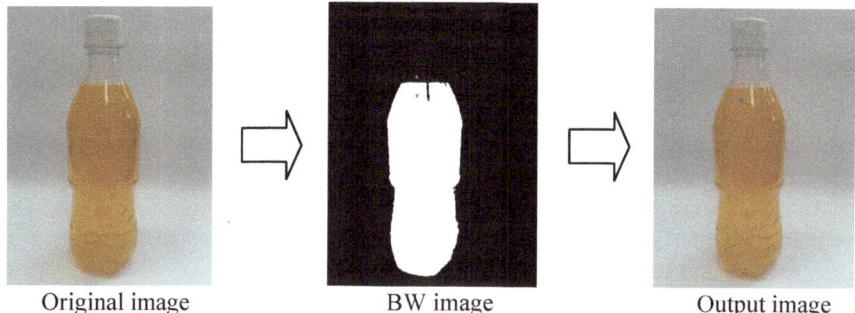

　　Original image　　　　　　　BW image　　　　　　　Output image

Fig. 5 Level pass

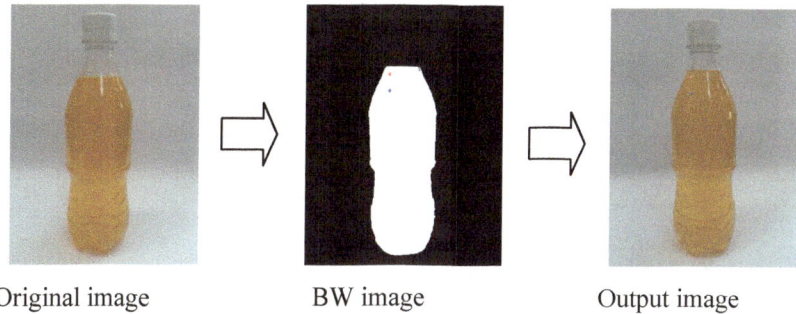

　　Original image　　　　　　　BW image　　　　　　　Output image

Fig. 6 Level overfill

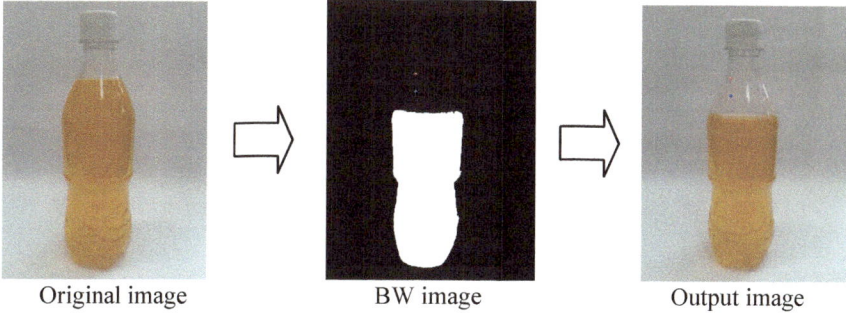

　　Original image　　　　　　　BW image　　　　　　　Output image

Fig. 7 Level underfill

that if the water is above from point 1, the level is overfill. Then, if the water is below from point 2, the level is underfill. The level is passed when the water is between at point 1 and point 2. This process is shown in Figs. 5, 6 and 7.

3 Results

Following are the results of the experiment that has been carried out. Figure 8 shows five types of reference images which are used as the template in classification stage. Figure 9 shows an example of beverages for color classification failed.

Figure 10 and Fig. 11 shows the histogram distributions for color intensity of sample in Fig. 8b and Fig. 9b, respectively.

Afterwards, the 8-bit RGB histograms are grouped into 10-bin histogram. The results are shown in Figs. 12, 13 and 14 for red, green and blue component of reference image strawberry in Fig. 8b, respectively.

(a) Tropical   (b) Strawberry   (c) Orange

(d) Rootbeer   (e) Grape

Fig. 8 Five types products of reference image

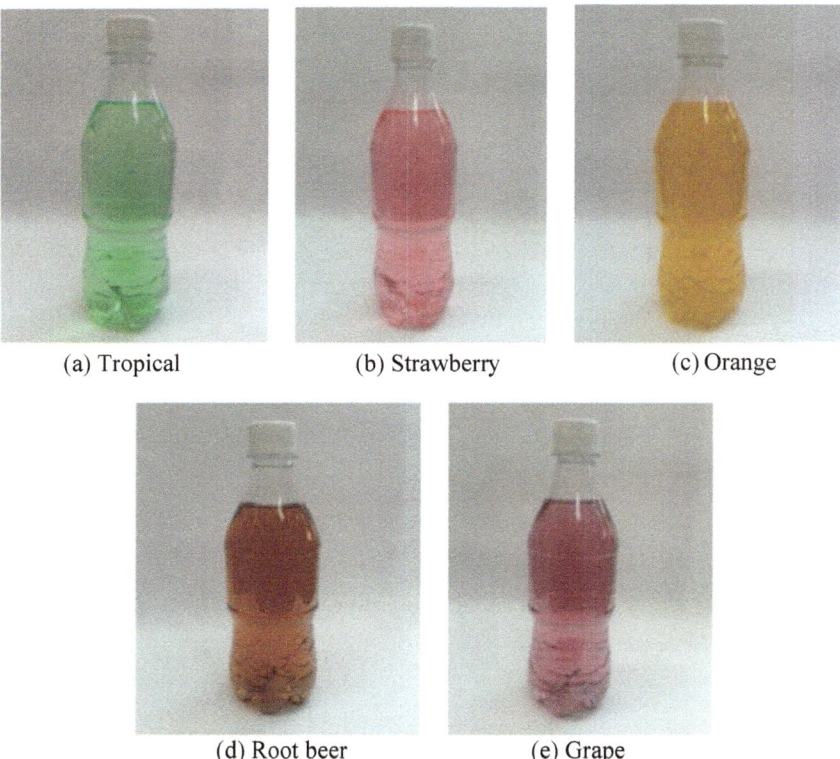

(a) Tropical   (b) Strawberry   (c) Orange

(d) Root beer   (e) Grape

Fig. 9 Five types examples of color fail

Fig. 10 RGB combine graph of reference image

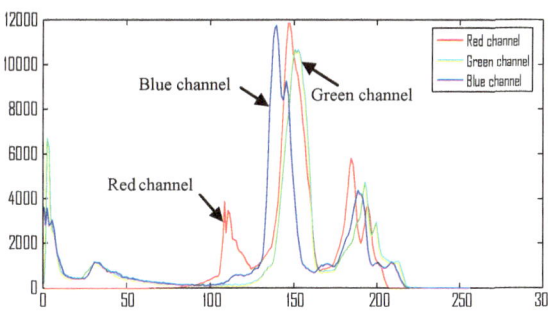

Bins first of histogram show the value of background image which is black image and the rest bins shown the value of color images. The background image was ignored and data from bin 2–10 was taken for analysis. These histograms are then combined into one variable for the classification process. This is shown in Figs. 15 and 16.

Fig. 11 RGB combine graph of test image

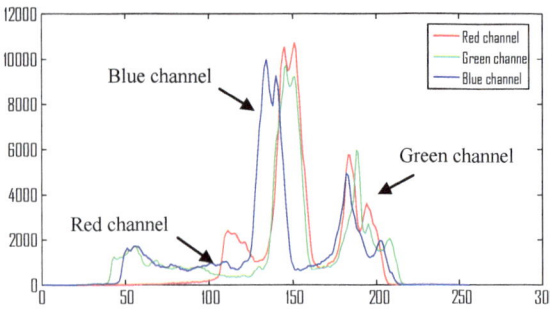

Fig. 12 Red histogram of reference image

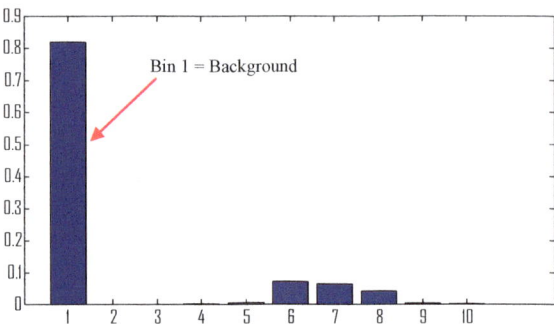

Fig. 13 Green histogram of reference image

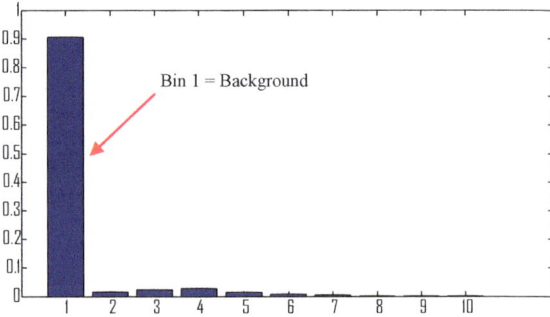

Value of red component is from bin 1–9, green component bin 10–18 and the blue component bin 19–27. The rest three bins are ignored and not taken because there is from the background image. Only value from bin 1–27 are used to calculate the distance and difference of two color images which are image color pass (reference image) and image fail (test image).

The data of each bin from two RGB histogram of the reference image, $P_1(i)$ and test image $P_2(i)$ are calculated by using the quadratic distance formula as shown in Eq. 1 to get distance value of two color images. This distance value can differentiate the color of two images. Figure 17 shows the classification process.

Fig. 14 Blue histogram of reference image

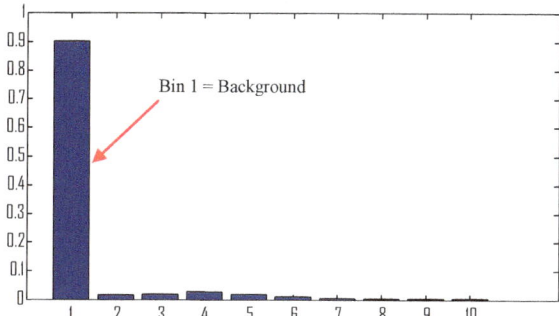

Fig. 15 Combination of RGB histogram for reference image

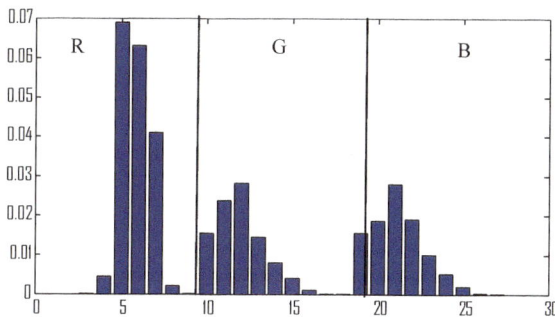

Fig. 16 Combination of RGB histogram for test image

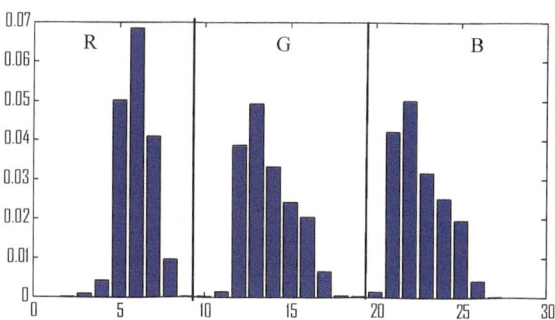

For all sample results, the threshold value is set by 0.2 which is for color PASS when the threshold value less than 0.2 and if more than 0.2 the color of the images were REJECT. Figures 18 and 19 show the complete graphical user interface for the proposed system.

The performance of the system is verified based on the accuracy of the system to classified color concentration and level of water. Figure 20 shows the percentage of each sample image that is tested in real-time system, 98% accuracy is achieved during this testing.

$$Vthresh = \sqrt{\sum_{i=1}^{30}\left[P_1(i) - P_2(i)\right]^2} \quad = 0.5091 \implies \boxed{\text{Color = Reject}}$$

$$\boxed{\begin{array}{l} \text{Threshold} = 0.2 \\ > 0.2 = \text{REJECT} \\ < 0.2 = \text{PASS} \end{array}}$$

Fig. 17 Block diagram of color classification process

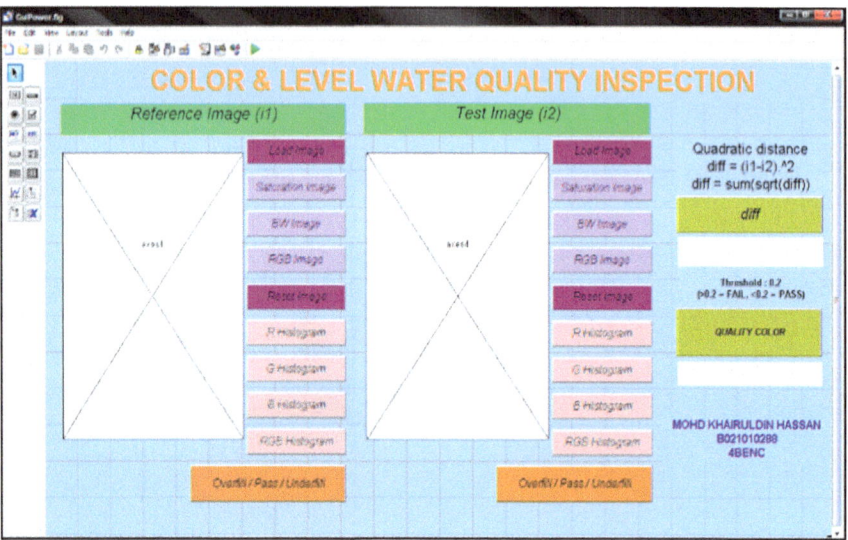

Fig. 18 GUI layout

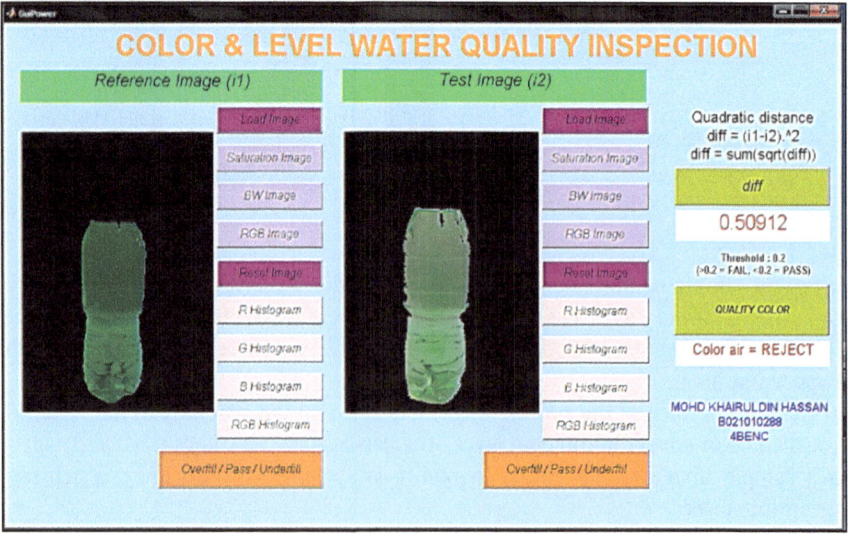

Fig. 19 Complete system of GUI

Fig. 20 System accuracy graph

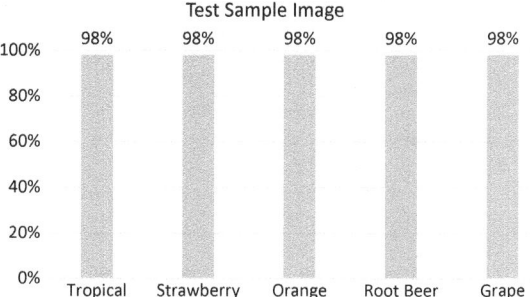

4 Conclusion

This paper has presented a system of beverage quality inspection process. The quality inspection was divided into three main processes; which are a color classification process, range level of water inspection and graphical user interface for the system. Color classification analysis introduces two proposed techniques which are Otsu' method and Quadratic distance classifier. Images of sample are taken by digital color camera for experimental setup. In color analysis, original image is converted to HSV and take S component which is saturation component to process images. After that, it uses Otsu method to threshold and become black and white image, and then ROI is converted for graphical analysis. The quadratic distance classifier technique is used to classify the color image and find the distance value between two color images based on the histogram. The coordinates of the image are set to find the range of pass, overfill and underfill of water level in the bottle. The GUI is designed by using user interface object that available in Matlab software. Protocol (IP) camera is used while validate the performance of the system. The result shows that the proposed system shows 98% accuracy using 246 samples.

Acknowledgements The authors would like to thank to the Universiti Teknikal Malaysia Melaka (UTeM), UTeM Zamalah Scheme, Rehabilitation Engineering & Assistive Technology (REAT) under Center for Robotics & Industrial Automation (CeRIA), Advanced Digital Signal Processing (ADSP) Research Laboratory and Ministry of Higher Education (MOHE), Malaysia for sponsoring this work under project GLuar/STEVIA/2016/FKE-CeRIA/l00009 and the use of the existing facilities to complete this project.

References

1. S.H. Huang, Y.C. Pan, Automated visual inspection in the semiconductor industry: a survey. Comput. Ind. **66**, 1–10 (2015)
2. A.R. Rababaah, Y. Demi-Ejegi, Automatic visual inspection system for stamped sheet metals (AVIS 3 M), in *2012 IEEE International Conference on Computer Science and Automation Engineering (CSAE)*, May 2012, vol. 2 (IEEE), pp. 661–665

3. A.M. Tuates Ir, A.R. Ligisan, Development of PHilMech computer vision system (CVS) for quality analysis of rice and corn. Int. J. Adv. Sci. Eng. Inf. Technol. **6**(6), 1060–1066 (2016)
4. I. Siena, K. Adi, R. Gernowo, N. Mirnasari, Development of algorithm tuberculosis bacteria identification using color segmentation and neural networks. Int. J. Video Image Process. Netw. Secur. IJVIPNS-IJENS **12**(4), 9–13 (2012)
5. S.A. Daramola, M.A. Adefunminiyi, Text Content Dependent Writer Identification (2016), pp. 45–49
6. K. Abou-Moustafa, F.P. Ferrie, Local generalized quadratic distance metrics: application to the k-nearest neighbors classifier, in *Advances in Data Analysis and Classification* (2017), pp. 1–23
7. D.P. Hutabarat, D. Patria, S. Budijono, R. Saleh, Human tracking application in a certain closed area using RFID sensors and IP camera, in *2016 3rd International Conference on Information Technology, Computer, and Electrical Engineering (ICITACEE)*, Oct 2016 (IEEE), pp. 11–16
8. F. Nie, P. Zhang, Fuzzy partition and correlation for image segmentation with differential evolution. IAENG Int. J. Comput. Sci. **40**(3), 164–172 (2013)
9. C.-S. Cho, B.-M. Chung, Development of real-time vision-based fabric inspection system. IEEE Trans. Ind. Electron. **52**(4) (2005)
10. X. Wang, Y. Xue, Fast HEVC intra coding algorithm based on Otsu's method and gradient, in *2016 IEEE International Symposium on Broadband Multimedia Systems and Broadcasting (BMSB)*, June 2016 (IEEE), pp. 1–5
11. K.A. Panetta, S. Nercessian, S. Agaian, *Methods and Apparatus for Image Processing and Analysis,* Trustees of Tufts College, 2016, U.S. Patent 9,299,130
12. N. Mohd Saad, N.N.S.A. Rahman, A.R Abdullah, A.R. Syafeeza, N.S.M. Noor, Quadratic distance and level classifier for product quality inspection system, in *Proceedings of the International MultiConference of Engineers and Computer Scientists 2017, IMECS 2017,* 15–17 Mar 2017, Hong Kong. Lecture Notes in Engineering and Computer Science, pp. 386–390

Use of Computed Tomography and Radiography Imaging in Person Identification

Thi Thi Zin, Ryudo Ishigami, Norihiro Shinkawa and Ryuichi Nishii

Abstract After a large scale of a natural or manmade disaster or fatal accident is hit all victims have to be immediately and accurately identified for the sake of relatives or for judicial aspects. Also it is not ethical for human being to lose their identities after death. Therefore, the identification of a person after or before death is a big issue in any society. In most commonly used methods for person identification includes utilization of different biometric modalities such as Finger-print, Iris, Hand-Veins, Dental biometrics etc. to identify humans. However only a little has been known the chest X-Ray biometric which was very powerful method for identification especially during the mass disasters in which most of other biometrics are unidentifiable. Therefore, in this paper, we propose an identification method which utilizes a fusion of computed tomography and radiography imaging processes to identify human body after death based on chest radiograph database taken prior to death. To confirm the validity of the proposed approach we exhibit some experimental results by using real life dataset. The outcomes are more promising than most of existing methods.

Keywords Computed tomography · Degree of similarity · HOG feature
Identity verification · MFV · Natural disaster · Radiograph
Ranking

T. T. Zin (✉) · R. Ishigami
Faculty of Engineering, University of Miyazaki, Miyazaki, Japan
e-mail: thithi@cc.miyazaki-u.ac.jp

N. Shinkawa
Faculty of Medicine, Department of Radiology, University of Miyazaki,
Miyazaki, Japan

R. Nishii
Department of Diagnostic Imaging Program, Molecular Imaging Center,
National Institute of Radiological Sciences, Chiba, Japan

© Springer Nature Singapore Pte Ltd. 2018
S.-I. Ao et al. (eds.), *Transactions on Engineering Technologies*,
https://doi.org/10.1007/978-981-10-7488-2_23

1 Introduction

In last decade the world has been hit with many big natural disasters such as Hurricane Katrina in USA, Earthquakes in Haiti and Asia, Great East Coast Earthquake and Tsunami in Japan. A natural disaster makes a lot of lives lost and economy as well. Similarly, manmade disasters like Twin Tower Terrorist attack know as 9/11 left many victims unidentified as in natural disasters [1]. We also have some knowledge about the case of the Japan Airlines 123 flight crash in 1985. Fortunately, the corpses of the victim of this case were identified by using methods of tooth treatment medical records, fingerprint collation, DNA matching etc. [2]. However, as the Great East Japan Earthquake happens, if the damage becomes larger and more extensive, corruption and damage caused by delays in the discovery of the bodies become significant, and it is difficult to verify dental findings, fingerprint verification, and identity in DNA collation. Specifically, in the case of the Great East Japan Earthquake, about 500 hospitals, etc. were dispatched from the national police force at the maximum (about 1,500 people in total), the number of dead bodies housed in three Tohoku prefectures and examined, etc. was 15,824, and 15,749 entities have been identified as being identifiable (as of March 11, 2008) [3].

Although it is possible to take various measures such as fingerprints, DNA appraisal, tooth treatment record, etc. to identify unidentified bodies, in modern times it is a very difficult task. The main reason is that when dealing with the corpse, naturally any mistake cannot be allowed so that more unidentified people appear in various forms in front of the doctors and the coroner. For example, if you do not have treatment with a dentist while you were alive, a tooth treatment record method could not be used for identification after the death. On the other hand, in case of clumsy body and cloth and all belongs have gone it is also difficult to narrow down the verification destination of the DNA appraisal the identification could not be confirmed.

At the time of Japanese earthquake disaster, the works of identification of unknown human body have been is drawn public attentions. Various types of methods have been applied for identification including the following activities.

(a) Checking process of blood donation information and making confirmation through the Japanese Red Cross Society whether blood specimens are preserved. If stored, receipt of blood specimen and verification.
(b) Confirmation process of documents directly related to himself, such as umbilical cord, diary, tooth treatment record etc., fingerprint detection, DNA type inspection, dental chart preparation.
(c) Decision making process by combining with other information such as collection of oral cavity cells, DNA type examination, physical characteristics, clothing, personal belongings and the like by related persons by parent-child appraisal method.

Finally, through the processes (a)–(c), the identification task is comprehensively confirmed for the identity. In the case such as the Great East Japan Earthquake that unidentified bodies become extensive and numerous, it had taken a tremendous effort for the identification processes. However, there is a dilemma between identification of dead bodies and rescuing living ones.

In this chapter, we propose a matching method using radiograph and CT scan image after death by using image processing technology. Since most of people in their life time, one time or another, they have made a medical checkup, chest X-ray photographs are taken in their life time. These data are stored in medical database. It is very rare not to have such a data during a person life time. In addition, it is bone which is the most robust and corruption-resistant part of the body that can be obtained with X-ray photograph and CT scan image. Thus, identification of an identity becomes possible even if identity confirmation by tooth treatment record is impossible. It can also be expected to prevent human error when visually checking for a long time. As an operation method when this program is actually completed, a CT scan image is performed on the unknown body to obtain a query image. Then, using the obtained data, a database is created from a hospital within a range where it is considered that an unknown identity body had been living beforehand, and a matching process is performed between the database and the query.

The purpose of this study is to develop an algorithm that creates a database using chest X-ray and matches with CT scan image after death and creates ranking with similar chest radiograph.

This rest of this chapter is organized as follows. The Sect. 2 describes some related works followed by in Sect. 3, the overview of proposed method along with the technical details. Some experimental results are to be shown in Sect. 4 before we conclude the chapter in Sect. 5.

2 Some Related Works

Several researchers have shown valuable methods in a variety of forensics applications including identification process when the common used methods have failed to perform good jobs [4, 5]. In this concern, radiology techniques such as X-Ray and CT images matching techniques have been played important roles. Chest X-ray images have also been studied for visual monitoring system and surveillance by other researchers [6]. Two of the prior studies relied on rule-based keyword search approaches. In another study, chest X-ray images are applied supervised classification to identify chest X-ray reports consistent with visual surveillance [7, 8]. It is also worthwhile to note that the image processing techniques are not only for diagnosis purposes in medical imaging but also they are reliable and accurate for identification of human body during alive or death. In this aspect, we refer some works described in the reference [9–11]. The commonly used method is based on the dental status and the comparison of ante-mortem and postmortem on radiographs. Since the computed tomography is invented the comparison of CT scans of postmortem and the X-ray image of ante-mortem becomes an efficient tool for identification especially when other conventional methods failed [12, 13].

In order to compare two images, the most popular technique is matching the features of images. Among the image features, HOG (Histograms of Oriented Gradients) feature [14, 15], the Gray-Level-Co-Occurrence Matrix (GLCM) feature [16], and the BoW (Bag-of-Visual Words) [17], Markov Feature Vector [18], Scale Invariant Feature Transform [19] are dominant in the literature. These features will be employed in this chapter along with the classification method of SVM (Support Vector Machine) [20] and K-Mean classifiers [21].

In our proposed work, different from prior research, we proposed a fully statistical approach where (1) the content of chest X-ray images are represented by extracted features (2) statistical feature selection was applied to select the most informative features, and (3) matching analysis was used to enrich the extracted features.

3 Overview of Proposed Method

The architecture of proposed human body identification system based on the matching algorithm the X-ray image of prior to death and CT scan image after death is described in Fig. 1a and b. The system contains three steps to identify the bodies by using X-ray and CT scan images. They are (i) Preprocessing Step (ii) Feature Extraction Step and (iii) Matching and Ranking Step.

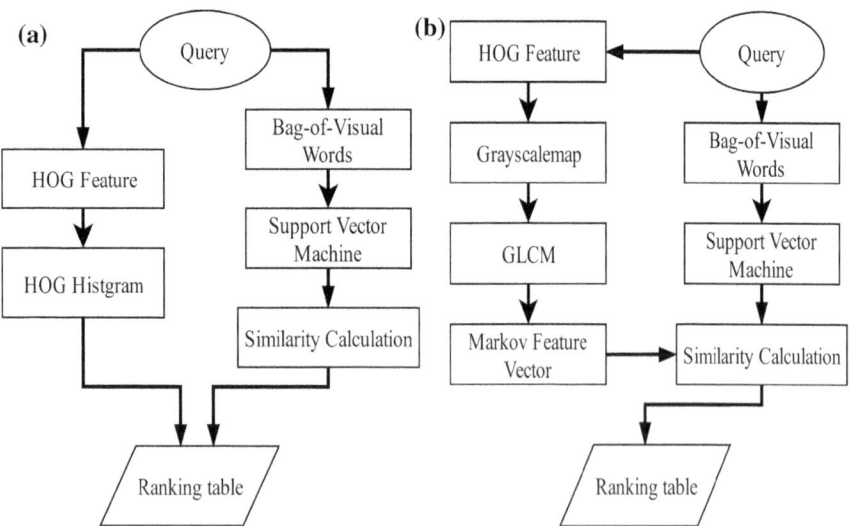

Fig. 1 Overview of proposed method

(i) Preprocessing Step

In this step, all images are to be resized manually. In order to so, first, a marker is placed at the junction of the sternum and the rib of each of the chest radiograph and the CT scan image, then the midpoint two markers is projected into the outermost part of each rib. After this an affine transformation is to be performed with the first rib and the second rib as the target image. The reason for using only the first rib and the second rib as a reference at this time is that when all the markers of the ribs are used, the influence of swelling of the lung accompanying respiration, aging, chest contraction, will occur. So, we chose the first rib and the second rib as the least affected part. On the basis of other parts, for example, the backbone, it is influenced by the curvature of the waist due to aging, and if the collarbone is used as a reference, if the chest X-rays are mutual, the attitude is roughly unified so that it may not be a serious problem, Eq. 1 shows the mathematical formulas used for affine transformation, and Fig. 2 shows an example before and after transformation.

$$\begin{pmatrix} x' \\ y' \end{pmatrix} = \begin{pmatrix} a & b \\ c & d \end{pmatrix} \begin{pmatrix} x \\ y \end{pmatrix} + \begin{pmatrix} +x \\ +y \end{pmatrix} \tag{1}$$

(ii) Feature Extraction Step

In this step, two levels of features will be extracted. First we will extract Histogram of Orientations of Gradients (HOG) feature and Bag of Words (BoW) feature from all X-ray images in the database. In this feature extraction process the luminance gradient and the luminance intensity for each local region (Cell) are extracted and the intensity for each direction is expressed in a histogram for each block, so that the rough shape of the object can be extracted. By using this feature quantity, we can expect extraction of features related to the slope of the ribs. In general, a normalization processing is performed in each block, and the shape of the histogram is adjusted, to characterize local illumination fluctuation. The image of HOG feature quantity is shown in Fig. 3. Similarly, the features are extracted from the query image too.

(a) **(b)**

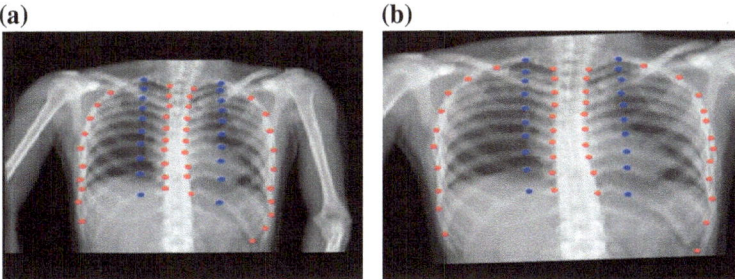

Fig. 2 Example before and after transformation

Fig. 3 Image of HOG feature

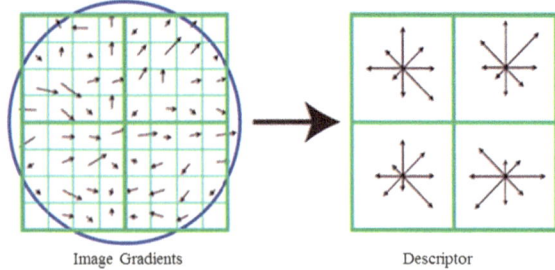

It is worthwhile to note that during extracting the HOG features process due the use the brightness gradient, large noise occurs if the entire photograph is whitish and blurred or shadows appear large. In order to improve the Level 1 feature extraction process, we move forward to proceed to Level 2 process by adding Markov Feature Vector developed for object image retrieval.

In this Level 2, after finding the HOG feature amount, a grayscale map of nine gradations (0–8) was generated using the maximum bin value in each block without obtaining the HOG histogram. A grayscale image is an image represented by only 256 gradations per pixel, whereas a normal photograph is represented by 256 gradations of each of Red, Green and Blue three colors. Figure 4 shows an example of a full-color image and a grayscale image.

We then form a Gray Level Co-Occurrence Matrix (GLCM) by considering the spatial relationship of pixels, which is also called a gray level spatial dependence matrix. Figure 5 shows the calculated value of GLCM (0-degree direction) of the 4-row and 5-column image with the density co-occurrence matrix. In this case, the value 1 is stored in the element (1, 1) of the GLCM. This is because there is only one instance in the image having the values 1 and 1 in two horizontally adjacent pixels. Similarly, 0 is stored in the element (2, 1) of GLCM by calculation. This is because there are no instances in the image that have two values horizontally adjacent and two and one. Do this in all eight directions. In the proposed method using the HOG histogram alone, it was learnt that it is difficult to capture stable feature quantities for X-ray photographs and CT scan images photographed under

Fig. 4 Examples of full-color images and grayscale images

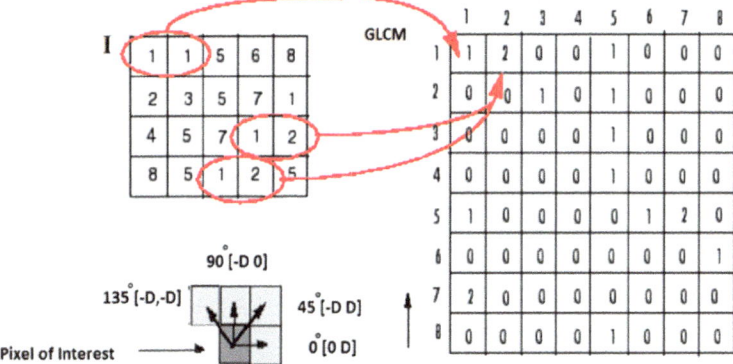

Fig. 5 Calculation example of GLCM

various conditions. Therefore, we employ the Markov Feature Vector similar to one used in [22].

From the grayscale map, it can be seen that the shape of the original data is blurredly captured and then Markov Feature Vector (MFV) is generated as follows. Let P (k × k) be the GLCM at an angle, then P in one direction is given by the equation.

$$C = C_{i,j}, \quad P_{(i,j)} = \frac{C_{(i,j)}}{\sum_{i=1}^{k} C_{(i,j)}}$$

GLCM in all 8 directions can be expressed as $\pi = \pi P$ where π is defined as MFV.

(iii) Matching and Ranking Step

In this step, two sub processes will be performed. One is finding the Euclidean distance between two feature vectors and the other is the ranking process according to the Euclidean distance measures.

Euclidean distance

We calculate similarity using Euclidean distances between features extracted from each chest radiograph stored in the database using MFV and BoW score obtained from the query image. In this chapter, we used an expression where MFV of query image data is x_m, image data in database is y_m, BoW score of query image data is x_b, image data in database is y_b, and distance is D.

$$D = \sqrt{200(x_m - y_m)^2 + 50(x_b - y_b)^2}$$

Ranking Evaluation method

Here, the ranking evaluation method of the matching result will be described. The evaluation method of similarity ranking includes a method of finding the Bull's eye score and relevance rate, a method of finding the recall rate, etc. However, due to the characteristics of this experiment, multiple chest radiographs and CT scan images are collected per person.

4 Some Experimental Results

In this section, we present some experimental works to illustrate the validity of the proposed method by using the medical information database of Miyazaki University School of Medicine. In the data base we investigate 27 subjects for whom pre-living data are available so that post-mortem data is to be extracted from data stored in the database. Experimental data was saved in DICOM format, and processing was performed on captured images of each data. Figures 6 and 7 show actual images. The chest radiographs were all 2470×2470 in resolution, and the CT scanned images were all 256×256 in resolution. ID1, ID2, ID3, … are allocated in order from the upper left in Figs. 6 and 7, and ID corresponds to the same person before and after death respectively.

Fig. 6 Chest radiograph before death

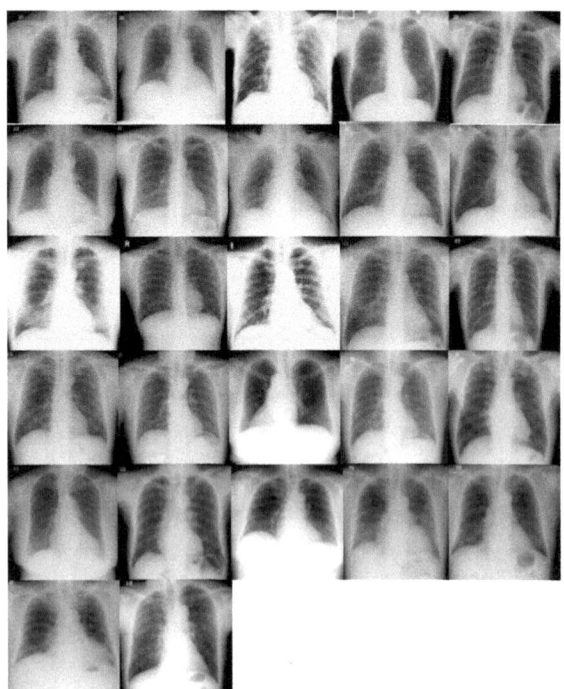

Fig. 7 CT scan image after death

Initially, we created a database using all images described in Fig. 6 and then present an input image from the Fig. 7 as query to the system to produce an outcome result. The output result shows that 16 out of 27 are ranked within top 10. Although there was no example where the correct answer image was displayed at the lowest position among all the data, the result with the lowest ranking was the 24th in ID9.

Here, we have shown both best and lowest rank output results. In the output data, the 15 highest scores of 27 people are displayed in ascending order. Figures 8 and 9 show examples of best output results and Fig. 10 shows an example in which the correct answer image could not be represented in the top 15 places. In addition, when the experiment was carried out using only the portion of the sixth rib or more after trimming the stable part, the accuracy was 59.3%. The accuracy when using the stable regions MFV and BoW which were the most accurate in each of the first and second feature amounts was the highest, 63.0%. Table 1 shows the accuracy and overall result of each feature amount for each proposed method.

Discussions on the Experimental Results

It is worthwhile to discuss about the experimental results presented in this chapter. In this experiment we used the data of 27 people. We found that the final correct answer rate was about 60%. The reason is that all the features used in here are feature quantities dependent on luminance, so it is clear that light and darkness

Fig. 8 Result of ID23

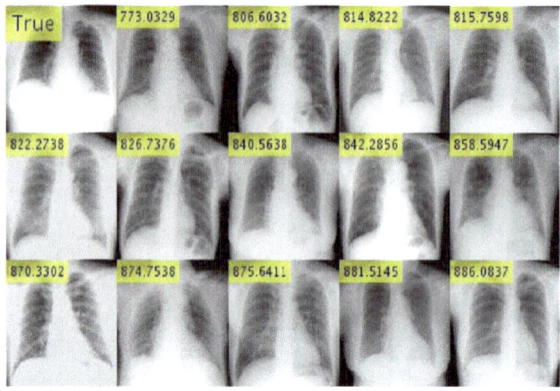

Fig. 9 Result of ID27

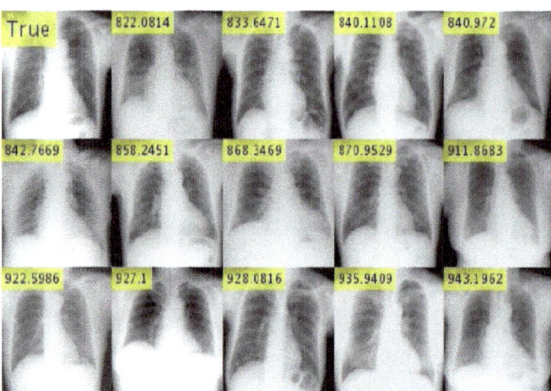

Fig. 10 Result of ID19

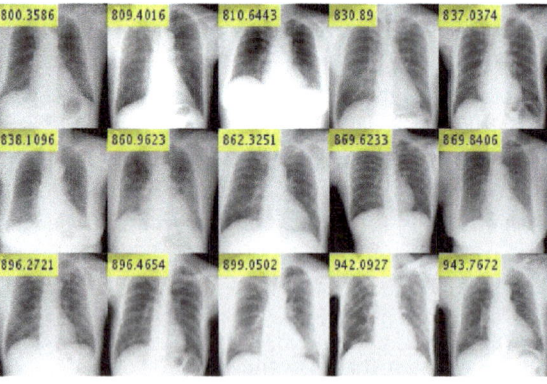

appears definitely which leads to firmly extracting feature quantities such as rib tilt and spine unevenness. As a result, it seems that an improvement of the ranking could be made. In addition, as a feature of the data with low ranking, it is a case

Table 1 Accuracy and overall accuracy

Method	First feature (%)	Second feature (%)	Accuracy (%)
[HOG histogram, BoW]	18.5	51.9	44.4
[MFV, BoW]	48.1	51.9	59.3
[Steady MFV, Steady BoW]	59.3	40.7	59.3
[Steady MFV, BoW]	59.3	51.9	63.0

where an overall X-rayed picture is used as a database. In this case as well, it is considered that the feature amount used greatly depends on luminance. When using each feature quantity alone, the accuracy was 48.1% in the case of MFV alone, and 51.9% in the case where BoW alone. It turns out that combining these two leads to an improvement in accuracy of improvement 110% of higher for MFV. We noticed that the highest is achieved in the case of the combination of MFV and BoW features. This is also due to the part cut out images may differ little by little. We expect that there is a possibility that further improvements in accuracy.

5 Conclusion and Future Work

In this study, we have proposed a method to score and output the similarity score based on the chest radiograph of the living as in a database and CT scan image of the corpus as a query to identify the human body as an extended version of the conference paper [23]. We have conducted an experiment to confirm its effectiveness. In the proposed method using features mainly dependent on luminance information, good results were obtained for cases where light and dark appear clearly, but overall whitish data is used for database creation. Since the DICOM format is a RAW image, it is possible to adjust the contrast and luminance, so if the parameter adjustment can be automated, the variation in the result due to the difference in contrast may be suppressed. In addition, since all the resizing images are performed at the beginning of the experiment manually, it is considered that there is a possibility that the change in ranking may be caused by the resizing, so it is necessary to implement an automatic resizing method. Again, we realized that the purpose of resizing is due to the difference in resolution and scale between the CT scanned image and X-ray photo. When CT scan of the body is performed the pose and the size as the reference at the time of photographing such as the position of the arm and the size of the rib may not be uniform. If it can be unified, we think that it will not be necessary to resize, which will lead to an improvement in accuracy.

Future Works

In future we expect to perform automation of image resizing which is currently performed and automatic setting of parameters such as luminance contrast information, experiments using the proposed method for larger scale data, and extraction of new feature quantities. It is observed that automating the resizing of images is essential when conducting experiments with larger data, but it is sometimes made mistakes to read information such as ribs and spine from radiographs and CT scan images with experts' eyes. This is a challenging task. Experiments on larger data sets suggest that it is sufficient for this purpose, which assumes large-scale disasters and large-scale accidents alone based on the data for 27 people only. Although it is necessary to experiment with more data than ever, it is not easy to prepare 10000 pieces of data due to the characteristic of dealing with human death data. It is conceivable that experiments are carried out after considering X-ray photographs as dead, processing to the prepared data, and trying to increase the apparent number of data. Also, at present, only one piece of post-mortem data is used for each person, but as for chest X-ray photographs, there are many cases where multiple sheets were stored before death. If it becomes possible to use all, there is a possibility of leading to an improvement in the average ranking. Finally, although the new feature makes an accuracy of less than 60%, the combined features achieved, the final accuracy is improved up to 60% or more. With respect to practical point of view, it is insufficient for practical use. Furthermore, by increasing the number of feature quantities, it is also expected that countermeasures will be taken in the case where one feature amount is greatly influenced by noise and the overall accuracy is greatly lowered.

Acknowledgements This work was partially supported by JSPS KAKENHI Grant Number 15K15457.

References

1. D. Pushpalal, J. Rhyner, V. Hossini, The great east japan earthquake 11 March 2011: lessons learned and research questions, Proofreading: J. Kandel, S. Yi, S. Stoyanova, and Print: D. Paffenholz, Bonn, Germany, (UN Campus, Bonn, 11 Mar 2013)
2. T.D. Ruder, M. Kraehenbuehl, W.F. Gotsmy, S. Mathier, L.C. Ebert, M.J. Thali, G.M. Hatch, Radiologic identification of disaster victims: a simple and reliable method using CT of the paranasal sinuses. Eur. J. Radiol. **81**(2), e132–e138, 29 Feb 2012
3. Police Agency: Police procedure accompanying the great east Japan great earthquake (Japanese). https://www.npa.go.jp/archive/keibi/biki/keisatsusoti/zentaiban.pdf, 25 Jan 2017
4. R.W. Byard, K. Both, E. Simpson, The identification of submerged skeletonized remains. Am. J. Forensic Med. Pathol. **29**(1), 69–71, 1 Mar 2008
5. Y. Bilge, P.S. Kedici, Y.D. Alakoç et al., The identification of a dismembered human body: a multidisciplinary approach. Forensic Sci. Int. **2**(137), 141–146, 26 Nov 2003
6. G.A.M. Loaiza, A.F.O. Daza, G.A. Archila, Applications of conventional radiology in the medical forensic field. Rev. Colomb. Radiol. **4**(24), 3805–3817 (2003)
7. C.A. Bejan, L. Vanderwende, F. Xia, M. Yetisgen-Yildiz, Assertion modeling and its role in clinical phenotype identification. J. Biomed. Inform. **46**(1), 68–74 (2003)
8. C.A. Bejan, F. Xia, L. Vanderwende, M. Wurfel, Y.M. Yildiz, Pneumonia identification using statistical feature selection. J. Am. Med. Inform. Assoc. **5**(19), 817–823, 26 Apr 2012
9. B.G. Brogdon, *Forensic Radiology* (CRC Press LLC, Boca Rakon, Florida, 2002)

10. D.R. Smith, K.G. Limbird, J.M. Hofman, Identification of human skeletal remains by comparison of body details of the cranium using computerized tomographic (CT) scans. J. Forensic Sci. **47**(5), 937–939, 1 Sept 2002

11. T. Riepert, D. Ulmcke, F. Schweden, B. Nafe, Identification of unknown bodies by X-ray image comparison of the skull using X-ray simulation program FoXSIS. Forensic Sci. Int. **1** (117), 89–98, 1 Mar 2001

12. M. Pfaeffli, P. Vock, R. Dirnhofer, M. Braun, S.A. Bolliger, M.J. Thali, Post-mortem radiological CT identification based on classical ante-mortem X-ray examinations. Forensic Sci. Int. **2**(171), 111–117 (2007)

13. G.M. Hatch, F. Dedouit, A.M. Christensen, M.J. Thali, T.D. Ruder, RADid: a pictorial review of radiologic identification using postmortem CT. J. Forensic Radiol. Im. **2**(2), 52–59, 30 Apr 2014

14. N. Dalal, B. Triggs, Histograms of oriented gradients for human detection, in *Proceedings of IEEE Conference on Computer Vision and Pattern Recognition (CVPR)*, vol. 1, pp. 886–893, 25 Jun 2005

15. F. Behnood, HOG (Histogram of Oriented Gradients) with matlab implementation. http:// farshbafdoustar.blogspot.jp/2011/09/hog-with-matlab-implementation.html, 27 Sept 2011

16. Concentration co-occurrence matrix (Japanese). https://jp.mathworks.com/help/images/ref/ graycomatrix.html, 25 Jan 2017

17. D.G. Lowe, Local feature view clustering for 3D object recognition, in *Proceedings of 2001 IEEE Conference on Computer Vision and Pattern Recognition*, vol. 1 (Kauai, Hawaii, 2001), p. 1

18. H. Bay, A. Ess, T. Tuytelaars, L. Van Gool, SURF: Speeded up robust features. Comput. Vis. Im. Understand. (CVIU) **3**(110), 346–359, 30 Jun 2008

19. B.E. Boser, I.M. Guyon, V.N. Vapnik, A training algorithm for optimal margin classifiers, in *Proceedings of the Fifth Annual Workshop on Computational Learning theory (COLT)*, pp. 144–152, 1 Jul 1992

20. J.B. MacQueen, Some methods for classification and analysis of multivariate observations, in *Proceedings of 5th Berkeley Symposium on Mathematical Statistics and Probability*, vol. 1, No. 14, pp. 281–297 (1967)

21. J. Sivic, A. Zisserman, Efficient visual search of videos cast as text retrieval. IEEE Trans. Pattern Anal. Mach. Intel. **4**(31), 591–605 (2009)

22. T.T. Zin, P. Tin, T. Toriu, H. Hama, Dominant color embedded markov chain model for object image retrieval, in *Intelligent Information Hiding and Multimedia Signal Processing*, pp. 186–189, 12 Sept 2009

23. R. Ishigami, T.T. Zin, N. Shinkawa, R. Nishii, Human identification using X-Ray image matching, in *Lecture Notes in Engineering and Computer Science: Proceedings of the International Multi Conference of Engineers and Computer Scientists 2017* (Hong Kong, 15–17 March, 2017), pp. 415–418

Improvement Method for Topic-Based Path Model by Using Word2vec

Ryosuke Saga and Shoji Nohara

Abstract Studying purchasing factor for product developers in the market place is important. Using text data, such as comments from consumers, for factor analysis is a valid method. However, previous research show that generating a stable model for factor analysis using text data is difficult. We assume that if the target text data are handled well, then the analysis can progress smoothly. This study proposes pre-processing text data by word2vec for factor analysis to improve the analysis. Word2vec regards words as vectors in text. Our proposed process is effective, because variables are expressed as the frequency of words in the analysis model. Experiment results also show that our proposed method is helpful in generating an analytical model.

Keywords Causal analysis · Data ming · Text mining · Topic model
Structural equation modeling · Word2vec

1 Introduction

Product developers in many companies gather customer opinions, especially focusing on text data from reviews or questionnaires. Developing a brand name or evaluating merchandise by using text data is beneficial. However, this approach cannot easily handle massive text data. In recent years, text mining has been used as an important method in market research when dealing with huge amounts of text data.

R. Saga (✉)
Graduate School of Humanities and Sustainable System Sciences,
Osaka Prefecture University, 1-1 Gakuen-cho, Naka-ku, Sakai, Osaka, Japan
e-mail: saga@cs.oskaafu-u.ac.jp

S. Nohara
Graduate School of Engineering, Osaka Prefecture Univesrity, 1-1 Gakuen-cho,
Naka-ku, Sakai, Osaka, Japan

© Springer Nature Singapore Pte Ltd. 2018
S.-I. Ao et al. (eds.), *Transactions on Engineering Technologies*,
https://doi.org/10.1007/978-981-10-7488-2_24

Topic models are a general method in the field of data-mining. Topic modelling is a machine learning technique that clarifies the structure of a document group by estimating words. The words constitute a topic based on the premise that each document group comprising the corpus belongs to that specific topic. Several studies have analyzed various consumer situations using topic models. For example, Kawanaka et al. proposed a method for analyzing competitive relations of brands using latent semantic analysis (LSA), which is a kind of modelling method [1]. Wajima et al. proposed the identification of a negative factor using latent Dirichlet allocation (LDA), which is effective for many applications [2]. These related studies indicate that factor analysis using topic modelling is an effective and valid approach in handling text corpus (i.e., sets of electronic documents). However, both LSA and LDA cannot define relationships among topics in an analysis model. Kunimoto et al. successfully analyzed the gaming software market by using structural equation modelling (SEM), which is a factor analysis method [3]. They proposed a path model generation process for SEM using hierarchical LDA (hLDA). Meanwhile, Saga et al. proposed using SEM with hLDA targeting Crowdfunding [4]. They also attempted to combine numeric and text data to explain how the identified relationships influence the decision to invest in a crowdfunding project [4]. Although these proposed methods employ visual and quantitative analyses, it remains difficult to generate a model. In addition, the significance level is not mentioned or not high enough to interpret in these works.

In the current study, we propose a pre-processing method that uses a novel technique, Word2vec, for pre-processing target text data [5]. Word2vec is a two-layer neural network that, after obtaining information from the text corpus, outputs the feature vectors of words in the text corpus. Word2vec is also able to compute the similarity of words as a similarity of vectors, thus allowing it to group words based on similarities. By using this technique, we can extract words that are not keywords but are related to keywords. Results show that higher evaluation values (such as the score of GFI) can be gained by using the analysis model.

2 Topic-Based Path Model Constructed for Structural Equation Modeling

SEM analyzes various relationships among several factors, i.e., latent and observed variables. A latent variable is an invisible concept that is used for target analysis. An observed variable is an observable item from the target analysis, and is used to estimate a latent variable. These variables have "causal" and "co-occurrence" relationships. SEM can quantify the influence and strength of these relationships.

A path model is used to understand the relationships among the variables. This model visualizes the factors and the relationships among them, as shown in Fig. 1. In the path model, the observed and latent variables are denoted by a rectangle and an ellipse, respectively. The relationships among the variables are expressed by the

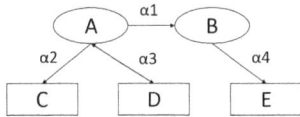

Fig. 1 Path Model

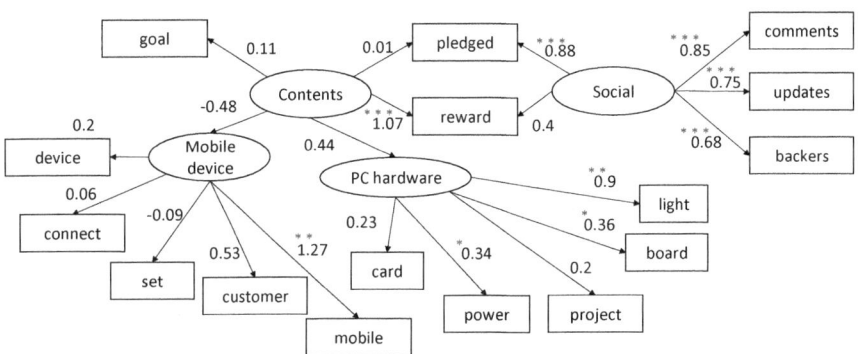

Fig. 2 Five keywords without word2vec (previous model [4])

unidirectional and bidirectional arrows, which respectively correspond to causal and co-occurrence relationships.

The path model shown in Fig. 2 consists of three observed variables (C, D, and E) and two latent variables (A and B). The relationship between A and D, denoted as α3, is co-occurrence, whereas the other relationships, denoted as α1, α2, and α5, are causal relationships.

Two methods can be used to build the models: the data-driven approach (exploratory factor analysis, EFA) and the hypothesis-based approach (confirmatory factor analysis, CFA). Especially for the former methods, Saga et al. proposed a method that can build the model using text data based on an LSA-approach [6]. Furthermore, Saga et al. [7] and Kunimoto et al. [3] have developed the approach of using multinominal topics based on LDA and hLDA, respectively. By using hLDA, we can automatically extract not only the topics but also the hierarchical structure of these topics from the text data, thus allowing us objectively and understandably construct the path model of SEM with hLDA.

However, this approach faces several challenges. The first of these is the identification problem. As the approach utilizes the term frequency of keywords, the zero-frequency problem may occur, because some keywords appear less frequently in some documents. Another problem, which is related to the first one, is the significance level for paths, that is, the fewer the term frequency, the harder the estimations of the path coefficients. Therefore, the generated models consist of insignificant paths that are difficult to explain and not at all reliable.

The root of the abovementioned problems lies in the low frequency of each keyword. As a solution, we aim to increase term frequency artificially. In doing so, the ontology approach is a useful technique, because it semantically regards equivalent words as the same words. However, manually constructing an ontology from scratch, especially for new domains like Kickstarter, can be quite costly. Thus, we utilize the word2vec which automatically creates a similarity structure among words from text data.

3 Pre-processing by Using Word2vec

3.1 Word2vec

Word2vec is a simple neural network composed of two layers: hidden and output layers [8, 9]. By grouping similar words, distributed representations of words in a vector space help learning algorithms achieve better performance in natural language processing tasks. The neural network of Word2vec includes two architectures: Continuous Bag-of-Words Model (CBOW) and Skip–Gram. The former uses continuous distributed representations of the context. The best performance on the task introduced in the next section is obtained by building a log-linear classifier with four future and four history words as inputs, where the training criterion is to correctly classify the current (middle) word. Using the Skip–Gram model, Mikolov et al. introduced an efficient method for learning high-quality vector representations of words from large amounts of unstructured text data. Unlike many of the previously used neural network architectures for learning word vectors, the Skip–Gram model does not involve dense matrix multiplications.

Another important technique that is used to derive word embedding is called negative sampling. While negative-sampling is based on the Skip–Gram model, it is in fact, optimizing a different objective. What follows is the derivation of the negative-sampling objective.

The word representations computed using neural networks are unique, because the learned vectors explicitly encode many linguistic regularities and patterns. Furthermore, many of these patterns can be represented as linear translations. For example, the result of a vector calculation vec("Tokyo") − vec("Japan") + vec ("France") is closer to vec("Paris") than to any other word vector.

3.2 Data Pre-processing

This section describes our proposed text pre-processing using word2vec. First, for learning the word2vec knowledge model, we obtain the corpus as word2vec import text data. After acquiring the knowledge model, the word2vec vector computing is ready for use. Next, word2vec is used to compute the feature vector for every word in the target text corpus, thereby comprising the text data for analysis. At this point, we perform the pre-processing of the target text data. Word2vec can compute word similarities as cosine similarities, and the similarity scores range between 0 and 1. If the score is closer to 1, this indicates that the similarity is the higher. Focusing on target text data, we compute the similarity between two words, and if the similarity exceeds the threshold score, we equalize the compared words.

Next, we obtain the modified term frequency that artificially affects similarity as term frequency, and is denoted by tf'_i

$$tf'_i = tf_i + \sum_{j \in S} sim(i,j)tf_j \qquad (1)$$

where tf_i is the term frequency of word i appearing in the text, and $sim(i, j)$ shows the similarity between words i and j. In addition, S is a set of words similar to word i, which exceed the threshold value. For example, for three words "A," "B," and "C," if the similarity between "A" and "B" is over the threshold score, and "C" is not similar to either "A" or "B," then "B" is converted to word "A" times by $sim(A, B)$ and "C" is ignored. Note that, as the threshold value becomes lower, many words may be regarded as the same. This means that the low threshold value transforms the model into more abstracted one.

4 Experiment

4.1 Dataset and Experiment Process

We perform the experiment to confirm that our proposed approach improves model fitness and significance level of paths. In this experiment, we collected from Kickstarter 754 live project data from the technology genre on a specific date (April 28, 2015) [10]. The data include the "backers," which indicate the number of persons who invested in a project; "pledged," which show the investment amount regarded as funded; "update," which is the number of updates of a project; "comments," which indicate the number of interaction with investors; and "reward information" to build the social variable. Regarding the reward information, we collected the smallest amount of money required to obtain a product or an item or to receive the first reward. In this study, we focused on 11 genres: 3D printer, Apps, DIY, Flight, Gadgets, Hardware, Robots, Sound, Wearables, Space Exploration and Web categories for analysis.

For the learning corpus for word2vec, we extracted 10,000 pages of text data from a Google web search with each category word ("hardware," "3d printer" and so on). To maintain generality, we performed additional learning by Text8 corpus (the corpus was the word sample text data in the genism package).

The models were evaluated based on the GFI, AGFI, and RMSEA indices. GFI and AGFI have values between 0 and 1; the higher the value of the model is, the better the model. In general, AGFI is lower than GFI. Meanwhile, RMSEA should be lower than 0.10; if the value is lower than 0.5, then the model is considered a good model and is between compatibility and information quantity. For topic extraction using hLDA, we used the method employed by Mallet [11]. For SEM analysis, we used the SEM package (3.1–5) provided in R (3.2.0) [12, 13]. For word2vec, we used the genism package (0.12.4) in Python (3.5.2) [14]. As the parameter of word2vec, we changed the threshold score between 0.5 and 0.9 per 0.1.

4.2 Result and Discussion

Table 1 shows the result of the analysis changing the number of embedded keywords in model about hardware category. The evaluation measure is better when the threshold score is 0.8. Models 2 and 6 have higher GFI and AGFI indices and lower RMSEA scores compared with the other models. Therefore, our proposed process of using the SEM with hLDA analysis model is useful.

As can be seen, Models 6, 9, and 11 have better scores in the number of variables exceeding the significance level. The results of the conventional and the proposed models in Fig. 2 and Fig. 3, respectively, are compared. The asterisks *, **, and *** indicate significance at the 10%, 5%, and 1% levels, respectively. In Model 5, the number of variables that reach the significance level increases.

Table 1 Result for each parameter in hardware category

Model number	Number of keywords	Threshold value	GFI	AGFI	RMSEA	Number of variable exceeding significance level
1	3	Without wd2vc	0.871	0.792	0.0836	11
2	3	0.9	0.893	0.827	0.0682	10
3	3	0.8	0.894	0.829	0.0669	11
4	3	0.7	0.881	0.807	0.0772	11
5	3	0.6	0.842	0.745	0.1007	10
6	3	0.5	0.833	0.731	0.127	15
7	5	Without wd2vc	0.86	0.805	0.0678	9
8	5	0.9	0.889	0.845	0.0472	11
9	5	0.8	0.89	0.845	0.0468	12
10	5	0.7	0.864	0.809	0.0659	9
11	5	0.6	0.844	0.781	0.0774	12
12	5	0.5	0.841	0.777	0.079	5

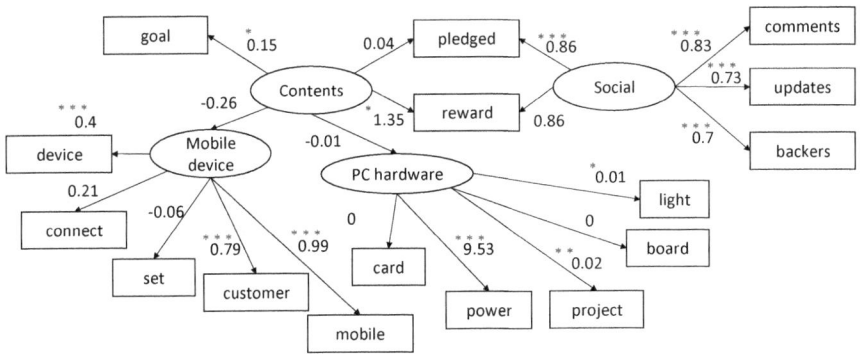

Fig. 3 Five keywords with a threshold value of 0.8

One common problem encountered by previous research is that it is difficult to construct a stable analysis model. This suggests that some analysis models do not have good evaluations in terms of GFI and AGFI scores. In comparison, our proposed method successfully increased the evaluation value. In addition, Figs. 2 and 3 show there exist optimal similarities with which to equalize words. When the similarity is low, the words that do not have a strong relationship are regarded as keywords. Hence, the observed variables lose touch with the models, because words that comprise the topic change significantly, as in the case with the models with a threshold value 0.9, which is worse than 0.8. Hence, it is difficult to distinguish a model's threshold value 0.9 from the conventional models (without word2vec).

From Table 1, we found that GFI is better than when the number of keywords was three but when we consider not only GFI and AGFI and RMSEA, our model could output when the number of keywords was five. Therefore, we output models and evaluate them for other categories in the same way.

Table 2 shows the results of analyses. From the table, we found the same trends among almost models to the case of hardware category. That is, each model has a maximum value and as the threshold decreases, evaluation values are worse. Therefore, there is a high possibility that our method can improve the SEM model.

However, sometimes the method does not work well. For example, GFI decreases on the threshold around 0.7 in 3D printer category. This can be guessed that there are a lot of keywords with similarity 0.7–0.8 related to 3D printer categories in the corpus, which influence to evaluation. In Web category, the previous work (without Word2vec) is the best. As a reason, we can guess that the learning of Word2vec is not enough because "web" is nowadays general word so that the general words related to web is not appropriate for data to explain technical items in Kickstarter.

Table 2 Result for each category with five keywords

Genre	Model number	Threshold value	GFI	AGFI	RMSEA	Number of variable exceeding significance level
3D Printer	1	Without wd2vc	0.633	0.462	0.163	10
	2	0.9	0.643	0.477	0.160	14
	3	0.8	0.647	0.484	0.159	16
	4	0.7	0.600	0.414	0.172	13
	5	0.6	0.630	0.459	0.164	14
	6	0.5	0.558	0.353	0.183	12
Apps	1	Without wd2vc	0.668	0.596	0.078	26
	2	0.9	0.716	0.655	0.065	27
	3	0.8	0.722	0.661	0.063	26
	4	0.7	0.721	0.661	0.064	28
	5	0.6	0.687	0.619	0.074	23
	6	0.5	0.664	0.591	0.079	24
DIY	1	Without wd2vc	0.672	0.540	0.142	13
	2	0.9	0.702	0.582	0.133	13
	3	0.8	0.700	0.580	0.134	9
	4	0.7	0.696	0.573	0.135	6
	5	0.6	0.667	0.533	0.143	11
	6	0.5	–	–	–	–
Flight	1	Without wd2vc	–	–	–	–
	2	0.9	0.698	0.501	0.186	9
	3	0.8	0.693	0.490	0.188	10
	4	0.7	0.703	0.510	0.184	9
	5	0.6	0.661	0.440	0.199	10
	6	0.5	0.675	0.464	0.194	8
Gadgets	1	Without wd2vc	0.880	0.827	0.062	9
	2	0.9	0.885	0.834	0.059	10
	3	0.8	0.885	0.834	0.059	10
	4	0.7	0.897	0.852	0.049	13
	5	0.6	0.907	0.866	0.038	13
	6	0.5	0.844	0.774	0.085	13
Hardware	1	Without wd2vc	0.860	0.805	0.068	9
	2	0.9	0.889	0.845	0.047	11
	3	0.8	0.890	0.845	0.047	12
	4	0.7	0.864	0.809	0.066	9

(continued)

Table 2 (continued)

Genre	Model number	Threshold value	GFI	AGFI	RMSEA	Number of variable exceeding significance level
	5	0.6			0.077	12
	6	0.5	0.841	0.777	0.079	5
Robots	1	Without wd2vc	0.721	0.540	0.177	10
	2	0.9	0.721	0.540	0.177	10
	3	0.8	0.721	0.540	0.177	10
	4	0.7	0.730	0.554	0.174	7
	5	0.6	0.725	0.546	0.176	5
	6	0.5	0.722	0.541	0.177	8
Sound	1	Without wd2vc	–	–	–	–
	2	0.9	0.456	0.102	0.261	4
	3	0.8	–	–	–	–
	4	0.7	0.417	0.038	0.271	5
	5	0.6	0.493	0.164	0.251	9
	6	0.5	0.474	0.132	0.256	6
Space exploration	1	without wd2vc	–	–	–	–
	2	0.9	0.459	0.108	0.256	12
	3	0.8	–	–	–	–
	4	0.7	–	–	–	–
	5	0.6	–	–	–	–
	6	0.5	–	–	–	–
Wearables	1	Without wd2vc	–	–	–	–
	2	0.9	0.703	0.566	0.142	5
	3	0.8	0.703	0.566	0.142	5
	4	0.7	–	–	–	–
	5	0.6	0.715	0.583	0.138	7
	6	0.5	0.660	0.503	0.155	11
Web	1	Without wd2vc	0.808	0.766	0.026	19
	2	0.9	0.797	0.753	0.033	7
	3	0.8	0.796	0.752	0.033	20
	4	0.7	0.797	0.752	0.033	19
	5	0.6	0.766	0.715	0.048	19
	6	0.5	–	–	–	–

5 Conclusion

This study proposed a method of pre-processing for text-based analysis with hLDA and SEM by using word2vec. The basic idea of the proposed process is that pre-processing text data as corpus for topic modelling is an effective approach in performing factor analysis by using SEM. Previous research suffered from such problems as an unstable construction analysis model and low significance levels. Here, we proposed the use of word2vec to achieve keyword flexibility in the text corpus for topic modelling. The results demonstrate that our proposed approach successfully resolved the abovementioned problems.

For our future work, we plan to combine word2vec with topic modelling. In other words, we will combine neural-network and LDA. Moreover, we aim to improve the proposed process so that word2vec can adjust to many kinds of special text data, such as Twitter posts and customer reviews.

Acknowledgements This work was supported by KAKENHI 25240049.

References

1. S. Kawanaka, A. Miyata, R. Higashinaka, T. Hoshide, K. Fujimura, Computer analysis of consumer situations utilizing topic model, in *25th Annual Conference of the Japanese Society for Article Intelligence* (2011)
2. K. Wajima, T. Ogawa, T. Furukawa, S. Shimoda, *Specific Negative Factors Using Latent Dirichlet Allocation*, DEIM Forum, A9–3 (2014)
3. R. Kunimoto, H. Kobayashi, R. Saga, Factor analysis for game software using structural equation modeling with hierarchical latent Dirichlet allocation in user's review comments. Int. J. Knowl. Eng. **1**(1), 54–58 (2015)
4. R. Saga, S. Nohara, Factor analysis of investment judgment in crowdfunding using structural equation modeling, in *The Fourth Asian Conference on Information Systems* (2015)
5. S. Nohara, R. Saga, Preprocessing method topic-based path model by using Word2vec, in *Proceedings of The International MultiConference of Engineers and Computer Scientists 2017*. Lecture Notes in Engineering and Computer Science, pp. 15–17, Mar 2017, Hong Kong, pp. 317–320
6. R. Saga, T. Fujita, K. Kitami, K. Matsumoto, Improvement of factor model with text information based on factor model construction process, in *IIMSS*, 2013, pp 222–230
7. R. Saga, R. Kunimoto, LDA-based path model construction process for structure equation modeling. Artif. Life Robot. **21**(2), 155–159 (2016)
8. T. Mikolov, K. Chen, G.S. Corrado, J. Dean, Efficient Estimation of Word Representations in Vector Space, CoRR (2013). arXiv:1301.3781
9. T. Mikolov, I. Sutskever, K. Chen, G.S. Corrado, J. Dean, Districted representations of words and phrases and their compositionality, in *27th Annual Conference on Neural Information Processing Systems*. Advances in Neural Information Processing Systems 26. Proceeding of a meeting held December 5–8, Lake Tahoe, Nevada, United States (2013), pp. 3111–3119
10. Kickstarter, https://www.kickstarter.com/
11. MALLET: A Machine Learning for Language Toolkit, http://mallet.cs.umass.edu
12. The R Project for Statistical Computing, http://www.r-project.org/

13. J. Fox, Structural equation modeling with the SEM package in R. Struct. Equ. Model. **13**, 465–486 (2006)
14. Genism: A Topic Modeling Free Python Library, https://radimrehurek.com/gensim/index. html

How to Find Similar Users in Order to Develop a Cosmetics Recommender System

Yuki Matsunami, Mayumi Ueda and Shinsuke Nakajima

Abstract Most shopping web sites allow users to provide product reviews. It has been observed that reviews have a profound effect on item conversion rates. In particular, reviews of cosmetic products have significant impact on purchasing decisions because of the personal nature of such products, and also because of the potential for skin irritation caused by unsuitable items, which is a major consumer concern. In this study, we develop a method for user similarity calculation for a cosmetic review recommender system. To realize such a recommender system, we propose a method for the automatic scoring of various aspects of cosmetic item review texts based on an evaluation expression dictionary curated from a corpus of real-world online reviews. Furthermore, we consider how to calculate user similarity of cosmetic review sites.

Keywords Automatic review rating · Clustering cosmetic products · Evaluation expression dictionary · Explanation of reviews · Recommender system · Text mining · User similarity

1 Introduction

In recent years, most shopping web sites have added support for user-provided reviews. These are very useful in helping consumers decide whether to buy a product, and they have been shown to have a significant impact on conversion rates. In particular, consumers tend to be cautious while making choices about cosmetics because unsuitable items frequently cause skin irritations. "@cosme" [1] is a

Y. Matsunami (✉) · S. Nakajima
Kyoto Sangyo University, Motoyama, Kamigamo, Kita-ku, Kyoto 603-8555, Japan
e-mail: i1788223@cc.kyoto-su.ac.jp

S. Nakajima
e-mail: nakajima@cc.kyoto-su.ac.jp

M. Ueda
University of Marketing and Distribution Sciences, 3-1, Gakuen-nishimachi, Nishi-ku, Kobe 651-2188, Japan
e-mail: Mayumi_Ueda@red.umds.ac.jp

© Springer Nature Singapore Pte Ltd. 2018
S.-I. Ao et al. (eds.), *Transactions on Engineering Technologies*,
https://doi.org/10.1007/978-981-10-7488-2_25

337

cosmetics review site that is very popular among young Japanese women. While the site can be helpful to them in their decision making, it is not easy to find genuinely suitable cosmetic items because of the lack of explanation and poor granularity in user's reviews and score systems of cosmetic items. As an example, there is no guarantee that a cosmetic item mentioned by a single user as good for dry skin is always suitable for people who have this skin type. As the compatibilities between skin and cosmetic products differ among users, we believe that it is important to identify users who share similar preferences in cosmetic items, and to share reviews among those niche communities. In order to realize the proposed approach, we design and evaluate a collaborative recommender system for cosmetic items, which incorporates opinions of similar-minded users and automatically scores fine-grained aspects of item reviews.

To develop the review recommender system, scores by similar users are adopted. We have proposed a basic concept of an automatic scoring method for various aspects of cosmetic items in our previous work [2].

Therefore, in this paper we propose the user similarity calculating method for the cosmetic review recommender system. In particular, we realize this system by using the following method (see Fig. 1).

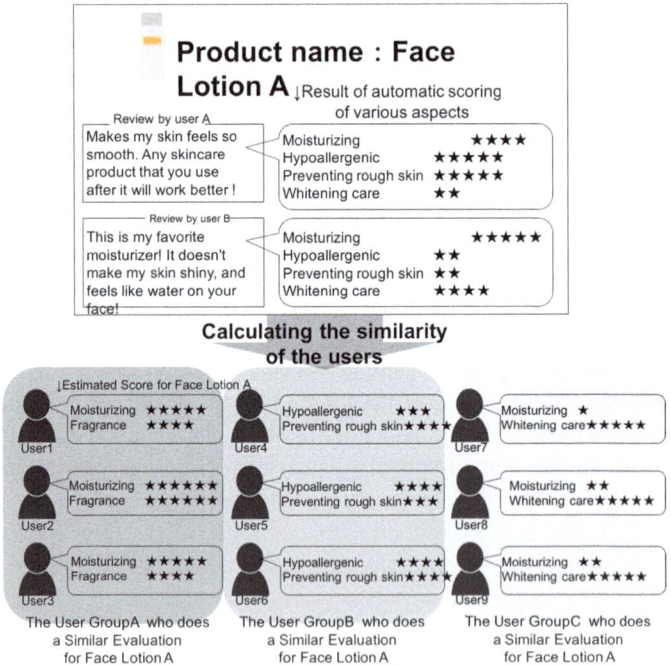

Fig. 1 Example of automatic scoring and user clustering

1. Automatic scoring of various aspects of cosmetic items (Sect. 3)
2. User similarity calculating method considering various aspects of cosmetic items (Sect. 4)

This paper is a revised version of the conference paper that we presented at IMECS 2017 [3].

The remainder of this paper is organized as follows. The related work is presented in Sect. 2. In Sect. 3 we describe the method for the automatic scoring of various aspects of cosmetic item review texts based on an evaluation expression dictionary. In Sect. 4, we describe the user similarity calculating method. We present our conclusions in Sect. 5.

2 Related Work

There are many websites which provide reviews written by consumers. For example, Amazon.com [4] and Priceprice.com [5] are popular Internet shopping sites which provide consumers reviews of their merchandise. Another such website, which is popular in Japan, is "Tabelog". This website does not sell products; it only provides restaurant information and reviews. In addition to the algorithmic aspects, researchers have recently focused on the presentation aspects of review data [6]. In recent years, "@cosme" has become very popular among Japanese young women. This website is a portal for beauty and cosmetic products, and it provides various information, such as reviews and shopping information. According to the report by istyle Inc., the company which operates that portal, in October 2016 the number of monthly page views was 280 million, the number of members was 3.9 million, and the total number of reviews was 1300 million [7]. The report also states that many women exchange information about beauty and cosmetics through the @cosme service. They provide a lot of information about cosmetic products of various brands. Hence, users can compare cosmetic products belonging to various brands. Reviews are composed of data such as user-written text, scores, and tag about effects. Furthermore, the system has user profile data that includes information about age and skin type, entered by users when they register as members. Therefore, users who want to browse reviews can filter them based on their own purposes, for example, reviews sorted by scores or focused on one specific effect.

Along with the popularization of these review services, several studies involving the analysis of reviews have been conducted in the past. For example, O'Donovan et al. evaluated their Auction Rules algorithma dictionary-based scoring mechanism for eBay reviews of Egyptian antiques. They showed that their approach was scalable and that a small amount of domain knowledge can considerably improve prediction accuracy compared to traditional instance- based learning approaches. In our previous study, we analyzed reviews of cosmetic items [8]. In order to determine if the review is a positive or a negative one, we make dictionaries for Japanese language morphological analysis, containing positive and negative expressions for cosmetic products. The previous research was aimed at developing the system to provide reviews that take into account the users profile, and then the system tries to retrieve information from blogs and SNS and merge them into the same format. The final goal of our current study is to develop a method for an automatic scoring of review texts, according to various aspects of cosmetic products.

Nihongi et al. proposed a method for extracting the evaluation expression from the review texts, and they developed the product retrieval system using evaluation expressions [9]. Our research focuses on the analysis of reviews for cosmetic products, and our aim is to find users with similar preferences and feelings in order to recommend truly useful reviews.

Yao et al. administered a questionnaire to examine the impact of reviews on purchasing behavior [10]. In order to investigate what kind of reviews are considered trustworthy, they conducted a hypothesis verification. The results obtained from this investigation were contrary to their intention, but the results obtained from the variance analysis showed that reviews do have an impact on the purchasing behavior.

Titov et al. proposed a statistical model for sentiment summarization [11], which is a joint model of text and aspect ratings. In order to discover the corresponding topics, it uses aspect ratings, and therefore it is able to extract textual evidence from reviews without the need of annotated data.

Nakatsuji et al. analyzed the frequency of the description expression depending on the reviewed product and calculated the similarity of users considering the tendency of that description expression [12]. It is difficult to apply this approach to calculate the similarity of users of cosmetic products because the compatibilities between skin and cosmetic products differ from one user to another.

It is difficult to apply this approach to calculate the similarity of users of cosmetic products because the compatibilities between skin and cosmetic products differ from one user to another.

3 Automatic Scoring of Various Aspects of Cosmetic Item Review Texts

In this section we describe a method for automatically scoring various aspects of cosmetic item review texts based on an evaluation expression dictionary. First, in Sect. 3.1. we describe a brief overview of our proposed method, while the actual method is given in Sect. 3.2. In Sect. 3.3 we describe an experimental evaluation of our method using real review data.

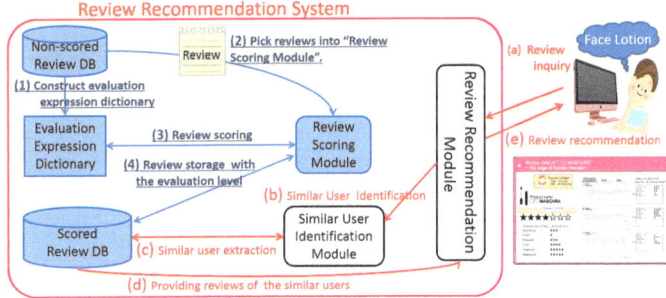

Fig. 2 Conceptual diagram of the cosmetic product review recommender system

3.1 Overview of Proposed Method

Our final goal is to develop a cosmetic item review recommender system which can recommend truly useful reviews to a target user. It operates in a manner similar to collaborative filtering, using a set of users who are deemed similar and have similar preferences and feedback on their experience regarding cosmetic products.

In order to make the significance of our study clear, in Fig. 2, we present a conceptual diagram of our final goal, the cosmetic item review recommender system. Numbers (1)(4) written in blue correspond to the procedure of the cosmetic review automatic scoring process, and Latin characters (a)(e) written in red correspond to the procedure of the review recommendation process. Further details are explained below:

Automatic Scoring

(1) Construct the evaluation expression dictionary which includes pairs of evaluation expression and their score by analyzing reviews sampled from non-scored DB.
(2) Take reviews from the non-scored DB to score them.
(3) Automatically score the reviews taken in step (2) using the evaluation expression dictionary constructed in step (1).
(4) Insert the reviews scored in step (3) into scored review DB.

Review Recommendation Process

(a) The user provides the name of a cosmetic item in which she is interested.
(b) The system refers to the "similar user extraction module" in order to extract users similar to the target user considered in step (a).
(c) The "similar user extraction module" obtains the information about reviews and reviewers, and identifies users similar to the target.
(d) Reviews of the users identified as similar in step (c) to the "review recommendation module" are provided.
(e) The system recommends suitable reviews to the target user.

This paper focuses on the development of a dictionary-based approach. Developing the live review recommendation method is part of our future work.

3.2 Automatic Scoring Based on Evaluation Expression Dictionary

We describe our automatic scoring based on an evaluation expression dictionary in this section. In Fig. 3, a conceptual diagram for the construction of the co-occurrence keyword-based dictionary is shown. The procedure for constructing such dictionary is as follows:

1. Analyze phrasal evaluation expressions extracted from reviews.
2. Divide the phrasal expressions into aspect keywords, feature words, and degree words.
3. Construct the dictionary by assembling co-occurrence relations and evaluation scores.

The procedure of automatic scoring reviews based on the evaluation expression dictionary is described below:

1. Acquire a review text data.
2. Read a review text from the outside.
3. Evaluate the end of the sentence using punctuation marks during the morphological analysis.
4. Extract the sentence of the k unit from one review text.
5. Evaluate the sentence which includes the keyword.
6. Pick up the sentence which satisfies co-occurrence condition from the sentences which include the keyword.

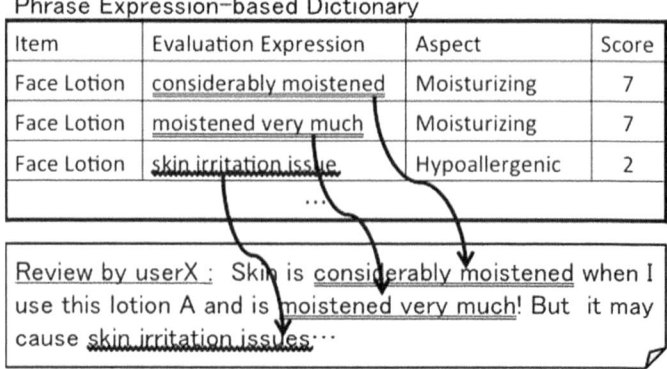

Phrase Expression–based Dictionary

Item	Evaluation Expression	Aspect	Score
Face Lotion	considerably moistened	Moisturizing	7
Face Lotion	moistened very much	Moisturizing	7
Face Lotion	skin irritation issue	Hypoallergenic	2

...

Review by userX : Skin is considerably moistened when I use this lotion A and is moistened very much! But it may cause skin irritation issues...

Fig. 3 Review scoring using a phrase expression-based dictionary

7. Survey the evaluation expression, score the corresponding co-occurrence condition and calculate the score for any evaluation expression to review texts.

3.3 Experimental Evaluation of Automatic Scoring Using Real Review Data

We examine an experimental evaluation of the automatic scoring method using real review data in order to verify the effectiveness of our proposed method. As a first step, we analyze 5,000 reviews randomly extracted from review data for "face lotion" posted at @cosme, to understand characteristics of the data.

3.3.1 Procedure of Experimental Evaluation

In this experiment, we use 10 review data for "face lotion" randomly selected from 5,000 reviews as described above, and then we compare the results by the following methods:

- Manual scoring method without the dictionary (as true data for comparison).
- Automatic scoring method based on the evaluation expression dictionary (proposed method).

In the case of the manual scoring, evaluators actually read review texts and score them on a scale from 0 to 7 stars for 10 aspects of "face lotion" set in advance. The evaluators are 30 females, aged 20–50. In the case of automatic scoring, the method scores review texts between 0 and 7 stars for the 10 aspects based on the co-occurrence keyword-based dictionary. The 10 aspects for "face lotion" set in advance for the experiment are: Cost performance, Moisturizing, Whitening care, Exfoliation and Pore care/Cleansing effect, Refreshing feeling/Sebum shine prevention, Refreshing-Thickening, Hypoallergenic, Rough skin prevention, Aging care, and Fragrance.

3.3.2 Result of the Experiment

In Fig. 4 we show the results of review scoring for 10 reviews based both on the co-occurrence keyword-based dictionary and the manual scoring by evaluators.

The contents of the 10 reviews are different from each other, so the detected aspects are different. The average score (# of stars) by manual scoring of all the aspects is 4.76, and the average score by our proposed method is 4.32. The mean absolute error (MAE) is 1.2.

Scores by the manual method tend to be a little higher than those by our proposed automatic method. However, the range of the scores is from 0 to 7 and MAE is 1.2, so

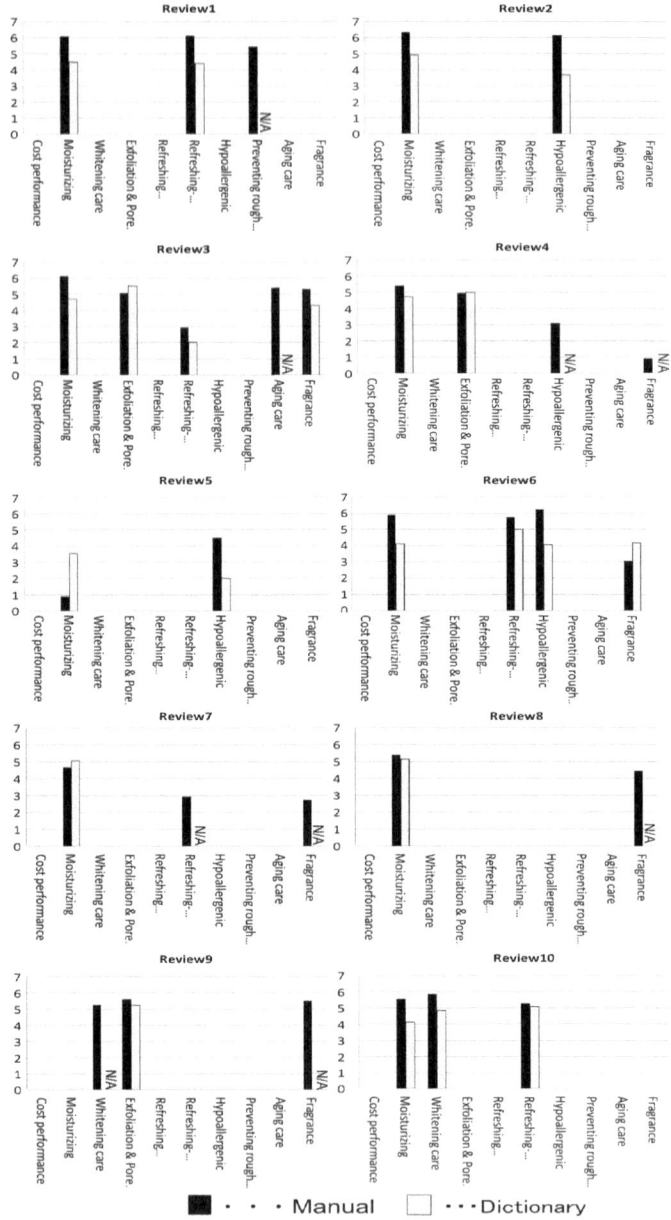

Fig. 4 Result of the review scoring based on the co-occurrence keyword-based dictionary and the manual scoring

Fig. 5 MAE of review scoring for each aspect (except N/A) before an additional experiment

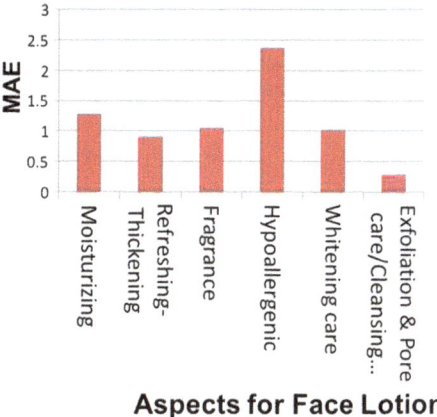

Aspects for Face Lotion

we may say that the results of the proposed automatic method are quite close to the results of the manual method used as true data for comparison. The total number of detected aspects is 31 by manual scoring (true data for comparison) and 22 by automatic scoring (proposed method). Therefore, the achievement rate of our proposed method versus the manual scoring is about 71%.

There are several "N/A" in Fig. 4 related to the automatic scoring method. However, there is room for improving the result of aspect detection for reviews by updating the dictionary. In a future work of ours, we will try to analyze a larger number of reviews, and then to improve and tune the dictionary. Moreover, we will develop a review recommender system for cosmetic products to further evaluate our novel scoring method.

Next, we focus on the MAE of the automatic review scoring. In Fig. 5 we show the results of review scoring for each evaluation aspect. The range of the score is from 0 to 7. Most MAE scores achieve results under 1.3, but the one for "Hypoallergenic" achieves a result higher than 2.0. This is caused by expressions related to problems of the users skin like "My skin easily feels stimulation because I have a sensitive skin, but this lotion is good for me!". Hence, we will try to remove the noise data of evaluation expressions using a correlative conjunction.

3.4 Additional Experiment

We conducted an additional experiment in order to improve the evaluation expression dictionary. The research participants (evaluators) were 37 females with an age between 19 and 25 years. In this experiment, 15 kinds of face lotions were prepared and the participants used all of them and then gave reviews and scores for all aspects of the considered products. Based on the reviews and scores, we tuned up the

Fig. 6 MAE of review scoring for each aspect (except N/A) after an additional experiment

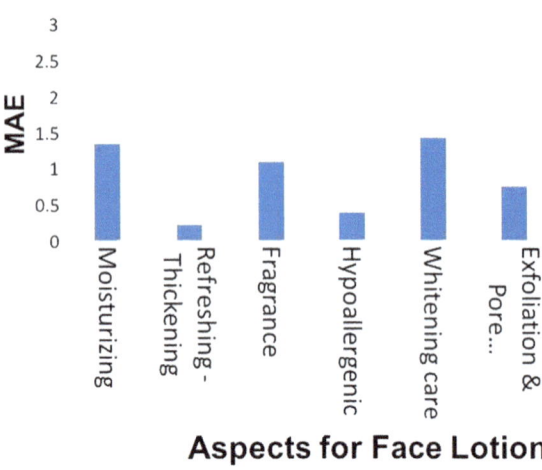

Aspects for Face Lotion

evaluation expression dictionary in order to improve the precision of the automatic scoring system.

After tuning up the dictionary, we calculated the MAE of the automatic review scoring. In Fig. 6 we show the results of review scoring for each evaluation aspect. The range of the score is from 0 to 7 as it is in Fig. 5. Basically, MAE scores in Fig. 6 achieve better results than MAE scores in Fig. 5. The mean of the scoring level based on the new evaluation expression dictionary after the tune-up was 4.56 whereas the mean of the scoring level by manual operations of participants, used as the correct answer data, was 4.76. The achievement rate of the automatic scoring after tuning was 81% whereas the rate before tuning was 71%. In addition, the mean absolute error after tuning was 0.9 versus the value of 1.2 obtained before tuning. In Table 1 we summarize the results before and after the tuning based on the additional experiment.

As mentioned above, the results indicate that we can improve the accuracy of the automatic scoring based on tuning the evaluation expression dictionary. In a future work, we will try to analyze a larger number of reviews, and also to improve the expression dictionary. Moreover, we will develop a review recommender system for cosmetic products with the construction of evaluation expression dictionaries for other products than face lotion.

Table 1 Comparing experimental results between before and after an additional experiment

	Before	After
Average score	4.32	4.56
Mean absolute error	1.2	0.9
Achievement rate of scoring (%)	71	81

4 User Similarity Calculating Method for the Cosmetic Review Recommender System

In this section, we describe the method for calculating the similarities of the users. First, in Sect. 4.1., we give an overview of the proposed method. Then, in Sect. 4.2, we describe the method for clustering cosmetic products based on their rating similarity. Finally, in Sect. 4.3, we explain the user similarity calculating method.

4.1 Overview of the Method

The purpose of our research is to develop a cosmetic review recommender system to provide helpful reviews for each user. In order to realize such recommender system, our process has to estimate the users who are deemed to be similar in order to extract reliable ratings for the products. In this section, we describe the method to estimate the similar users using the rating for similar item clusters. By using existing review providing services, users can get reviews by consumers with similar skin type or age. However, these reviews are not truly helpful for everyone.

For example, suppose that "Item B" has been highly appreciated by user1, user2 and user3 (Fig. 7). User1 and user2 post a high score for the "Moisture" aspect of Item B, and a low score for the "Hypoallergenic" aspect. User3 posts a high score for the "Hypoallergenic" aspect, and a low score for the "Moisture" aspect. Because of the similarity of these aspects, we consider that it is desirable to recommend Item A to user2, because it received high scores by similar users.

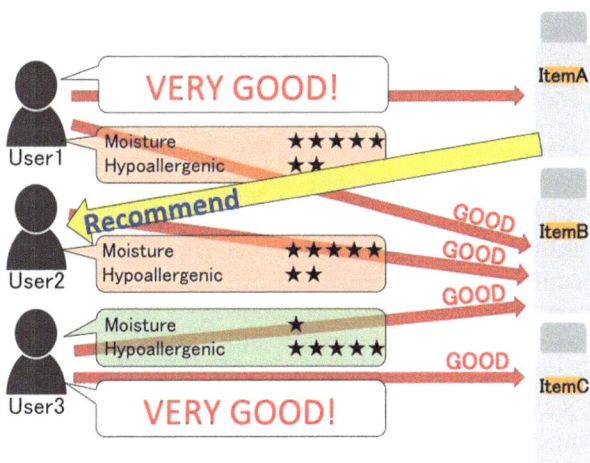

Fig. 7 Conceptual diagram of our recommendation method

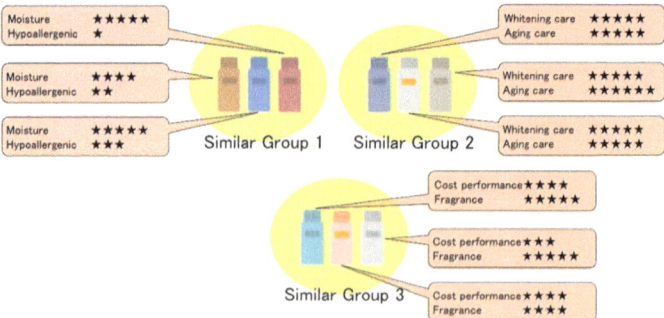

Fig. 8 Clustering cosmetic products based on their rating similarities

4.2 Clustering Cosmetic Products Based on Their Rating Similarities

We consider two users to be similar if they post similar ratings for the same products. However, because there is a considerable number of cosmetic products, similar users have difficulty in finding products of their choice, and may not choose the same product always. Therefore, we make clusters of similar products, and then we consider the users who post similar ratings for the products inside the cluster to be similar. The clustering method is as follow (Fig. 8).

1. Give scores of various aspects for the reviews.
2. Calculate the average scores for each item and each aspect.
3. Make clusters of cosmetic products by the similarity of the scores.

In order to cluster cosmetic products, we adopt the k-means clustering method.

4.3 User Similarity Calculating Method

In this section, we describe a method for calculating user-similarities. In this section we describe a method for calculating user-similarities. In most instances of a recommender system using collaborative filtering, to understand the relationships between users and products the ratings matrices are utilized. If we denote with m and n the number of products and users, respectively, the size of the ratings matrix equals $m \times n$, and therefore it is possible to calculate user similarities based on the Pearson correlation coefficient using this matrix.

However, in our method each product has not just one single score but multiple scores, as different aspects of the products are considered. Note that we proposed 10 aspects for face lotion. Thus, if we denote with p and n the number of scores and users, respectively, the size of the ratings matrix equals $p \times n$. In our case,

p corresponds to ten times m. By using the ratings matrix $p \times n$, in order to calculate user similarity our method can consider a correlation coefficient based not on just a single comprehensive evaluation score but on several evaluation scores for each item. As a result, we believe that our method can achieve a higher recommendation efficiency than conventional methods.

5   Conclusions

In this paper, we propose a system that can recommend truly effective reviews of cosmetic products for each user. In order to realize such a recommender system, first we proposed a method for automatically scoring various aspects of cosmetic item review texts based on an evaluation expression dictionary. To develop this automatic scoring method, we constructed a dictionary using co-occurrence keyword-based evaluation expressions. Then, we proposed a user similarity calculating method for cosmetic review recommender system using the rating of similar cosmetic item clusters.

Acknowledgements This work was supported in part by istyle Inc. which provided review data for cosmetic items, and the MEXT Grant-in Aid for Scientific Research(C)(#16K00425, #26330351).

References

1. @cosme, http://www.cosme.net/. Accessed 8 Jan 2017
2. Y. Matsunami, M. Ueda, S. Nakajima, T. Hashikami, S. Iwasaki, J. O'Donovan, B. Kang, Explaining item ratings in cosmetic product reviews, in *International MultiConference of Engineers and Computer Scientists ICICWS 2016*, (2016), pp. 392–397
3. Y. Matsunami, A. Okuda, M. Ueda, S. Nakajima, User Similarity calculating method for cosmetic review recommender system, lecture notes in engineering and computer science, in *Proceedings of The International MultiConference of Engineers and Computer Scientists 2017*, Hong Kong, 15–17 Mar 2017, pp. 312–316
4. Amazon.com, http://www.amazon.com/. Accessed 8 Jan 2017
5. Priceprice.com, http://ph.priceprice.com/. Accessed 8 Jan 2017
6. B. Kang, N. Tintarev, J. O'Donovan, Inspection mechanisms for community-based content discovery in microblogs, in *IntRS15 Joint Workshop on Interfaces and Human Decision Making for Recommender Systems* (http://recex.ist.tugraz.at/intrs2015/) *at ACM Recommender Systems 2015*, Vienna, Austria, Sep 2015
7. istyle Inc., Site data of @cosme, http://www.istyle.co.jp/business/uploads/sitedata.pdf (in Japanese). Accessed 8 Jan 2017
8. Y. Hamaoka, M. Ueda, S. Nakajima, Extraction of evaluation aspects for each cosmetics item to develop the reputation portal site, in *IEICE WI2-2012–15*, Feb 2012 (in Japanese), pp. 45–46
9. T. Nihongi, K. Sumita, Analysis and retrieval of the word-of-mouth estimation by structurizing sentences. in *Proceeding of the Interaction 2002*, (2002) (in Japanese), pp. 175–176
10. J. Yao, H. Idota, A. Harada, The affect of the internet's word of mouth on buying behavior: from the questionnair survey pf the cosmetics purchase of the women students, in *Proceeding of National Conference of JASMIN Spring 2014*, Kanagawa, Japan, May 2014 (in Japanese), pp. 231–232

11. I. Titov, R. McDonald, A joint model of text and aspect ratings for sentiment summarization, in *46th Meeting of Association for Computational Linguistics (ACL-08)*, Columbus, USA, 2008, pp. 308–316
12. M. Nakatsuji, M. Kondo, A. Tanaka, T. Uchiyama, Measuring similarity of users using sentimental tags for items, in *The 24th Annual Conference of the Japanese Society for Artificial Intelligence, 2010*

Applied Finite Automata and Quadtree Technique for Thai Sign Language Translation

Jirawat Tumsri and Warangkhana Kimpan

Abstract In this article, algorithm for transforming sign language into Thai alphabets is presented to fill the gap in communication between the hearing impairment and normal people. Leap Motion Controller was applied to detect 5-finger-tip position and palm center in the form of X and Y axis. Then, decision tree was created by using the Quadtree technique and the research result on transforming Thai sign language into finite automata was applied to improve algorithm in creating finite automata of Thai alphabet sign language to increase efficiency and speed in processing sign language. The test result shows that it can discriminate 42-Thai alphabet sign language at 78.70% accuracy.

Keywords Communication · Finite automata · Hearing impairment people
Leap motion controller · Quadtree · Thai sign language

1 Introduction

In daily life, human beings always need interpersonal communication. Communication is, therefore, an essential basic factor to be in a society. Normal people communicate mainly through speaking but the hearing impairment people use the sign language. One of the barriers in communication between the hearing impairment people and the normal people is that the normal people do not understand the sign language thus making a little difference in a sign language expression can completely change the meaning in communication.

J. Tumsri · W. Kimpan (✉)
Department of Computer Science, Faculty of Science, King Mongkut's Institute of Technology Ladkrabang (KMITL), Chalongkrung Rd, Ladkrabang, Bangkok 10520, Thailand
e-mail: warangkhana.ki@kmitl.ac.th

J. Tumsri
e-mail: 57605074@kmitl.ac.th

© Springer Nature Singapore Pte Ltd. 2018
S.-I. Ao et al. (eds.), *Transactions on Engineering Technologies*,
https://doi.org/10.1007/978-981-10-7488-2_26

The statistics during 2012–2014 indicated that 51.0% of the hearing impairment people need help [1]. The public sector has set the policy in helping and promoting the hearing impairment people to be able to communicate with normal people by developing Thailand Telecommunication Relay Service (TTRS) machine or the communication machine for the hearing impairment people to fill the gap in communication [2]. However, there is still a problem of insufficient TTRS machine service as there are only 120 service points available in various places, moreover, there is an accessibility problem for the hearing impairment people. When using TTRS machine service, the hearing impairment people need provided sign language translators.

The survey on the performance of public sector during 2012–2014, shows that the average of the hearing impairment people's quality of life is at a stable level (35.3%) and that there are 37.3% of the hearing impairment people who still face problems and difficulties in living. This shows that the service of the public or private sectors is still insufficient for the hearing impairment people to improve their quality of life [1].

According to the problems mentioned above and in order to help hearing impairment people to be able to communicate with normal people, this research focuses in developing computer program that can translate the sign language into Thai alphabets by applying Leap motion controller. This equipment detects the fingers and then Quadtree and finite automata for classifying are used in order to transform the position of each finger into a Thai alphabet.

2 Theory and Related Works

2.1 Thai Sign Language

Communication can be divided into 2 types: verbal communication in which words and alphabets are commonly used in socialization and non-verbal communication in which actions, gestures, tunes, eyesight, objects, signals, environments, and other expressions are used in communication. Showing hand symbols or sign language is a type of non-verbal communication for hearing impairment people.

In 1956, Khun Ying Kamala Krailuek designed the standard Thai sign language (THSL) by applying the fingerspelling principle of American sign language (ASL) in single hand fingerspelling. THSL consists of 42 alphabets and 23 vowels [3] as shown in Fig. 1.

Figure 1 shows the Thai sign language in which the Thai alphabets are classified into 3 types according to the steps of movement as follows: Type (I) 1-step movement in fingerspelling, consisting of 15 alphabets, e.g., ก, ฑ, ส, and so on. Type (II) 2-step movement in fingerspelling, consisting of 24 alphabets, e.g., บ, ค, ฆ, and so on. Type (III) 3-step movement in fingerspelling, consisting of 3 alphabets, ธ, บ, ญ [4].

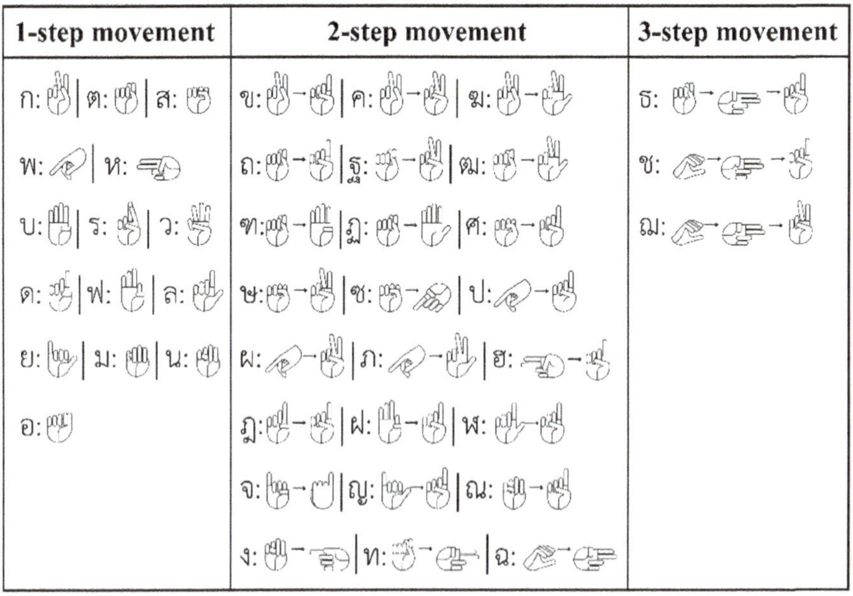

1-step movement	2-step movement	3-step movement
ก:	ข:	ต:	ส:	ช:	ค:	ฌ:	ร:
พ:	ห:	ฅ:	ฐ:	ฒ:	ช:	
บ:	ร:	ว:	ฑ:	ฏ:	ศ:	ฌ:
ด:	ฬ:	ล:	ษ:	ฆ:	บ:	
ย:	ม:	น:	ฝ:	ภ:	ฮ:	
อ:	ญ:	ผ:	ฟ:	
	จ:	ฌ:	ณ:	
	ง:	ท:	ถ:	

Fig. 1 Thai sign language diagram

2.2 Quadtree

Quadtree is a type of spatial data structures, similar to Q-tree in object-based image classification for solving the problem of extracting a large amount of data at the same time. Moreover, it can also reduce the number of irrelevant data extractions by continuously dividing the area in a two-dimension square into 4 equal parts in the form of plus sign, resulting in four and sixteen squares respectively. Then the decision tree is created from the results of the square division as shown in Fig. 2.

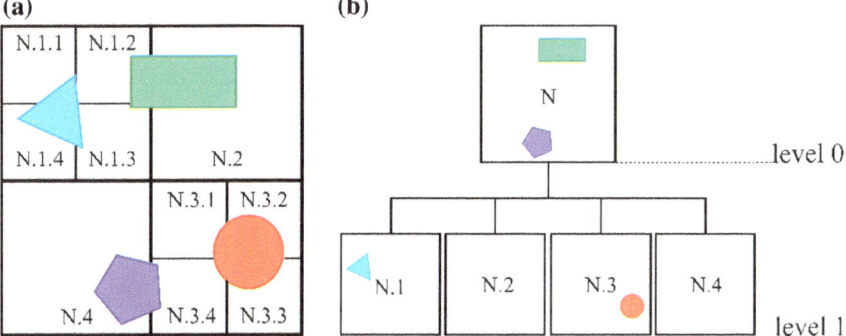

Fig. 2 Quadtree division for storing objects in decision tree

Figure 2a shows the square N before dividing. The decision tree creates 1 node in level 0, called "root node", which is the node N as shown in Fig. 2b. After dividing the square N into 4 parts and comparing it to the decision tree in Fig. 2b, it can be indicated that the node N in level 0 consists of 4 child nodes in level 1 which are N.1, N.2, N.3, and N.4. The division of nodes in level 1 is the last division, therefore the child nodes in this level are called "leaf node". Every node in the decision tree stores the objects which can be the member of only one node. However, in each node, many objects can be stored. The objects can be the members of node N when the criteria are as follows:

1. The objects are placed in N (triangle, rectangle, circle, and pentagon).
2. If the objects are placed in any child nodes of N, they are the members of those child nodes (triangle and circle).
3. If the objects are placed astride child nodes of N, they are the members of node N (rectangle and pentagon).

The Quadtree is mostly used in an object-based image classification, such as finding various positions on the map, providing spatial details of games, and so on. It is good for systematic data storage and reduction of the number of the other irrelevant space examinations [5].

2.3   Finite Automata

Finite automata is a model for studying the operation of computer. The overall elements of automata are divided into 3 parts consisting of input alphabet tape, processor, and data storage. The input tape stores alphabets and sends them to the processor, the processor controls internal state that changes according to the input, whereas the storage stores the outcome obtained during the calculation or processing [6].

Finite automata consists of 2 types: deterministic finite automata and non-deterministic finite automata. The definition of deterministic finite automata can be explained by substitution of 5 factors as shown in Eq. (1) [7].

$$M = (Q, \Sigma, \delta, q_0, F) \tag{1}$$

when

Q	is any state finite sets
Σ	is alphabet finite sets
$\delta: Q \times \Sigma \to Q$	is a transition function
q_0	is a initial state, for $q_0 \in Q$
F	is a set of acceptance state for $F \subseteq Q$ and F is a final state assuming that finite automata accepts the input alphabet tape

The transitive diagram definition of finite automata in each diagram is the graph that identifies the direction as followings:

1. There is Q as a set of all points and there are arrows pointing to the starting point.
2. There are 2 circles enclosing all points in F.
3. There is a line connecting point p to point q with the input symbol a, written as.

Figure 3 is a written example of finite automata. It is merely one-way. In each movement of pointer at the input tape, automata will always be in the set of some states. When the reading starts, the input is 0. Therefore, the direction is from the starting state (q_0) to state (q_1). Then, the pointer is moved to the right for 1 alphabet. The process is repeated until all the input is completely read. If the current state is in the set of acceptance state (q_3), it shows that the obtained outcome is the acceptance state.

From Eq. (1), it can be substituted as follows: $M = (\{q_0, q_1, q_2, q_3\}, \{0, 1, 2\}, q_0, q_3)$ which can be written in the table format showing the state of receiving the input of 0 and 2 as in Table 1.

2.4 Related Works

In 2007, Qutaishat Munib, Moussa Habeeb, Bayan Takruri, and Hiba Abed Al-Malik [8] presented the study on the development of automatic system for transforming American sign language by applying image processing to transform image into vector and comparing to the vector using neural network technique in learning the gesture set of sign language, which the rotation of gesture in the image

Fig. 3 Diagram of automata elements

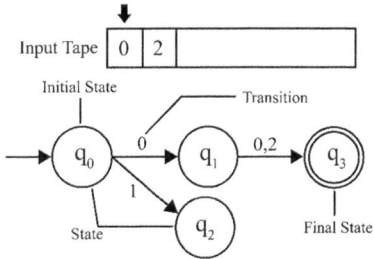

Table 1 State transition

State/Event	0	1	2
q_0	q_1	q_2	–
q_1	q_3	–	q_3
q_2	–	–	–
q_3	–	–	–

did not affect the data extraction. The result of system test, by using 300 sample images of hand signals, revealed that the system could perceive the sign language gesture with 92.3% accuracy.

In 2008, Suwannee Phitakwinai, Sansanee Auephanwiriyakul, and Nipon Theera-Umpon [9] proposed the study on Thai sign language translation using Fuzzy C-Means (FCM) and Scale Invariant Feature Transform (SIFT), which was the transform of sign language into 15 alphabets and 10 words by using camera as an input in processing. The algorithms used in this study were FCM and SIFT. The experiment result revealed that the system could transform sign language into alphabets and words with the accuracy of 82.19% and 55.08%, respectively. Whereas the nearby neighbor was at 3 and SIFT threshold was 0.7.

In 2014, Wisan Tangwongcharoen and Jirawat Tumsri [4] presented the study on decoding Thai sign language pattern using Quadtree and using Leap motion controller in detecting 5-fingertip positions. Those fingertip positions were analyzed by using Quadtree technique. The outcome consisted of the number codes for 44 alphabets, 3 digits for each one.

In 2015, Jirawat Tumsri and Warangkhana Kimpan [10] presented the study on code translation from the Thai sign language pattern to finite automata by creating finite automata from 1-step fingerspelling codes of 18 alphabets, one finger for 3-digit code. The obtained result was 1-step fingerspelling finite automata structure.

In this study, Quadtree technique [4] was applied in transforming finger positions into finger codes for 42 alphabets, and finite automata algorithm [10] was improved for supporting 3-step fingerspelling.

3 Proposed Method

3.1 Overview of the Method

The methods is divided into 2 main parts: creating finite automata prototype for Thai sign language as shown in Fig. 4a and transforming sign language into Thai alphabet pattern as shown in Fig. 4b.

According to Fig. 4a, the creation of finite automata for Thai sign language starts by collecting the samples of Thai sign language in the pattern of the position on X and Y axis. Then, the sample data values are adjusted into a proper pattern, then Thai sign language is decoded. The obtained result in each finger is a 3-digit number. Then, the 3-digit code is used to find out finger order and number for creating finite automata prototype for Thai sign language in order to reduce the number of data or fingers. After getting the finger order and number, the finite automata prototype for Thai sign language is created [12].

Figure 4b shows the steps in transforming sign language into Thai alphabet pattern. It starts by the user showing the sign language through Leap motion controller and then the data are sent to the program. The value of the data, which are

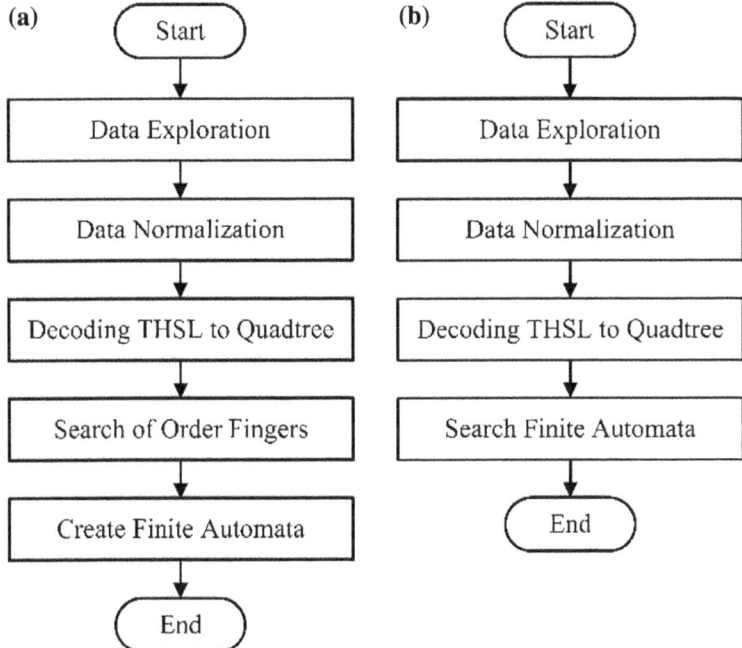

Fig. 4 Two main parts of overview method

in the pattern of positions on X and Y axis, is adjusted to be in a proper pattern. Then, the positions on X and Y axis with the adjusted value are decoded into Thai sign language and the outcome of 1 alphabet is found out from finite automata prototype for Thai sign language [12].

3.2 Data Exploration

In order to collect the prototype data or the transformation of sign language into Thai alphabet pattern, the hand gestures must be in the parallel position to Leap motion controller. First, hand gesture, open hand or waving hand, must be shown for 3 s so that the hand sizes of both the master and examiner can be adjusted. After that the position of X and Y axis on fingertip and palm center are stored for 3 s for each gesture after started. The data of positions on X and Y axis on fingertips which derived from Leap motion controller, are used in this study as follows:

1. The creation of finite automata prototype for Thai sign language consists of positions on X and Y axis of 5 fingertips and on X and Y axis of palm center.
2. For transforming sign language into Thai alphabet pattern, receiving data value of each finger relies on the outcome of finding out finger order and number in

the step of creating finite automata prototype for Thai sign language. For instance, the outcome of finding out finger order and number consists of thumb, index finger, and middle finger, while the position of X and Y axis are in the palm center.

3.3 Data Normalization

The adjustment of data value into the proper pattern is the master's and the examiner's hand size adjustment in order that their hands are in the same size, and determination of position on the same palm center is also done for accuracy in transforming sign language into Thai alphabet pattern.

The equation for finding the distance between 2 points in direct line or Euclidean distance [11] is shown in Eq. (2):

$$distance = \sqrt{(X_1 - X_2)^2 + (Y_1 - Y_2)^2} \qquad (2)$$

when

$distance$ is the length from middle fingertip to palm center
X_1, Y_1 is X and Y axis on middle fingertip
X_2, Y_2 is X and Y axis on palm center

After the distance between middle fingertip and palm center is obtained, the X and Y axis positions are used in finding out the ratio as shown in Eq. (3) and the outcome of Eq. (3) is used to adjust the X and the Y axis positions of each finger as shown in Eq. (4) and Eq. (5), respectively. This adjustment is conducted to solve the problem of different hand sizes of the master and the examiner and reduce the error in transforming sign language into the Thai alphabet pattern.

$$ratio = distance_1 / distance_2 \qquad (3)$$

When

$ratio$ is the ratio between sign language of the master and examination
$distance_1$ is the length from middle fingertip to palm center of the master
$distance_2$ is the length from middle fingertip to palm center of the examination

$$X_{post} = X_{pre} * ratio \qquad (4)$$

$$Y_{post} = Y_{pre} * ratio \qquad (5)$$

when

X_{post}, Y_{post} is X and Y axis positions after adjusting the ratio

X_{pre}, Y_{pre} is X and Y axis positions before adjusting the ratio

ratio   is the ratio between sign language of the master and Examination

In addition to the adjustment of finger positions, there is also the shift of finger positions presenting the proper Quadtree table, which the position on palm center is always set as a fixed point in referring to other fingers by setting the palm center position on the midpoint of X and Y axis in the Quadtree table. In doing so, it can cover all of the hand even though there is a shift position or changes of the positions of other fingers.

The shift of finger position is done by finding out the distance between the former palm center position (X_1, Y_1) and palm center position after being shifted to the midpoint of X and Y axis (X_2, Y_2) as shown in Eq. (2). After finding out the distance of the shift of palm center position, X and Y axis positions of each finger are adjusted as shown in Eqs. (6) and (7).

$$X_{post} = X_{pre} + distance \qquad (6)$$

$$Y_{post} = Y_{pre} + distance \qquad (7)$$

when

X_{post}, Y_{post} is X and Y axis positions after adjusting the ratio

X_{pre}, Y_{pre} is X and Y axis positions before adjusting the ratio

distance   is the length of shifting palm center position toward the midpoint of Quadtree table

The creation of finite automata prototype for THSL starts by finding out the value between middle fingertip and palm center as shown in Eq. (2). Then, X axis position is moved as shown in Eq. (6) and Y axis position as shown in Eq. (7) to transform sign language into the correct Thai alphabet pattern.

In adjusting data value into the proper pattern in the step of transforming sign language into Thai alphabet pattern, the values of X and Y axis positions on middle fingertip and on palm center are used in finding out the distance between 2 points as shown in Eq. (2). Then, the ratio of the distance is found out between middle fingertips and palm centers of the master and the examiner as shown in Eq. (3). After the ratio is obtained, the value of X axis position is adjusted as shown in Eq. (4) and of Y axis position as shown in Eq. (5). The next step is to move X axis position as shown in Eq. (6) and Y axis position as shown in Eq. (7).

3.4 Decoding THSL to Quadtree

Algorithm for decoding THSL by using Quadtree technique is a method for decoding the gesture of one finger. The input consists of X and Y axis positions on each finger, the number of alphabet codes, and the size of Quadtree square table,

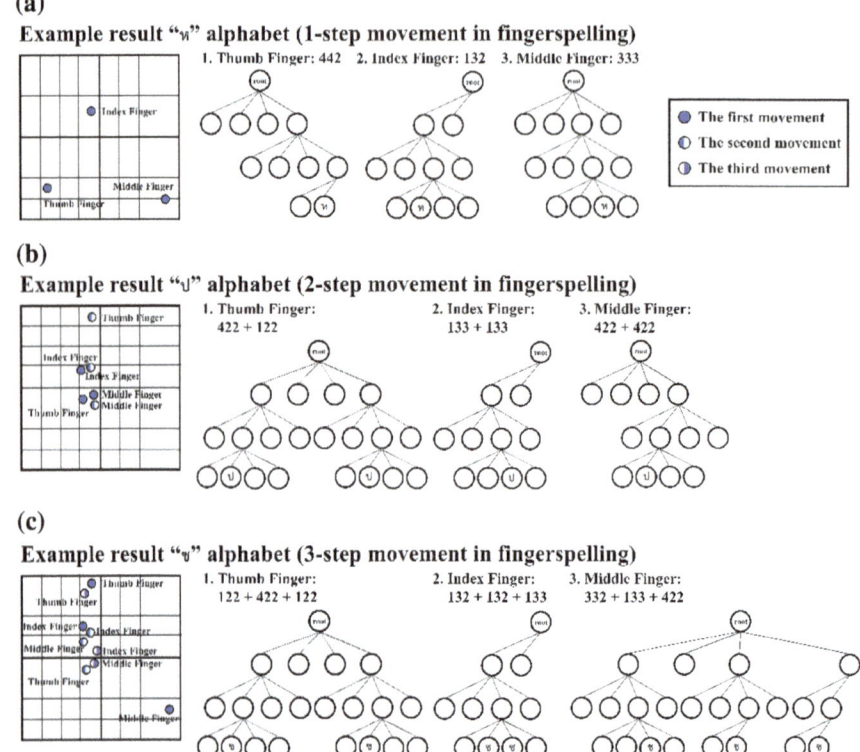

Fig. 5 Example of decision tree for the alphabet "ต", "ป", and "บ" of 3 fingers

whereas the obtained outcome consists of 3-digit number code of each finger that can be transformed into the decision tree pattern as shown in Fig. 5.

3.5 Search of Order Fingers

Finger order search is a step in creating a finite automata prototype of THSL to reduce data storage and increase processing efficiency in transforming sign language into Thai alphabets. The data on X and Y axis of 5 fingers be entered and the outcome will be obtained in form of the finger order as described in the following algorithm.

Algorithm: Search of Order Fingers
1. found duplicate = true, number of finger = 5, finger number i = 1
2. **While** found duplicate
3.   **Set** temp finger number = 1, minimum code = 42
4.   **While** temp finger number <= number of finger
5.   number duplicate finger array = find code of finger duplicate
6.   **If** minimum code > number duplicate finger array **Then**
7.   minimum code = number duplicate finger array
8.   finger number i = type of finger
9.   **End if**
10.   **Next** temp finger number
11.   **End while**
12.   **If** found duplicate code of finger array **Then**
13.   Remove code of finger in temp finger number
14.   number of finger = number of finger - 1
15.   **Next** i
16.   **Else**
17.   found duplicate = false
18.   **End if else**
19. **End while**
20. **Return** finger number

The above algorithm is the steps of finger order and number search. First, the first finger order is searched by selecting to store each finger code of duplicate alphabet. For example, the codes on a thumb finger for the first, second, and third alphabets are 244, 123, and 244, respectively. Therefore, 244 is stored as an alphabet code of the thumb finger. Secondly, after all alphabets are completely searched, the algorithm examines whether the initial finger has more finger codes in storage than the current one. If so, the number of duplicated code is stored and the current finger is set as the first finger. Finally, the stored finger codes are examined whether there are any duplicate codes. If there are, the next finger or the second finger is then searched. In the search of the second finger, only the stored duplicate code of the first finger is considered and the outcome of the first finger from the initial search will not be considered. For example, if the outcome of the first finger is the thumb finger, the second round is, therefore, to consider the codes of index finger, middle finger, ring finger, and little finger. This process kept repeating until the duplicate finger code is no longer found or all of 5 fingers are completely considered. For example, the final outcome reveals that there are 3 fingers: the first one is the thumb finger, the second is the index finger, and the third is the middle finger.

3.6 Create Finite Automata

From Eq. (1), the factors of finite automata are set to be the code numbers of sign language in each digit of each finger and final state is the set of one alphabet. Therefore, the alphabet is encoded into sign language. Then, each digit is decoded by starting at the outcome of finger order and number search, which each finger has 3-digit code.

In algorithm for creating finite automata prototype, the input is THSL prototype for decoding 42 alphabets. The outcome is finite automata structure of THSL. The algorithm for creating finite automata is shown below.

Algorithm: Create Finite Automata
1.   alphabet = 1, number of alphabet = 42
2.   **While** alphabet <= number of alphabet
3.   finger number = 1, current state = initial state
4.   **While** finger number <= size finger number
5.   Read code of finger number 3 integer
6.   **For** i = 1 to 3
7.   **Get** code of finger number integer i
8.   Find state code finger at current state
9.   **If** not found state code finger **Then**
10.   Create state finite automata
11.   **End if**
12.   **Next** i
13.   **End for**
14.   **Next** finger number
15.   **End while**
16.   state automata = final state
17.   **Next** alphabet
18.   **End while**
19.   **Return** finite automata

The above algorithm is finite automata creating algorithm. Firstly, the current state pointing at the initial state and receiving 1-alphabet code is set. Then, the 3-digit code is read from the first finger, the first digit code is read, the current state is searched whether there is a movement direction toward another state of the first digit code. If it is not found, the state is created with the movement direction toward another state to be the first digit and the current state is set to point to the created state. The process is repeated until all of three digits are completely done. Secondly, when the state is created for all 3 digits, the code of the next finger is read and repeated the same process as creation steps of automata of the first finger code. Finally, when the creation is completed according to finger order and number for one alphabet, the automata of next alphabets are created until all of 42 alphabets are completely done.

3.7 Search Finite Automata

The inputs of search finite automata for Thai alphabets are the examiner's and the master's codes of each finger, and code for decoding in THSL pattern. The obtained outcome consists of Thai alphabets which is described in the algorithm of search finite automata.

At the starting, the current state is set to point to the first state. Then, the value is read from the pointer that points to the input tape by considering all directions of transition of current state. If value on transition is equal to value read from input tape, the current state point to transition state in that direction. Then, the pointer of the input tape is moved to the right and the direction of the transition is considered as done before. This process is repeated until the end of the input tape pointer, then this algorithm outcome from the acceptance state of current state is considered. If it is the acceptance state, the alphabet of current state is retrieved. If it is not, it reveals that no Thai alphabet is found.

Algorithm: Search Finite Automata
1.   current state = first state of THSL finite automata
2.   **While** EOF of input tape
3.   found = false
4.   **Read** value at current pointer of input tape
5.   **While** consider all transition or not found
6.   **If** value on transition == value of input tape **Then**
7.   **Next** state of current transition
8.   found = true
9.   **Else**
10.   **Next** transition of current state
11.   **Next** i
12.   **End if else**
13.   **End while**
14.   **Next** pointer of input tape
15.   **End while**
16.   **If** state of current state == final state **Then**
17.   **Get** alphabet of current state
18.   **Return** alphabet

4 Experimental Results and Discussion

From the steps of THSL finite automata creation for 42 alphabets, the outcome from the step of decoding THSL to Quadtree was the code of 3-digit number of each alphabet on each finger. Then, the outcome from the step of decoding THSL to Quadtree was used for the search of finger order which reveals that three fingers namely thumb finger, index finger, and middle finger, can sequentially classify sign language of 42 Thai alphabets. The results of this step are shown in Table 2. Finally, these results are used for creation of finite automata.

Table 2 shows the code of 3-digit number of each alphabet on each finger. For example, the outcome in the second-movement group of alphabet "ป" was the

Table 2 Example result of search finite automata

Alphabet	Accuracy (%)	Alphabet	Accuracy (%)	Alphabet	Accuracy (%)
ต	80.00	ป	81.67	ธ	82.22
ห	83.33	ศ	85.00	บ	83.33
ม	56.67	ษ	85.00	ณ	80.00

Table 3 Result of decoding THSL to Quadtree method

Alphabet	Thumb finger	Index finger	Middle finger
ต	132	122	122
ห	422	132	333
ม	122	122	122
บ (พ + 1)	422 + 122	133 + 133	422 + 422
ศ (ส + 1)	122 + 122	422 + 133	122 + 422
ธ (ห + 1)	422 + 122	132 + 133	333 + 422
ธ (ท + 1)	(132 + 422) + 122	(122 + 132) + 133	(122 + 133) + 422
บ (ฉ + 1)	(122 + 422) + 122	(132 + 132) + 133	(332 + 133) + 422
ฌ (ฉ + 2)	(122 + 422) + 422	(132 + 132) + 112	(332 + 133) + 333

results from the mixture of alphabet "พ" (thumb: 422, index finger: 133, middle finger: 422) and the number "1" (thumb: 122, index finger: 133, middle finger: 422). Whereas the outcome in the third-movement group of alphabet "ธ" was the results from the mixture of alphabet "ท" in the second movement (the first movement, thumb: 132 + 422, index finger: 122 + 132, middle finger: 122 + 133) and the number "1" (thumb: 122, index finger: 133, middle finger: 422).

On the steps of transforming sign language into Thai alphabet pattern, the accurate outcome in transforming sign language into Thai alphabets is shown in Table 3.

5 Conclusion and Future Work

This research presented the transformation of sign language into Thai alphabets to help the hearing impairment people use to communicate with normal people. Two steps were applied in the study: (1) creation of the prototype of finite automata for sign language on 42 Thai alphabets which classified into 3 groups according to the finger movements and (2) transformation of sign language into Thai alphabets by using Quadtree and finite automata. The average score of obtained outcome accuracy in sign language transformation was 78.70%. The alphabets with the highest accuracy score, 85.00% were "ศ" and "ธ". Whereas the alphabets with the lowest accuracy score, 56.67% were "ม".

References

1. J. Bunyarattanasuntorn, Quality of Life of the Deft: Reflection of Monitoring and Evaluation of the 4th National Development Plan for Quality of Life of The Hearing Impairment People, 2015–2016, in *Proceeding Articles in 7th National Academic Conference on the Deft People*, vol. 2 (2015), pp. 33–58

2. Thai Telecommunication Relay Service. TTRS Machine for the hearing impairment people: Reduction of communication gap, http://ttrs.or.th/index.php/submenu1/news/156-it24hrs-ttrs. Accessed 22 Nov 2016

3. M. Thammasaeng, "Sign Language," Bangkok: Jong Chareon Printing, Sign Language, vol. 2 (1996)

4. W. Tangwongcharoen, J. Tumsri, Decoding thai sign language pattern using quad tree, in *Proceedings of the 37th Electrical Engineering Conference (EECON37)*, Khon Kaen, Thailand, vol. 2, pp. 873–876, 19–3 Nov 2014

5. Ajangz. 2015. Quadtree, http://ajangz.exteen.com/20150528/quad-tree. Accessed 10 Nov 2016

6. K. Jiaranaithanakit, *Theory of Computation*, 2nd edn. (Bangkok: Technology Promotion Association, Thailand-Japan, 2014), ch. 2, pp. 20–23

7. P. Linz, *An Introduction to formal languages and Automata*, 4th edn. (London, UK: Jones and Bartlett Publishers, 2006), ch. 1, pp. 26–28

8. Q. Munib, M. Habeeb, B. Takruri, H.A. Al-Malik, American sign language (ASL) recognition based on Hough transform and neural networks. Sci. Direct Expert Syst. Appl. **32**, 24–27 (2007)

9. S. Phitakwinai, S. Auephanwiriyakul, N. Theera-Umpon, Thai sign language translation using fuzzy c-means and scale invariant feature transform, in *ICCSA 2008*, pp. 1107–1119, 30 Jun.- 3 Jul. 2008

10. J. Tumsri, W. Kimpan, Code translation from Thai sign language pattern to finite automata, in *Proceedings of the 7th Conference of Electrical Engineering Network of Rajamangala University of Technology 2015 (EENET 2015)*, A-one The Royal Cruise Hotel, Pattaya, Chonburi, Thailand, pp. 125–128, 27–29 May 2015

11. M.M. Deza, E. Deza, *Encyclopedia of Distances* (Springer Science & Business Media, 2009), p. 94

12. J. Tumsri, W. Kimpan, Thai sign language translation using leap motion controller, in *Proceedings of The International MultiConference of Engineers and Computer Scientists 2017*. Lecture Notes in Engineering and Computer Science, 15–17 Mar 2017, Hong Kong, pp. 46–51

Historical Event-Based Story Creation Support System for Fostering Historical Thinking Skill

Yuta Miki and Tomoko Kojiri

Abstract Historical thinking skill is a reasoning skill to analyze the historical events and infer about events that will occur in the future. In order to infer the future events, learners should understand intentions and background of people in the historical events, consider the similarity between historical events and modern situation, and infer what will be happened if the similar events are happened in the modern situation. This research focuses on the inferring phases and proposes the learning method for fostering the inferring skill for future events. In order to improve the inferring skill, it is important to consider the various problems and their future situations that can apply the actions in the historical events. Thus, this research introduces the new learning method of creating story based on the given historical event so as to derive various problem situations by learners themselves. We have also developed the learning system to support this learning method. The experiment result indicated that our learning method has a potential effect to grasping historical event from abstract point of view and understanding the condition to apply the solution in the historical events.

Keywords History learning · Historical thinking · Learning support system
Reasoning skill · Situation change · Story creation

Y. Miki
Graduate School of Science and Engineering, Kansai University, 3-3-35,
Yamate-cho, Suita, Osaka, Japan
e-mail: k867858@kansai-u.ac.jp

T. Kojiri (✉)
Faculty of Engineering Science, Kansai University, 3-3-35, Yamate-cho, Suita,
Osaka, Japan
e-mail: kojiri@kansai-u.ac.jp

© Springer Nature Singapore Pte Ltd. 2018
S.-I. Ao et al. (eds.), *Transactions on Engineering Technologies*,
https://doi.org/10.1007/978-981-10-7488-2_27

1   Introduction

Historical thinking skill is a reasoning skill to analyze and explain the historical events [1, 2]. This skill is important for inferring about events that will occur in the future [3, 4]. Abbott and Adler reviewed about learning method of planning on urban design from history [5]. In order to infer the future events, learners should understand intentions and background of people in the history [6], consider the similarity between historical events and modern situation, and infer what will be happened if the similar events are happened in the modern situation. Such skill should be learned through the historical learning. However, especially in Japan, learners tend to memorize facts in historical learning and historical thinking skill is merely acquired.

Several researches try to foster the historical thinking skill. Seixas and Peck proposed learning method to have learners consider important events by comparing two historical events [7]. This research succeeded in making learners focus on backgrounds of historical events, such as life style, politics and technology. However, this method did not focus on applying historical events to the problem in the modern situation. Ikejiri developed card game-based teaching material that makes learners consider about differences and similarities between a historical event and a modern situation [8]. This teaching material only provides similar cards and whether derived differences and similarities are correct or not should be judged by learners subjectively. In addition, it does not focus on inferring the future events.

In the context of the history, the important actions in the historical events are regarded as knowledge to solve the problem situation. To inferring the future events corresponds to the application of the knowledge to the modern problem. For improving the skill for applying acquired knowledge to other problems, learning by problem/question generation is one of the hottest learning method [9]. Yu et al. developed an environment for posing the questions generated by learners and assessing questions of other learners [10]. This learning environment was proved as potential effective in improving cognitive skill to apply the solution in the problems. However, few researches apply this learning activity into history learning.

This study proposes a learning method based on problem generation for enhancing the inferring skill in historical learning. Generally, the problem consists of the problem sentence and the solution. Sometimes explanation sentences are also provided that indicate reason for applying the solution to the problem. Therefore, our learning method makes learners generate various situation that the action in the given historical event can be applied and the situation changes after applying it. This is regarded as a creation of new story based on a given historical event, thus, we call our learning activity as "story creation".

We have developed a system for supporting this story creation activity. Situation is represented by three components: characters, states of the characters, and relationship between characters. Our system provides an environment for creating these components individually. In addition, the system provides the support environment for creating the situation changes easily only by defining the actions in the events and indicating the order of applying them. By creating various stories of similar or different results with the given historical event using the system, learners can improve the ability for applying historical event to the modern situation.

Note that this paper is a revised version of the conference paper that we presented at IMECS2017 [11].

2 Applying Solution in Historical Event to Modern Problems

Historical events contain problem situation and solutions that were taken by historical people. In order to apply solutions to a modern situation, learners need to take following three steps:

Step1: abstracting the historical event,
Step2: understanding similarity between the abstraction and modern situation, and
Step3: specializing the abstraction so as to apply to the modern situation.

The relation between these steps are shown in Fig. 1.

Let's consider the Tokusei Edict of Kamakura period as historical event and a modern situation that "a friend is suffering from a student loan." The brief description of the Tokusei Edict is shown in Table 1. Tokusei Edict was a debt of samurai cancellation order issued by the Shogun. In this historical event, Tokusei Edict is the solution in the historical event. In the situation in which Tokusei Edict was applied, there were three types of characters, such as samurai, Doso, and Shogun. Economic situation of samurai was not good and lent money from Doso. Shogun had power over samurai and Doso.

In the step 1, problem situation and solution of Tokusei Edict are abstracted. As an example of the abstraction of the problem, samurai is regarded as a money

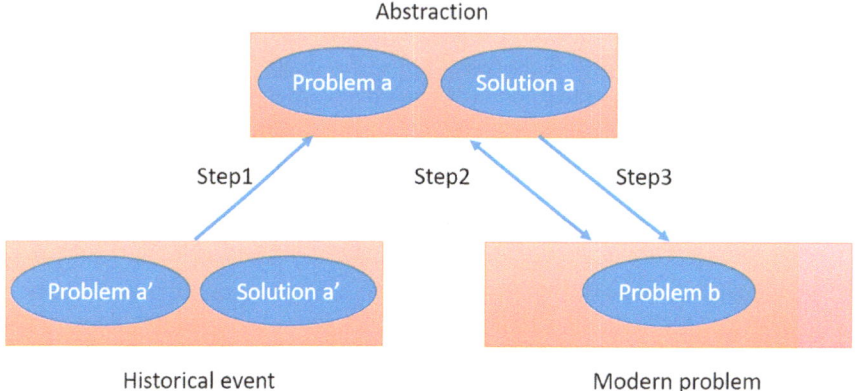

Fig. 1 Steps for applying solution in the historical event to the modern problem

Table 1 Kamakura Bakufu's Tokusei Edict

Samurai became impoverished in aftermath of the Mongol invasions and became indebted to moneylenders, Doso. With the goal of helping impoverished samurai, the Shogun issued Tokusei Edict cancelling samurai debt. Initially, this helped the samurai. Instead, Doso got angry because their loan was not paid

debtor and Doso is creditor. As for the abstraction for the solution, Tokusei Edict is changed to the law which makes contract of the lending and borrowing void. In the step 2, situation of money debtor (samurai in Tokusei Edict) and that of a friend in the modern problem are discovered as similar from the economical viewpoint. In addition, the dominator (Shogun in Tokusei Edict) and government in the modern situation also have similar situation that they both have a power to issue laws. In step 3, the law which makes contract of the lending and borrowing void is modified by replacing money debtor as friend, and the dominator as government.

Several researches try to support steps 1 and 2. For step 1, Knoblock insisted that abstracting is effective when people search for solutions of problems [12]. Our research group has already proposed the learning method for acquiring lessons from historical events by abstracting them [13]. As for supporting step 2, Ikejiri developed card game-based teaching material that makes learners consider about differences and similarities between a historical event and a modern situation [8].

There are few researches that focus on step 3. In order to specialize general solution to the specific situation, to consider whether conditions of the abstraction satisfies the problem situation of the modern problem. Therefore, in the learning method for supporting step 3, it is important to make learners be aware of the condition situation of the given solution in the abstraction and to create problem situation that satisfy the condition. This way of thinking corresponds to the thinking in the learning by problem/question generation, such as to create the problem based on the given solution.

As for the research of the learning by problem/question generation, Kojima et al. classified generated problems into four types according to the differences from given problem: (1) same situation and same solution, (2) same situation and different solution, (3) different situation and same solution, and (4) different situation and different solution [14]. This research insisted that changing the situation is rather easy than changing the solutions in the context of mathematics. Mathematical problems usually consist of conditions only necessary to apply the solution so that learners can easily find solution to change. On the contrary, in the history learning, situation contains several characters, states of characters and their relations, and most of them are not related to the solution. Therefore, it is difficult to change the situation because learners need to discriminate factors that relate to the condition to apply solution before changing the situation.

Our learning method gives learners opportunity to consider the conditions of the solution by creating a situation of the story.

3 Supporting for Generalization of Historical Event by Creating Story

Akaishi insisted that story composed of five components; world model, story, scene, event, and character [15]. World model is a set of stories and story is a sequence of scenes. Scene is a chunk of events. Events correspond to actions and characters are people. Characters are defined as a set of properties, such as name, gender, feeling, power, amount of money, and so on. Properties have values, which is called state. Also, characters have relations with other characters.

Historical events consist of the same components. However, history does not have end point, so world model cannot be defined. In the history, occurrence points of events are defined as the points at which the states of properties of characters are changed. Therefore, the scene is regarded as a set of state changes of characters.

Events occur not only by the intended actions by the characters but also the indirect effect from actions of others. For instance, when the citizen becomes rich, the country also becomes rich because citizen is a member of the country. This indirect effect is caused by the relation between characters. In above example, the government owns a citizen, which means when citizens money increases, the government's money is also increased.

In applying the historical event to the modern situation, it is important to compare characters, states of the characters, and their relations in the historical event with those of the modern situation. Also, in order to check if the applied solution is appropriate for the modern situation, state changes after applying the solution should be examined carefully. Therefore, in the story creation based on the historical event, character, properties of characters, relationship between characters, and event should be defined. In addition, the state transitions need to be represented according to the events or relationships.

Table 2 shows the components of the story that learners need to define. Character is a people and state of the character corresponds to the value of property that the character has. Properties of the character are determined according to the character's type. For example, "country" has "power", "land" and "money" as its property. Necessary properties to describe in the story are determined by the given historical event. That is, properties that are changed by events or relations in the given historical event should be included in the created story. For example, properties that are needed to deal story based on the Tokusei Edict is money since

Table 2 Components of story

Components	Explanation
Character	People appearing in story
State of character	Values of properties of character, indicating the kind of the character
Relationship between characters	Relationship between characters. State changes of one character affects to the state of the other
Event	Action that change the state of characters

money of the samurai is focused. Relationship between characters are described by the name of the relationship and the state of properties that the relationship affects to. Events are also defined by the state changes that the corresponding action derives. Figure 2 shows the example of the event, such as Tokusei Edict. This event changes the money and the land of samurai from little to normal. Figure 3 is the example of the relationship, such as debtors-creditors relationship. It defines that money of the creditor decreases the increase of the money of the debtor.

History can be represented by transition of states occurred by events or relationships. Creating story based on a given history means that creating the same state changes as the given history. For example, state changes of the Tokusei Edict are shown in Fig. 4 and example of story created by Tokusei Edict is shown in Fig. 5. In this story, the friend corresponds to samurai and the government corresponds to Doso. This story shows that because of the cancellation of loan by the government, the friend who lent a student loan has been survived while the government's economic situation got worse.

There are various stories that learners can create. Learners can add more characters or more events as long as they do not change the state of the given history. By considering the relations among relations, events, and characters' state changes, learners are able to be conscious of the conditions that the solution in the history can apply.

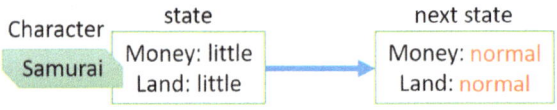

Fig. 2 Example of event "Tokusei Edict"

Fig. 3 Example of relationship "debtor and creditor"

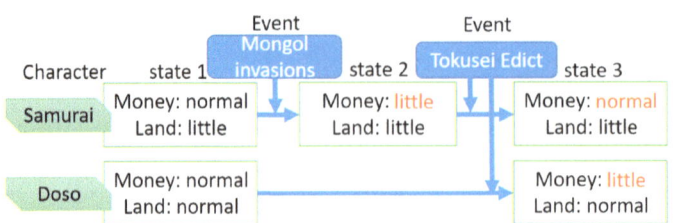

Fig. 4 State change of the Tokusei Edict (changed states are shown by orange words)

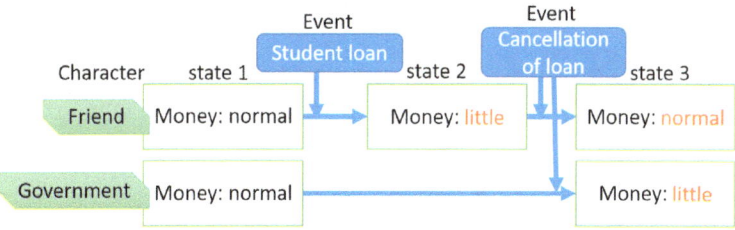

Fig. 5 Story referred to the Tokusei Edict

4 Supporting System for Creating Story

We have developed a support system for creating stories based on state changes of given history. It contains two databases. One is the history database which contains components of historical events. The other is story database that has information regarding to the story created by learners. In addition, the system consists of two interfaces. Story and history view interface shows selected story in the story database or history in the history database. Story creation support interface provides environment to create a story based on components of history. These interfaces are emerged when learners execute the system. In the story creation support interface, learners create characters, their properties, and their relations with other characters. They can also define initial states of the characters and represent state changes by creating events or applying relations. Stories created by learners are stored in the story database.

In the story and history view interface, after the learner selects historical event from a list, a text that explains the historical event and its state changes are shown in history view window (Fig. 6). Leaners are able to learn about the historical event by reading the text and observing the state changes. In addition, they can see the stories that they created before through this window.

Story creation support window is shown (Fig. 7). This window consists of four tabs. They correspond to the creation of characters, events, relationships, and scenes individually. Learners are able to create these components of the story by selecting each tabs and filling in the forms shown in the tabs. State changes that leaners create in the scene tabs are shown on the state change area. If the show story button is pushed, the new window has been emerged and the system transforms the input components to story sentences. For example, when Tom is created as characters and his type is selected as a human, the system shows "There is a person named Tom." If state of Tom's money is changed to "much" by an event named "part time job", the system shows "Tom's money increases by part-time job." Following shows the way to create each components.

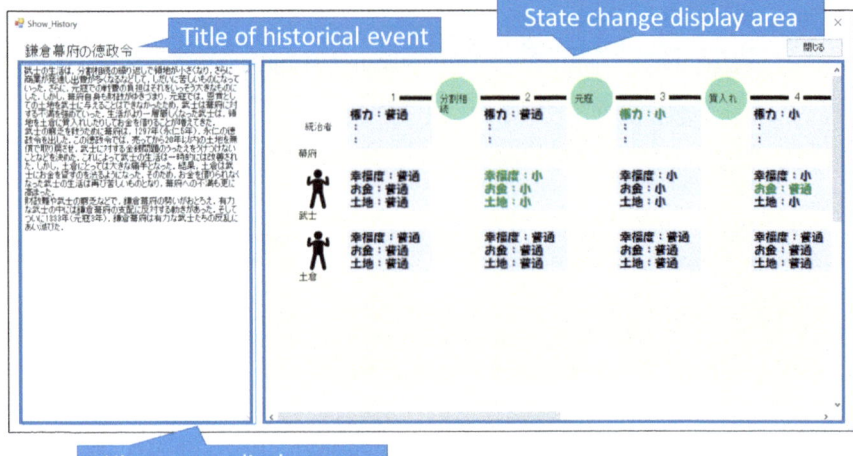

Fig. 6 Story and history view window

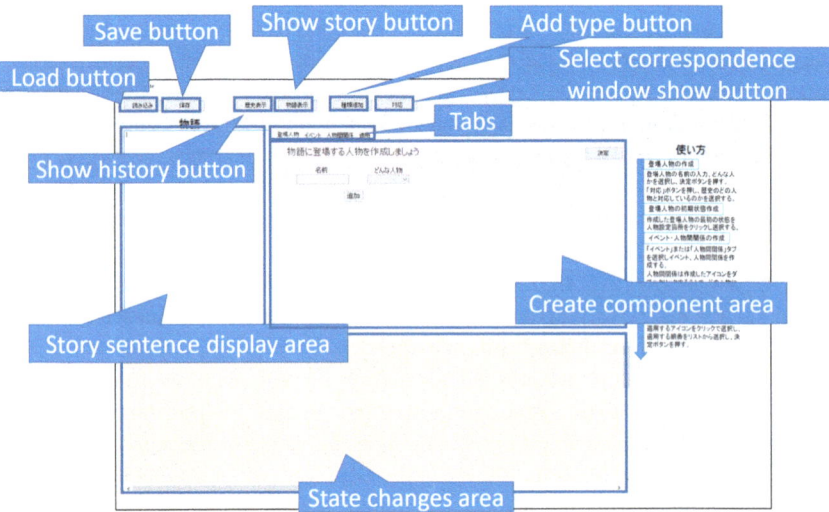

Fig. 7 Story creation support window

4.1 Character Tab

Story creation support window with character tab is shown on Fig. 9. Learners can input names of characters into character name textbox and determine their types by selecting types from character type list. They can create new characters by pushing

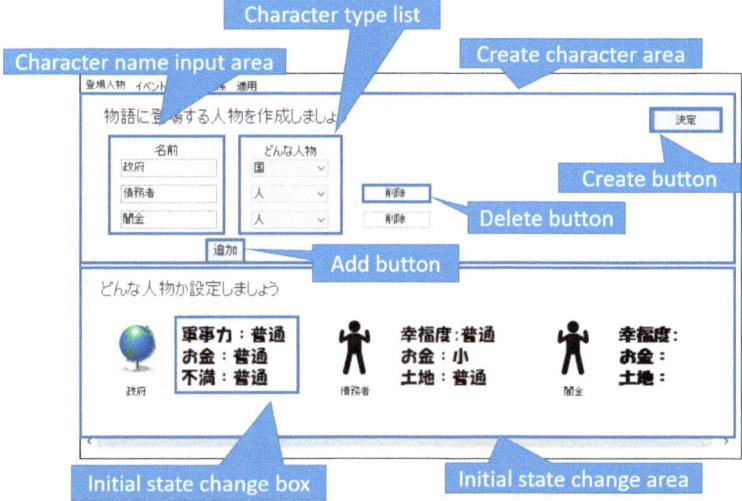

Fig. 8 Character tab

add button. Typical types of characters are prepared, such as human or country. Learners can also create new type by pushing add type button on Fig. 7. When a new type is created, created type is added on the character type list in Fig. 8. When learners push create button, new character is created and its initial state is appeared in the initial state change area. The states that property can take are much, normal and little. If learners are not satisfied with assigned initial state, they can change it by clicking an initial state change box.

4.2 Event Tab

Story creation support window with event tab is shown on Fig. 9. Learners can input event name into event name textbox. Events change the state of characters, so learners need to indicate whose and what properties are changed and how they are changed by the created event. After the event is created by pushing the add button, created event is shown on show created event area as a circle. Its contents are shown by locating mouse cursor on the circle. In addition, learners can delete created events by pushing delete button beside the circle.

4.3 Relationship Tab

Story creation support window with relationship tab is shown on Fig. 10. In create relationship area, learners first create the relationships between types and then

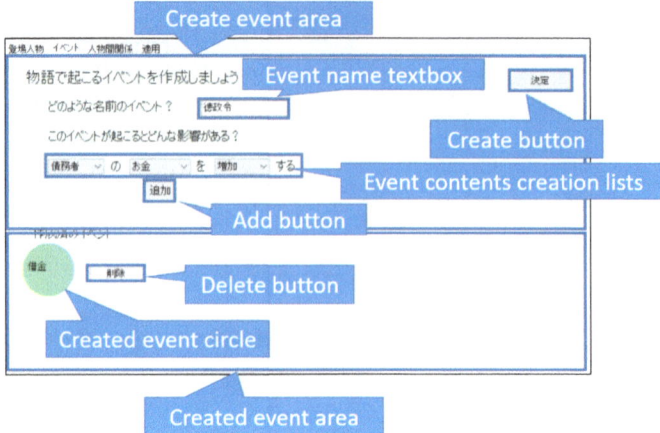

Fig. 9 Event tab

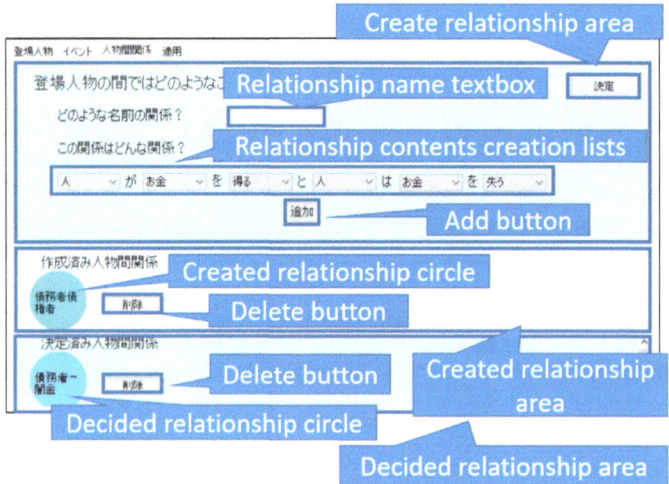

Fig. 10 Relationship tab

attach the created relations to the characters. In creating relationships, learners need to input relationship name and select the pairs of properties of character types. For example, as a creditor-lender relationship, "creditor's money increase and lender's money decrease" is indicated. When learners push create button, the system draws created relationship as a circle on created relationship area. Its contents are shown by locating mouse cursor on the circle. Created relationship is able to be deleted by pushing delete beside near the circle.

When clicking the circle, window for selecting characters is appeared (Fig. 11). Learners can assign the created relationships to the characters. Assigned

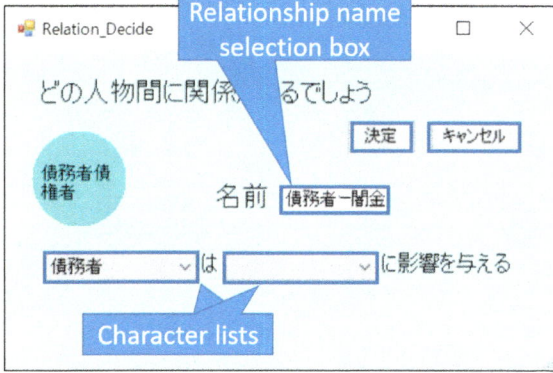

Fig. 11 Characters selecting window

relationships are appeared on the decided relationship area on Fig. 10 so that learners are able to grasp which relationships are not assigned.

4.4 Scene Tab

Story creation support window with scene tab is shown on Fig. 12. In this window, learners select events or relations that occur in the story and create state changes of

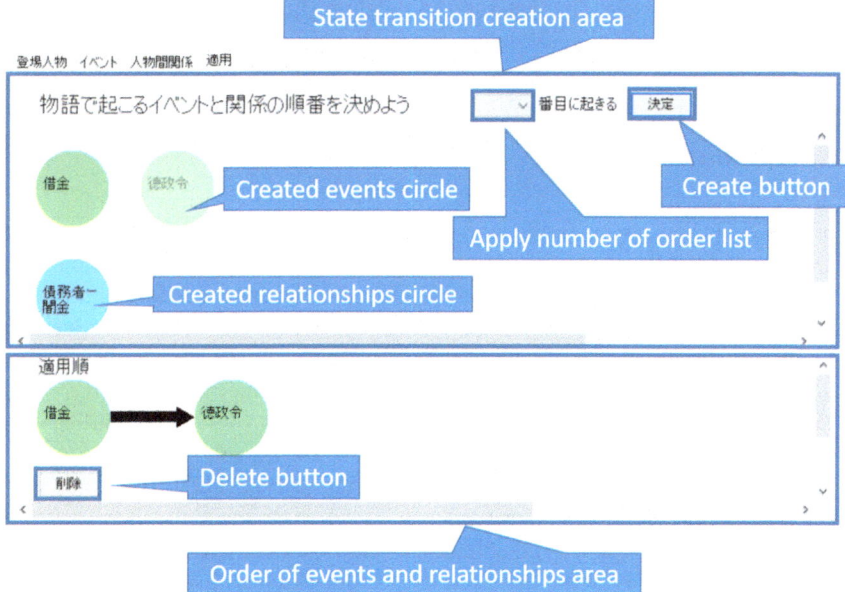

Fig. 12 Scene tab

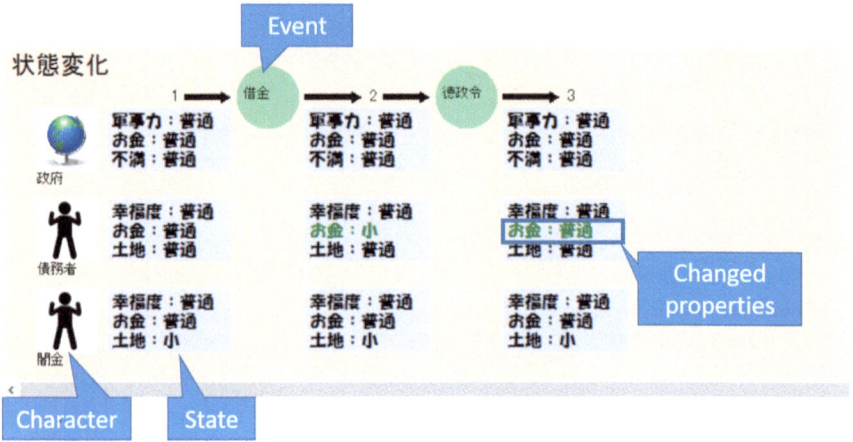

Fig. 13 Example of state change area

characters according to these events or relations. In the window, created events and relationships as green and blue circles on state transition creation area, respectively. Learners select events or relations by their applying order. When the events or relations are selected, states of characters are automatically changed according to the definition of the events or relations. Colors of the selected events or relations become thinner. The sequence of events or relations shown in the order of events and relationships area. Learners can check the state changes by watching state change area of the story creation support window (Fig. 7). In this area, changed properties are highlighted with green color as Fig. 13.

5 Experiment

We have evaluated potential effective of the learning method of the story creation based on historical event and the system for supporting story creation. The experiment especially focuses on evaluating whether the understanding of the given historical event, especially the condition to apply, is changed through the learning method of the story creation. Tokusei Edict of Kamakura period is used as a given historical event. In this historical event, "money" is the important property.

The experiment was done on 15 university students $(A - O)$ as participants. Figure 14 represents the process of the experiment. Participants were asked to write the condition for applying Tokusei Edict before the story creation. After they were asked to create the story using paper and pen. In this experiment, they were asked to create the story that applies the solution in the historical event but the result is different. That is, they were asked to create the story with the happy ending, because the Tokuse Edict ended up with the bad situation. Then, they were asked to

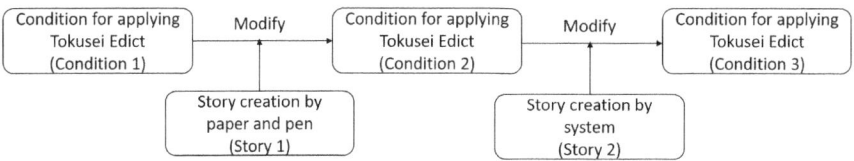

Fig. 14 Process of experiment 1

modify the condition. After that, they were asked to create the story using the system. In this time, they were allowed to change the story from story 1. Finally they were asked to modify the condition again. Also, they were asked to write the reason for answering each conditions and stories. In the end of the experiment, they were asked to answer questionnaires. Questionnaire items were as follows:

Question 1: Was "to create story" useful for thinking about conditions of success or failure of the solution in the given historical event?
Question 2: Was "to create story using the system" useful for thinking about condition of success or failure of the solution in the given historical event?
Question 3: Was the system easy to use?

These questions were rated on a four-point Likert scale. 1 is negative and 4 is positive.

Table 3 shows the change of conditions from 1 to 2 and from 2 and 3, and the change of story from 1 to 2. Y means that they had change and N means no. Through all experiment, 9 participants changed the conditions for applying Tokusei Edict, either from condition 1 to 2 or condition 2 to 3. Table 4 shows the conditions written by some participants. The underlined parts are the added description. Participant D added more condition after created the story. On the other hand, participant H embodied the "support" of condition 1 in condition 2. For these participants, the story creation was effective for understanding the condition of the given historical events. However, for other participants, the conditions are not changed.

For the story creation, 8 participants changed the story after using the system. This means that the system has some effects for deepen the created story. Table 5 is the examples of the created story. Underlined sentence is added description.

Table 3 Result of experiment 1: change of conditions and stories

Participants	A	B	C	D	E	F	G	H	I	J	K	L	M	N	O
Change of condition from 1 to 2	Y	N	N	Y	N	N	N	Y	Y	N	Y	N	N	N	Y
Change of condition from 2 to 3	N	N	Y	N	Y	N	N	N	Y	N	N	N	N	Y	N
Change of story from 1 to 2	Y	Y	Y	N	N	N	Y	N	Y	Y	Y	N	Y	N	N

Table 4 Examples of conditions 1 and 2: conditions for applying Tokusei Edit with its result success

Participant	Condition 1	Condition 2
D	If there are people who are exploited by others	If there are people who are exploited by others. When the Bakufu has enough money
H	If the creditor can get support afterward	If there are some other properties that the creditor can get benefit

Table 5 Example of created stories

Participant	Story 1	Story 2
A	The man does not have enough money. The company suggested that if he works for the company, he does not pay the money back	The man does not have enough money. The company who lend the money to the man meet him. The company suggested that if he works for the company, he does not pay the money back
G	One man borrowed money from a creditor. However, he was not able to give the money back. The government issued the law that if creditor canceled the debts, they would not need to pay the tax. After this law has been issued, the man and the creditor both get happy	One man lost his money and borrowed money from a creditor. However, he was not able to give the money back. The government issued the law that if creditor canceled the debts, they would not need to pay the tax. After this law has been issued, the man and the creditor both get happy

Participant *A* added the description about the relation between the man and the company. Participant *G* wrote the state change of the man. Both of them are factors of the story. Therefore, for these participants, to create story using the system was useful for create story in detail.

The results of the questionnaires are shown on Table 6. According to the result of the question 1, many participants could consider conditions of success or failure by creating story. For question 2, many participants answered that the system was useful for considering conditions. Some participants were commented that representing state changes makes them recognize the mistake of their understanding for the states the characters. This comment indicated that participants could grasp the state of characters more accurately through the story creation. Therefore, story creation has potential effect on fostering the reasoning skill in the historical thinking.

Table 6 Result of questionnaires

	Answer (Number of participants)			
	1	2	3	4
Question1	1	1	8	5
Question2	1	0	6	8
Question3	0	5	7	3

In question 3, 10 participants answered that they could use the system easily. However, 5 participants gave negative comments. One participant complained that the size of the characters and buttons are too small. Another participant requested to add modify function in the system. Current our system does not have a modify function. Participants should delete the inappropriate one and add new one in order to modify it. We need to revise the system according to these comments.

According to the result of the experiment, our system may have potential for supporting the learning activity of creating story. However, the number of the participants in the current experiment was small so that we need further experiment with more participants.

6 Conclusion and Future Work

In this study, we proposed the learning method for fostering historical thinking skill by creating story based on a given historical event. We have also developed the system for supporting the proposed learning method. History is composed of the state change of historical characters, so they need to consider states of characters, events and relationships by creating similar stories to a given history. This fosters learners to consider the historical events in detail and problem situation which is necessary to apply the solution taken in the historical event. According to the experimental result, our learning method was indicated to be effective for fostering historical thinking skill. Also, the system was well-designed to support the story creation.

Created stories are different among learners. Current experiment did not check the validity of the story. It may be estimated that the quality of the story reflects to the learning result. However, it is difficult for the system to check the validity of the created story automatically. One solution is to make introduce the collaborative evaluation function. By observing the stories of others, learners are may be able to consider if the condition situations which they think are valid. Thus, we plan to embed the collaborative evaluation function into the system and encourage learners to discuss the differences of their stories.

Acknowledgements The work was supported in part by JSPS KAKENHI Grant-in-Aid for challenging Exploratory Research (16K12563).

References

1. L. Elder, M. Gorzycki, R. Paul, *Student Guide to Historical Thinking* (Foundation for Critical Thinking, 2012)
2. R. Ikejiri, Y. Yamauchi, Categorize of historical thinking and effective training methods, in *28th National Convention of Japan Society for Educational Technology* (2012), pp. 495–496 (in Japanese)

3. P. Lee, Historical literacy: theory and research. Int. J. Hist. Learn. Teach. Res. **5**(1), 25–40 (2005)
4. R.J. Parkes, D. Donnelly, Changing conceptions of historical thinking in history education: an Australian case study. Revista Tempo e Argumento, Florianópolis **6**(11), 113–136 (2014)
5. C. Abbott, S. Adler, Historical analysis as a planning tool. J. Am. Plan. Assoc. **55**(4), 467–473 (1989)
6. C. van Boxtel, J. Van Drie, Historical reasoning: a comparison of how experts and novices contextualise historical sources. Int. J. Historical Learn. Teach. Res. **4**(2), 84–91 (2004)
7. P. Seixas, C. Peck, *Teaching Historical Thinking*, Challenges and Prospects for Canadian Social Studies (2004), pp. 109–117
8. R. Ikejiri, T. Fujimoto, M. Tsubakimoto, Y. Yamauchi, Designing and evaluating a card game to support high school students in applying their knowledge of world history to solve modern political issues, in *International Conference of Media Education* (2012)
9. E.A. Silver, J. Mamona-Downs, S.S. Leung, P.A. Kenney, Posing mathematical problems: an exploratory study. J. Res. Math. Educ. **27**(3), 293–309 (1996)
10. Yu. Fu-Yun, Y.-H. Liu, T.-W. Chan, A web-based learning system for question-posing and peer assessment. Innov. Educ. Teach. Int. **42**(4), 337–348 (2005)
11. Y. Miki, T. Kojiri, Story creation for fostering historical thinking skill and its support system, in *Proceedings of The International MultiConference of Engineers and Computer Scientists*, Lecture Notes in Engineering and Computer Science, 15–17 Mar 2017, Hong Kong (2017), pp. 40–45
12. C.A. Knoblock, Learning abstraction hierarchies for problem solving, in *Eighth National Conference on Artificial Intelligence*, vol. 2 (1990), pp. 923–928
13. T. Kojiri, Y. Nogami, K. Seta, Lesson discovery support based on generalization of historical events, in *Proceedings of 17th International Conference on Artificial Intelligence in Education, LNAI 9112* (2015), pp. 674–677
14. K. Kojima, K. Miwa, T. Matsui, Study on support of learning from examples in problem posing as a production task, in *Proceedings of the 17th International Conference on Computers in Education* (2009), pp. 75–82
15. M. Akaishi, A dynamic decomposition/recomposition framework for documents based on narrative structure model. Trans. Japan. Soc. Artif. Intell. **21**(5), 428–438 (2006) (in Japanese)

Japanese-Chinese Cross-Language Entity Linking Adapting to User's Language Ability

Fuminori Kimura, Jialiang Zhou and Akira Maeda

Abstract In this chapter, we propose a method to automatically discover valuable keyphrases in Japanese and link these keyphrases to related Chinese Wikipedia pages. Our proposed method has four stages. Firstly, we extract nouns from a Japanese document using a morphological analyzer and extract the candidates of keyphrases using a method called top consecutive nouns cohesion (TCNC) (Horita et al. Int. J. Comput. Theory Eng. 8(1):32–35, (2016) [1]). Secondly, we judge the degree of difficulty of the extracted keyphrases and tag them with different linguistic levels. Thirdly, we translate the extracted Japanese keyphrases into Chinese using a combination of three translation methods. Fourthly, we extract the corresponding Chinese Wikipedia articles of the translated keyphrases. Fifthly, we translate the original Japanese document into Chinese and make a vector of noun frequencies. Sixthly, we calculate the cosine similarities of the translated original document and candidate Chinese Wikipedia articles. Finally, we create links from the Japanese keyphrases to the top-ranking Chinese Wikipedia articles.

Keywords Cross-language link discovery · Entity disambiguation
Keyphrase extraction · Linguistic difficulty level estimation · Wikification
Wikipedia · Word2vec

F. Kimura (✉)
Faculty of Economics, Management and Information Science, Onomichi City University,
1600-2 Hisayamada-cho, Onomichi, Hirosima 722-8506, Japan
e-mail: f-kimura@onomichi-u.ac.jp

J. Zhou
Graduate School of Information Science and Engineering, Ritsumeikan University,
1-1-1 Noji-higashi, Kusatsu, Shiga 525-8577, Japan

A. Maeda
College of Information Science and Engineering, Ritsumeikan University,
1-1-1 Noji-higashi, Kusatsu, Shiga 525-8577, Japan
e-mail: amaeda@is.ritsumei.ac.jp

© Springer Nature Singapore Pte Ltd. 2018
S.-I. Ao et al. (eds.), *Transactions on Engineering Technologies*,
https://doi.org/10.1007/978-981-10-7488-2_28

383

1 Introduction

Recently, because of the wide use of tablets and smartphones, Internet services have become even more popular all over the world. An enormous amount of information is stored in a variety of languages. In addition, the hyperlinks on the Web provide links to many related pieces of information. However, related information is not always available in the native language of the user, which may make it difficult for the user to understand it.

To solve this problem, it is desirable to automatically find potential links from documents written in their native languages (Fig. 1). Therefore, we propose a method to obtain Chinese encyclopedia articles that give the meaning of the key-phrases in a Japanese document.

Such a mechanism will be useful for Chinese students studying Japanese and could further enhance the utility of online encyclopedias, such as Wikipedia. Our proposed method aims to support knowledge discovery using an online encyclopedia as a learning tool for Chinese students studying Japanese. Considering the difference of foreign students' language proficiencies, the system ought to recommend appropriate keywords. Therefore, the purpose of this research is to perform cross-language entity linking and provide appropriate information, adapting to each user's language ability.

Fig. 1 Example of cross-language entity linking

2   Related Work

Wikipedia is a multilingual online encyclopedia that anyone can use and edit on a Web browser. Articles for the same topic in different languages are usually linked via inter-language links. However, for some articles in some languages, there are no appropriate articles in different languages. A variety of previous researches have been performed so far that deal with the same problem.

There are studies about Wikification [2] at the NTCIR-9 and NTCIR-10 conferences [3, 4] that aim to reuse Wikipedia resources effectively. According to Horita et al. [1], Wikification is a method for automatically extracting keyphrases from a document and linking them to an appropriate Wikipedia article. From NTCIR-9, one of the related researches of Wikification is called CrossLink. CrossLink is a task aiming to automatically find potential links between online documents in different languages [2]. The linked text is called "anchor text" in the CrossLink task. This task mainly focuses on extracting anchor texts from English Wikipedia and linking them to appropriate Wikipedia articles in Japanese, Chinese, or Korean.

In the Text Analysis Conference (TAC) [5], the task of cross-language entity linking (CLEL) was being performed. The purpose of this task is to extract PER (person), ORG (organization), and GPE (geopolitical entity) from Chinese or Spanish documents. Then, they link them to appropriate English documents. In our proposed method, the target entities are not limited to places or personal names, which is the main difference between our proposed method and this task. Wang et al. [6] proposed a cross-lingual knowledge linking approach for building cross-lingual links across Wikipedia knowledge bases. Their approach uses only language-independent features of articles and a graph model to predict new cross-lingual links.

Chen et al. [7] proposed an approach in which the first step is extracting n-grams from the query source documents as potential anchors. The next step is anchor expansion and ranking. The final step of the anchor selection process is to re-rank anchors by computing the similarity between the title of the current query Wikipedia page and each element in the vector of expanded potential anchors using Wikipedia Miner [8]. Unlike our work, they only use Google Translate to translate the potential anchors. In our proposed method, we translate the keyphrase using three methods and extract all the Chinese articles of the translated keyphrase. Finally, we make a ranking using a cosine similarity comparison of the Japanese document and Chinese Wikipedia articles.

Liu et al. [9] divided their cross-language link discovery task into three steps: anchor mining, cross-lingual linking to related articles, and disambiguation. Like with Chen et al.'s work, they chose Google Translate as the anchor translation tool. In our proposed method, we use three translation methods and make a ranking using a cosine similarity comparison of the Japanese document and Chinese Wikipedia articles [10]. The difference between our approach and their work is that they use two different ways, which are Dice coefficient based and LDA model based measures, to calculate the similarity of keyphrases. Furthermore, they apply the POS (part of speech) tag analysis module.

3   Proposed Method

In this section, we describe our method to detect an appropriate Chinese Wikipedia article for a keyphrase in a Japanese document. The proposed method consists of four processes: 1. keyphrase extraction, 2. translation, 3. obtaining Chinese articles, and 4. ranking Chinese candidate articles. We used the top consecutive nouns cohesion (TCNC) method [1] for extracting keyphrase candidates. In this method, when consecutive nouns appear in a sentence, we adopt all possible binding patterns starting from the first noun.

Figure 2 shows the overview of the proposed method. When extracting the keyphrase candidates, to extract new compound words, we use MeCab [11] to analyze the Japanese document. By considering the surrounding context of the keyphrase, we can create links among nouns.

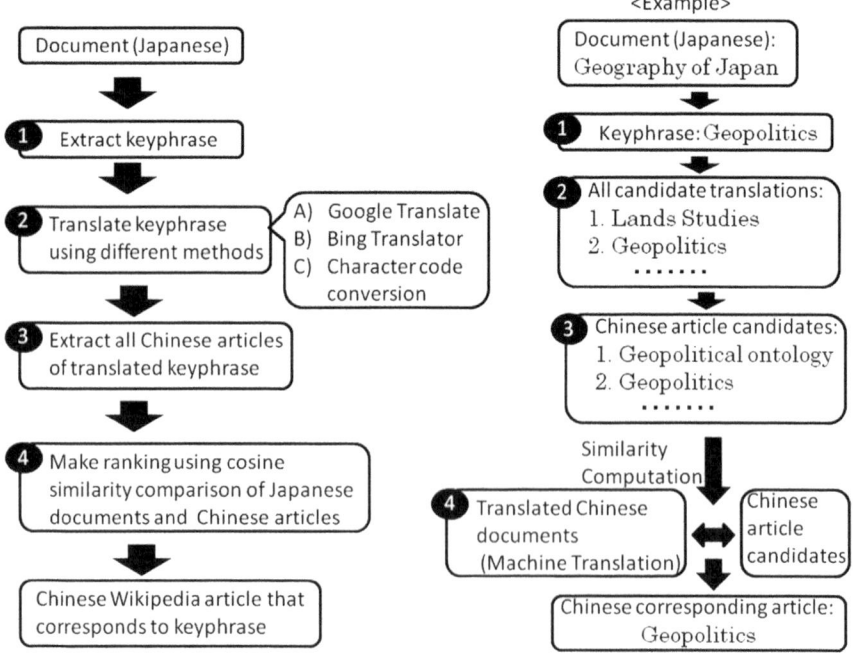

Fig. 2 Overview of proposed method

3.1 Keyphrase Extraction

In this section, we describe keyphrase extraction. It consists of three processes. First, we extract keyphrase candidates from sentences in a document. Second, we attach the linguistic difficulty level to the extracted keyphrase candidates. Third, we choose keyphrases with an appropriate linguistic difficulty level for the user. The user inputs his/her own linguistic difficulty level.

3.1.1 Keyphrase Candidate Extraction

We adopt the top consecutive nouns cohesion (TCNC) method [1] to extract keyphrase candidates from sentences in a document. Horita et al. [1] proposed a method for keyphrase extraction with two steps. Firstly, they conduct a morphological analysis and extract keyphrase candidates by the top consecutive nouns cohesion (TCNC) method, which means combining two characters and treating them as one compound word. Secondly, they rank the keyphrase candidates by Dice coefficient and keyphraseness measures [2] (Fig. 3). Therefore, we can extract all keyphrase candidates from consecutive nouns appearing in a sentence.

3.1.2 Attaching Linguistic Difficulty Level to Extracted Keyphrase Candidates

In this chapter, we propose a method to extract appropriate keyphrases for users, considering the user's Japanese proficiency. It is pointless and burdensome for users to show the potential links of easy phrases. We use the vocabulary of the Japanese-Language Proficiency Test (JLPT) [12] as a criteria of linguistic difficulty level. The JLPT categorizes Japanese words into five levels: N1 to N5. The words for N1 are the most difficult, and the words for N5 are the easiest.

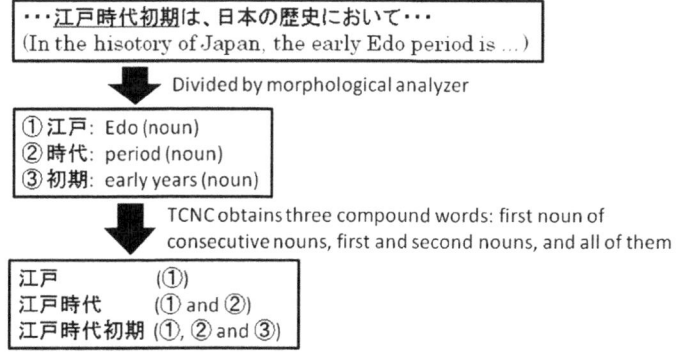

Fig. 3 Example of TCNC

In this method, we first refer to the JLPT vocabulary list to find out whether the keyphrase candidates exist in the list. If it exists in the list, we judge the degree of difficulty of the keyphrase by its level and attach its difficulty to the keyphrase (Fig. 4).

When it does not exist in the list, its linguistic difficulty level is estimated from the JLPT vocabulary list. We calculate the relevancies between the keyphrase and each word in the list using word2vec [13]. We choose the most relevant word in the list and attach its linguistic difficulty level to the keyphrase.

In word2vec, a feature of each word is represented by a word vector. This feature of the word is estimated from the words that appear before and after the word. Word2vec can estimate the probability that each word appears around a certain word. It is calculated by softmax function.

$$p(\omega_0|\omega_i) = \frac{\exp\left(v_{\omega_0}^T \cdot v_{\omega_i}\right)}{\sum_{\omega_v \in V} \exp\left(v_{\omega_v}^T \cdot v_{\omega_i}\right)}$$

where ω_i is the keyword, ω_0 is the word that appears around ω_i, v_{ω_i} is the feature vector of ω_i, v_{ω_0} is the feature vector of ω_0, and V is the JLPT vocabulary list.

For example, we extract the word "トランプ (Trump)" from a news article; to give it a tag, we use words from the JLPT vocabulary list. By calculating word vector similarities, we can find out the words that have the most similar meanings to the word "トランプ (Trump)". These are "大統領 (President)" and "国家元首 (head of state)" with a score of 0.624 and 0.596, respectively. Then, we know that Trump is the president of the United States of America. Thus, we can label the word "トランプ (Trump)" with the same language proficiency level tag as the N2 level.

3.2 Translation Using Multiple Methods

In the proposed method, we translate the Japanese keyphrases into Chinese using a combination of three translation methods: two machine translation services (Google

Fig. 4 Example of estimating linguistic level of keyphrase

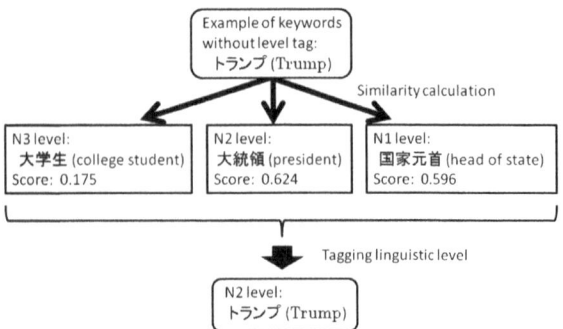

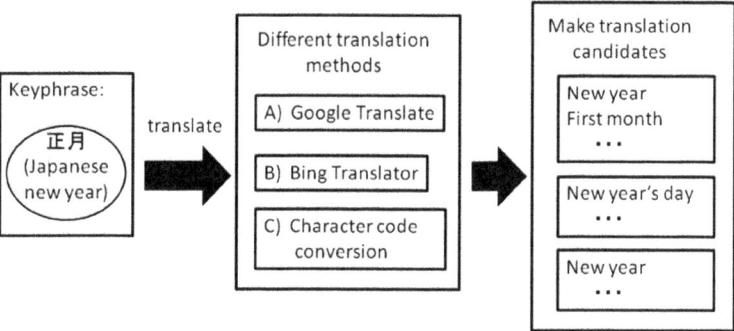

Fig. 5 Example of keyphrase translation using different translation methods

Translate [14], Bing Translator [15]) and the character code conversion method [16, 17] (Fig. 5). We consider all the obtained Chinese translations from all of the three translation methods (including incorrect translations) as the translation candidates for a Japanese keyphrase. The reason for using all the translation methods is to prevent missing the appropriate translation for the Japanese keyphrase. Incorrect translations will be eliminated in the ranking process, which will be explained in Sect. 3.4.

3.3 Obtaining Chinese Articles

In this process, we obtain the corresponding Chinese Wikipedia articles for each translation candidate obtained in the previous process. We use partial string matching between each obtained translation candidate and the titles of Chinese Wikipedia articles [17].

The reason for performing partial string matching is that some compound words might be partially translated into appropriate translation, although all of them cannot be translated (Fig. 6). Finally, we obtain the Chinese Wikipedia articles whose titles partially match each translation candidate.

3.4 Ranking of Chinese Candidate Articles

In this procedure, we rank the obtained Chinese Wikipedia articles to figure out the most appropriate Chinese article for the Japanese keyphrase. Figure 7 shows the flow of this ranking procedure. First, we translate the original Japanese document of the keyphrase into Chinese. Second, we extract nouns from the translated Japanese document and obtain Chinese Wikipedia articles. Third, the frequencies of these nouns are regarded as a vector, and we calculate the cosine similarities of the

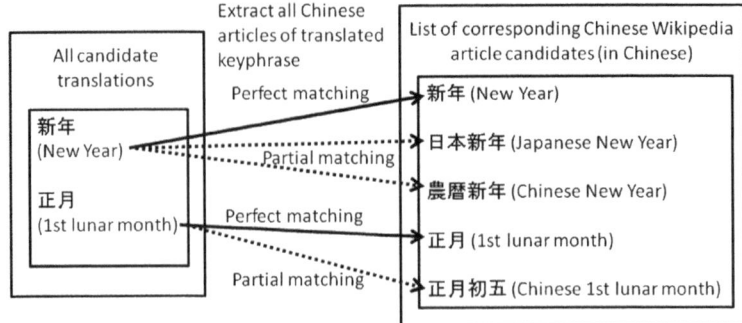

Fig. 6 Example of partial string matching of translation candidates and Chinese Wikipedia article titles

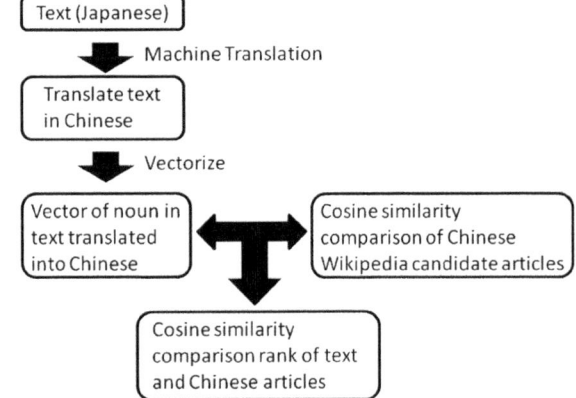

Fig. 7 Process flow of ranking of Chinese candidate articles

original document and all the Chinese Wikipedia articles. Fourth, we use the highest ranked article as the corresponding Chinese article for the Japanese keyphrase. In this way, we can find the corresponding Chinese article for the Japanese keyphrase even if we obtain many Chinese articles as the candidates.

4   Experiments

In this section, we describe the experiments to evaluate the proposed cross-language entity linking methods. We used 10 randomly selected news articles from the Mainichi Shimbun newspaper in 2014, which are the same as the experimental data used in Horita et al.'s study [1]. Table 1 shows some examples of the titles and genres of the articles used for the experiments.

Table 1 Examples of news articles from each genre

Article #	Title of news articles	Genre
1	[エアバッグ] タカタ, 不信払拭できず…米公聴会	経済
	(Airbag: Takata unable to dispel distrust – US hearing)	(Business)
5	[エボラ出血熱] リベリア, ギニア駐日大使が支援訴え	国際
	(Ebola hemorrhagic fever: Liberia and Guinea Ambassadors to Japan appeal for support)	(International)
6	［アベノミクス］景気, 期待か実感か 与野党が有権者に訴え	政治 (Politics)
	(Abenomics: Hope or realization of economic growth - ruling and opposition parties appeal to voters)	
9	[テニス]錦織「生まれ変われた」目標に４大大会優勝明言	スポーツ
	(Tennis: Nishikori says "I was able to be reborn" - sets his goal to win 4 major tournaments)	(Sports)
10	［大阪市議会］地下鉄・バス民営化条例案 野党反対で否決	社会 (Society)
	(Osaka municipal assembly: Regulation proposal to privatize subway and bus systems denied by opposition parties)	

Section 4.1 describes the procedure for creating the ground truth keyphrase data, which are necessary for evaluating the effectiveness of the cross-language entity linking methods. Section 4.2 describes the experiment to evaluate the proposed cross-language entity linking method.

4.1 Creation of Ground Truth Keyphrases

To create the ground truth keyphrases, we conducted a questionnaire survey with 18 Chinese students whose Japanese proficiency is N2 or N1 level. We used the 10 news articles explained in the previous section and asked them to pick out words or phrases that they did not know or had difficulty understanding. From these words and phrases, we selected ones that at least half of the students had picked out as the ground truth keyphrases.

The resulting keyphrases for N1 and N2 are shown in Table 2 and Table 3, respectively. The number of selected keyphrases is 23 for N1 and 34 for N2. The number for N2 is larger than that for N1 due to the difference of language ability in N1 and N2. From Tables 2 and 3, we observe that many proper nouns are selected that do not exist in the JLPT vocabulary list.

4.2 Experiments of Cross-Language Entity Linking

In this section, we describe the experiments of the proposed cross-language entity linking method. Section 4.2.1 describes the experiment of keyphrase extraction, Sect. 4.2.2 describes the experiment for obtaining corresponding Chinese articles,

Table 2 Ground truth keyphrases for N1

Article #	Ground truth keyphrases
1	タカタ (Takata) , リック・ショステク (Rick Szostek), エアバッグ (airbag), ホンダ (Honda)
2	入札 (bid)
3	グリーンスパン (Greenspan), リーマン・ショック (bankruptcy of Lehman Brothers), レーダ (radar)
4	ホワイトハウス (White House), 野党 (opposition party)
5	エボラ (Ebola), テレウォダ (Telewoda), リベリア (Liberia), ギニア (Guinea)
6	アベノミクス (Abenomics), 野党 (opposition party), リーマン・ショック (bankruptcy of Lehman Brothers)
7	ベストミックス (best mix)
8	マニフェスト (manifesto)
9	ATPツアー・ファイナル (ATP Finals), ジョコビッチ (Đoković)
10	腐心 (to make a great effort), 橋下徹 (Tōru Hashimoto)

Table 3 Ground truth keyphrases for N2

Article #	Ground truth keyphrases
1	大手 (major company), タカタ (Takata), エアバッグ (airbag), リック・ショステック(Rick Szostek), ドライバー (driver), リコール (recall), ホンダ (Honda), スバル (Subaru)
2	入札 (bid), 応札 (to make a bid)
3	リーマン・ショック (bankruptcy of Lehman Brothers), シンガポール (Singapore), グリーンスパン (Greenspan), ワシントン (Washington)
4	ホワイトハウス (White House), 野党 (opposition party), ベイナー (Boehner), オバマ (Obama), 恩赦 (amnesty), フェイスブック (Facebook)
5	エボラ (Ebola), テレウォダ (Telewoda), リベリア (Liberia), ギニア (Guinea), シラ (Sylla)
6	アベノミクス (Abenomics), 野党 (opposition party), マイナス (minus)
7	原発 (nuclear power plant), 稼働 (operation)
8	マニフェスト (manifesto)
9	錦織圭 (Kei Nishikori) ジョコビッチ (Đoković), シングルス (singles)
10	腐心 (to make a great effort), リセット (reset)

and Sect. 4.2.3 describes the experiment with the method for ranking the candidate articles to be linked.

4.2.1 Keyphrase Extraction

In this section, we evaluate to what degree the keyphrases obtained from the proposed method match the ground truth ones created in Sect. 4.1. The levels of

Table 4 Keyphrases obtained by proposed method

Article #	N1	N2
1	同社 (that company), スバル (Subaru), タカタ (Takata), 無償 (free of charge), ホンダ (Honda), リコール (recall), エアバッグ (airbag), 公聴会 (public hearing)	全米 (all of America), いんぺい (concealment)
2	構図 (composition), 発電所 (power plant), 入札 (bid)	長崎 (Nagasaki), 製造業 (manufacturing industry)
3	公示 (public announcement), グリーンスパン (Greenspan), 指標 (indicator)	述懐 (recollection), 不動産バブル (real property bubble)
4	ホワイトハウス (White House), フェイスブック (Facebook), 野党 (opposition party)	恩赦 (amnesty), オバマ (Obama), 合法 (legal)
5	シラ(Sylla), 言及 (mention), エボラ (Ebola), 終息 (eradication), 内戦 (civil war), ギニア (Guinea), リベリア (Liberia)	n/a
6	民主党 (Democratic Party), アベノミクス (Abenomics), 野党 (opposition party)	根拠 (ground/reason)
7	明示 (to specify), 自民党 (Liberal Democratic Party), ベストミックス (best mix), コスト (cost), 原発 (nuclear power plant), 発電所 (power plant)	稼働 (operation)
8	民主党 (Democratic Party), マニフェスト (manifesto), 菅義偉 (Yoshihide Suga)	衆院 (House of Representatives)
9	シングルス (singles), 錦織圭 (Kei Nishikori), 明言 (to declare), 準決勝 (semifinal), ジョコビッチ(Ðoković)	今季 (the present season)
10	自民 (Liberal Democratic), 橋下徹(Tōru Hashimoto), 腐心 (to make a great effort)	リセット (reset)

keyphrases are given by consulting the JLPT vocabulary list. The results of giving the levels of keyphrases are shown in Table 4.

Next, we compare the effectiveness of the proposed keyphrase extraction method with our previous method [1]. Our previous method measures the relation of each candidate keyphrase and the candidate articles to be linked, then ranks them by combining them with the *keyphraseness* measure [2], and selects the keyphrases that exceed the threshold.

We use precision, recall, and F-measure for evaluating the effectiveness of the methods. Their formulas are shown below:

$$precision = \frac{\text{number of extracted and correct keyphrases}}{\text{number of extracted keyphrases}}$$

Table 5 Comparison of previous method and proposed method (N1)

	Previous method	Proposed method
Precision	0.10	0.39
Recall	0.52	0.74
F-measure	0.16	0.51

Table 6 Comparison of previous method and proposed method (N2)

	Previous method	Proposed method
Precision	0.16	0.47
Recall	0.65	0.79
F-measure	0.26	0.59

$$recall = \frac{\text{number of extracted and correct keyphrases}}{\text{number of ground truth keyphrases}}$$

$$F\text{-}measure = \frac{2}{\frac{1}{precision} + \frac{1}{recall}}$$

The precision, recall, and F-measure of keyphrase extraction for proficiency level N1 and N2 are shown in Table 5 and Table 6, respectively. From the results, we observe that the proposed method successfully extracted more correct keyphrases than the previous method, thus improving recall. 27 out of 38 (including both N1 and N2) correct keyphrases were extracted by the proposed method. The drawback of the proposed method is that since it extracts all the keyphrases by consulting the JLPT vocabulary list, it resulted in relatively poor precision of 0.39 and 0.47, respectively.

4.2.2 Obtaining Corresponding Chinese Articles

In this section, we describe the experiments of obtaining corresponding Chinese articles for entity linking from the 27 extracted keyphrases in the previous section. One of the authors manually compared the keyphrases and the corresponding Chinese articles and decided whether the article describes the meaning of the keyphrase. For one keyphrase, the maximum number of candidate Chinese articles was 42, average was 11, and some keyphrases had no candidates. To evaluate the results, we classified them into four categories based on the results of the translation and the acquired corresponding article candidates, as shown in Table 7.

The results of the correct ratio of acquiring the relevant Chinese candidate articles for 27 Japanese keyphrases by the proposed method are shown in Table 8. The results are classified into one of the categories in Table 7. By using the combination of the three translation methods, the correct articles were acquired for 22 (19 in category 1 and 3 in category 3) out of 27 keyphrases, and the accuracy was 0.81.

Table 7 Classification of results of acquired translation and article candidates

Category	Whether correct translation was extracted	Whether correct article was acquired
1	Yes	Yes
2	Yes	No
3	No	Yes
4	No	No

Table 8 Experimental results of acquiring Chinese corresponding articles from Japanese keyphrases

Translation methods	1	2	3	4
MT (Bing Translator)	0.51 (14/27)	0.07 (2/27)	0.11 (3/27)	0.29 (8/27)
MT (Google Translate)	0.55 (15/27)	0.14 (4/27)	0.11 (3/27)	0.18 (5/27)
Character code conversion	0.07 (2/27)	0.00 (0/27)	0.00 (0/27)	0.92 (25/27)
Combination of three translation methods	0.70 (19/27)	0.14 (4/27)	0.11 (3/27)	0.03 (1/27)

Table 9 Examples of ranking candidate Chinese articles for some keyphrases

Keyphrase	Candidate Chinese article	Similarity score	Rank
	本田技研工業	0.29	1
	本田车队	0.27	2
ホンダ (Honda)	廣汽本田汽車	0.20	3
	本田飞度	0.17	4
	本田雅阁	0.17	5
	白宫 (莫斯科)	0.32	1
	白宫	0.29	2
ホワイトハウス (White House)	白宫办公厅	0.27	3
	白宫地图室	0.21	4
	白宫 (热那亚)	0.15	5
	利比里亚国旗	0.18	1
	利比里亚历史	0.18	2
リベリア (Liberia)	利比里亚地理	0.15	3
	利比里亚	0.15	4
	利比里亚国徽	0.14	5

Table 10 Summary of effectiveness of ranking candidate articles

Rank	Top 1	Top 10	Top 20
No. correct/no. keyphrase	16/22	21/22	22/22
Accuracy	0.72	0.95	1.00

4.2.3 Ranking Candidate Articles to Be Linked

In this section, we describe the results of the experiments for ranking the candidate Chinese articles using the cosine similarity between the Japanese document translated into Chinese and the candidate Chinese encyclopedia articles, which are explained in Sect. 3.4.

Examples of the ranked Chinese articles for some keyphrases are shown in Table 9. The articles names in boldface are the correct article to be linked from the keyphrase. The effectiveness of the ranking is summarized in Table 10.

5 Conclusion

In this chapter, we proposed a method for keyphrase extraction from Japanese documents considering the user's language ability in cross-language entity linking from Japanese to Chinese.

From the questionnaire survey, we found that Chinese students have difficulty in understanding keyphrases from popular phrases and topics in Japanese news articles. This makes it harder to find Chinese Wikipedia articles that are suitable for students with different linguistic proficiency levels. Further research is needed to solve this problem in the future. Furthermore, machine translation services are primarily designed for translating sentences rather than translating individual keyphrases as our current method does. We are planning to consider the surrounding context of the extracted keyphrase for resolving the word ambiguity problem.

Acknowledgements This work was supported in part by JSPS KAKENHI Grant Numbers 24500300, 16K00452, and the MEXT-Supported Program for the Strategic Research Foundation at Private Universities (S1511026).

References

1. K. Horita, F. Kimura, A. Maeda, Automatic keyword extraction for wikification of East Asian language documents. Int. J. Comput. Theory Eng. **8**(1), 32–35 (2016)
2. R. Mihalcea, A. Csomai, Wikify!: linking documents to encyclopedic knowledge, in *Proceedings of the 16th ACM Conference on Information and Knowledge Management*, pp. 233–242 (2007)

3. L.X. Tang, S. Geva, A. Trotman, Y. Xu, K.Y. Itakura, Overview of the NTCIR-9 crosslink task: cross-lingual link discovery, in *Proceedings of the 9th NTCIR Conference*, pp. 437–463 (2011)
4. L.X. Tang, I.S. Kang, F. Kimura, Y.H. Lee, A. Trotman, S. Geva, Y. Xu, Overview of the NTCIR-10 cross-lingual link discovery task, in *Proceedings of the 10th NTCIR Conference*, pp. 8–38 (2013)
5. J. Heng, J. Nothman, B. Hachey, Overview of TAC-KBP2014 entity discovery and linking tasks, in *Proceedings of TAC2014* (2014)
6. Z. Wang, J. Li, Z. Wang, J. Tang, Cross-lingual knowledge linking across wiki knowledge bases, in *Proceedings of the 21st International conference on World Wide Web*, pp. 459–468 (2012)
7. S. Chen, G.J.F. Jones, N.E. O'Connor, DCU at NTCIR-10 crosslingual link discovery (CrossLink-2) task, in *Proceedings of the 10th NTCIR Conference*, pp. 74–78 (2013)
8. D. Milne, I.W. Witten, An open-source toolkit for mining Wikipedia. Artif. Intell. **194**, 222–239 (2013)
9. Y. Liu, J. Boisson, J.S. Chang, NTHU at NTCIR-10 crosslink-2: an approach toward semantic features, in *Proceedings of the 10th NTCIR Conference*, pp. 62–68 (2013)
10. J. Zhou, F. Kimura, A. Maeda, Cross-language entity linking adapting to user's language ability, in *Proceedings of The International MultiConference of Engineers and Computer Scientists 2017*. Lecture Notes in Engineering and Computer Science, 15–17 Mar 2017, Hong Kong, pp. 24–29
11. MeCab: yet another part-of-speech and morphological analyzer (in Japanese), http://taku910.github.io/mecab/. Accessed 21 Aug 2017
12. JLPT Japanese-language proficiency test, http://www.jlpt.jp/e/index.html. Accessed 21 Aug 2017
13. T. Mikolov, I. Sutskever, K. Chen, G. Corrado, J. Dean, Distributed representations of words and phrases and their compositionality, in *Proceedings of Advances on Neural Information Processing Systems 26 (NIPS 2013)*, pp. 3111–3119 (2013)
14. Google Translate, https://translate.google.com. Accessed 21 Aug 2017
15. Bing Translator, http://www.bing.com/translator. Accessed 21 Aug 2017
16. Pinconv 4 (in Japanese), http://www.karak.jp/chinese/pinconv-4-00.html. Accessed 21 Aug 2017
17. J. Zhou, X. Song, F. Kimura, A. Maeda, A cross-language entity linking method using combination of multiple translation methods," in *Proceedings of the 4th ICT International Student Project Conference (ICT-ISPC2015)* (2015)

Author Index

© Springer Nature Singapore Pte Ltd. 2018
S.-I. Ao et al. (eds.), *Transactions on Engineering Technologies*,
https://doi.org/10.1007/978-981-10-7488-2

Subject Index

© Springer Nature Singapore Pte Ltd. 2018
S.-I. Ao et al. (eds.), *Transactions on Engineering Technologies*,
https://doi.org/10.1007/978-981-10-7488-2

Printed by Printforce, the Netherlands